Follicular Growth, Ovulation and Fertilization

Molecular and Clinical Basis

Other Books of Interest

Adaptation Biology and Medicine
Volume 1: Subcellular Basis (81–7319–111–5)
B.K. Sharma et al

Adaptation Biology and Medicine
Volume 2: Molecular Basis (81-7319-246-4)
K.B. Pandolf et al

Avian Endocrinology (1-84265-033-5)
Alistair Dawson and C.M. Chaturvedi

Biological Rhythms (1-84265-034-3)
Vinod Kumar

Brain Protection and Neural Trauma (81-7319-261-8)
V.K. Khosla et al

Diarrhoeal Diseases: Research Perspectives (81-7319-343-6)
N. Appaji Rao and N.K. Ganguly

Essentials of Clinical Toxicology (81-7319-219-7)
S.B. Lall

Immunomodulation (81-7319-118-2)
S.N. Upadhyay

Immunopharmacology (81-7319-241-3)
S.N. Upadhayay

An Introduction to Immunology (1-84265-035-1)
C. Vaman Rao

Liver and Environmental Xenobiotics (81-7319-121-2)
S.V.S. Rana and K. Taketa

Medical Diagnostic Techniques and Procedures (81-7319-350-9)
Megha Singh et al

Multi-Drug Resistance in Emerging and
Re-Emerging Diseases (81-7319-346-0)
R.C. Mahajan and Amu Therwath

Physiological Fluid Dynamics III (81-8519-856-X)
N.V.C. Swamy and M. Singh

Radiation Sensitizers (1-84265-053-X)
N.G. Huilgol

Radiobiological Concepts in Radiotherapy (81-7319-055-0)
D. Bhattacharjee and B.B. Singh

Editors: Anand Kumar □ Amal K. Mukhopadhyay

Follicular Growth, Ovulation and Fertilization

Molecular and Clinical Basis

Narosa Publishing House

New Delhi Chennai Mumbai Kolkata

Editors
Anand Kumar, MD
Department of Reproductive Biology and
Central Radioimmunoassay Facility
All India Institute of Medical Sciences
New Delhi-110 029, India

Amal K. Mukhopadhyay, PhD
Institute for Hormone and Fertility Research
University of Hamburg, Hamburg, Germany

Associate Editor
P.K. Chaturvedi, PhD
Department of Reproductive Biology
All India Institute of Medical Sciences
New Delhi-110 029, India

Editorial Assistants: **Urmil Vij** (PhD) and **Nishat Ahmad**
Department of Reproductive Biology
All India Institute of Medical Sciences
New Delhi-110 029, India

NAROSA PUBLISHING HOUSE

22 Daryaganj, Delhi Medical Association Road, New Delhi 110 002
35–36 Greams Road, Thousand Lights, Chennai 600 006
306 Shiv Centre, D.B.C. Sector 17, K.U. Bazar P.O., Navi Mumbai 400 705
2F–2G Shivam Chambers, 53 Syed Amir Ali Avenue, Kolkata 700 019

ISBN 81-7319-358-4

Published by N.K. Mehra for Narosa Publishing House, 22 Daryaganj, Delhi Medical Association Road, New Delhi 110 002 and printed at Replika Press Pvt. Ltd. Delhi 110 040 (India).

Foreword

Population control and maternal and child health have always been primary focus of the health planners which is reflected in frequent reviews of these programmes at national and international levels. The Sub-Committee on Reproductive Health Research Needs Assessment (RHRNA), set up in 1997 by the Government of India, has recommended research on fertility regulation including family planning and infertility, reproductive tract infections and old age. The National Population Policy of India, formulated in the year 2000 continues to promote research on reproductive health. All the abovementioned areas of reproductive health care are represented in this book.

The All India Institute of Medical Sciences has been a centre of excellence in India and has made significant contributions in the field of reproductive sciences. The Institute for Hormone and Fertility Research, University of Hamburg, Hamburg, is well known today as a premier institution in the discipline of reproductive biomedical sciences. I am happy that the two have joined hands to produce this book. Eminent researchers from different parts of the globe have contributed their latest findings. The book deals with the fundamental and applied aspects of reproductive biology and medicine including andrology. The papers on cell and molecular biology, physiology and pathophysiology of reproductive system can be considered timely and authoritative.

With the advent of *in vitro* fertilization we can reproduce asexually and can wilfully interfere in the evolutionary process of nature to improve our own stock. A good coverage is given to the biotechnological advances in assisted reproduction including *in vitro* fertilization.

I congratulate the editors, Dr Anand Kumar and Dr Amal K. Mukhopadhyay, for publishing this book on behalf of the All India Institute of Medical Sciences. I am happy to offer this book to all those interested in the subject.

PROF P.K. DAVE
Director
All India Institute of Medical Sciences
New Delhi, India

Prologue

India and the Federal Republic of Germany were major partners in the "International Symposium on Molecular and Clinical Aspects of the Regulation of Ovarian Function: Impact on the Reproductive Health of Women" and "Indo-German Satellite Workshop on Fertilization: Lessons from *In Vitro* Approaches". It was a matter of great pleasure for the scientists of these two countries and of other countries too to have deliberated in New Delhi on the recent developments in reproductive biology and medicine. India provided a historic destination for these events.

India is one of the few countries which has given a lead to the world in matters of sexual and reproductive health because of her liberal outlook since the time of Vatsyayana to date. The then president of the Indian National Congress, Subhas Chandra Bose had set up a National Planning Committee (NPC) under Jawaharlal Nehru in 1938 to prepare a plan for socio-economic development of India after Independence. Pandit Nehru set up a sub-committee under the chairmanship of noted economist Radhakamal Mukerjee to study the best approach to India's population problem. The NPC, in its final report in 1940, noted that "measures for the improvement of quality of population and limiting excessive population pressure .... were necessary" and that "in the interest of social economy, family happiness and national planning, family planning and a limitation of children are essential, and the State should adopt a policy to encourage them". India was the first country to make an official policy for birth control as early as 1952, immediately after Independence in 1947. Although it failed, but it laid the foundation for all future actions. In view of this history, Delhi was the most appropriate site for this crucial gathering of scientists from all over the world.

The Indo-German workshop provided an opportunity for a greater partnership to scientists from Federal Republic of Germany to travel to India and interact with their colleagues in India. Germany has a long history of research in reproductive biomedicine. The Institute for Hormone and Fertility Research at the University of Hamburg is a pioneering institution in the world today. It was started in 1983 and has now become a cradle of many innovative ideas leading to not only generation of high quality research publications but also to setting up of several start-up companies in the sector of biotechnology.

The collaborative workshop and the jointly organised International Symposium have resulted in compilation of this book which I think will be very useful as a reference volume for both students and scientists engaged in research in reproductive biology and medicine.

Dr Amal K. Mukhopadhyay and Dr Anand Kumar deserve my full appreciation for their pains-taking work in presenting this book.

PROF DR MED F.A.LEIDENBERGER
Former Director and Chairman, Governing Body
Institute for Hormone and Fertility Research
University of Hamburg, Hamburg, Germany

Preface

This book covers wide range of topics deliberated on at the "International Symposium on Molecular and Clinical Aspects of the Regulation of Ovarian Function: Impact on the Reproductive Health of Women" and "Indo-German Satellite Workshop on Fertilization: Lessons from *In Vitro* Approaches" held in October, 1999. These meetings were jointly organized by the All India Institute of Medical Sciences, New Delhi and the Institute for Hormone and Fertility Research, University of Hamburg, Hamburg. The topics on contraception, infertility management, environment, aging and basic understanding of reproductive processes are the highlights of the proceedings.

The world population which was 1.7 billion in 1950 grew to 4.9 billion in 2000. It is predicted that the population will soar to over 8 billion by the year 2050. This rapid rate of population growth, especially in the developing countries, has kept the social scientists, political thinkers and reproductive biologists busy in finding ways and means to contain the population explosion. The search for an ideal contraceptive which would be safe, hundred per cent effective, convenient, inexpensive, reversible and which does not interfere with normal intercourse, is still on.

Since the increase in population has largely occurred in the less developed countries whereas population has shown a trend of negative growth in some developed countries, it seems that the interests of developing and developed countries diverge on population growth management. It is clear that on the one hand there is a need to make available newer methods of fertility control, giving wider choice to the user depending upon individual need and on the other hand to take effective care of infertility–related problems, incidence of which has tended to increase in recent years. The problem of infertility cannot be ignored in the developing countries too because of the high absolute number of infertile couples in these countries. The number of infertile men and women in the world was estimated to be 50-80 million in 1990s. Assisted Reproductive Technologies have given a new hope in otherwise incurable conditions.

Although the role of environmental toxins in affecting reproductive performance in humans has not been conclusively proven, this may have a bearing on the incidence of infertility. However, in terms of future challenges, e.g. with respect to the impact of environmental disrupters and aging, the gap in the perspectives of developed and developing nations has already started narrowing. In future, there will be a need to invest considerable amounts of money and effort to understand the impact of environmental disrupters and aging on reproductive system. Thus biology and medicine of reproduction will continue to be a high-impact science for years to come.

Furthermore, global population is greying at an unprecedented pace, in which developing countries are no exception, thanks to an increase in longevity due to better healthcare systems. The world population of elderly (over 60 years of age) in 2030 is likely to be three times more than what it was in 1990. Most of this growth will take place in Asia, particularly India and China. In India alone the number of elderly will be over 136 million in the year 2021. In the USA, currently there are 35 million people above the age of 65 which is set to double by 2030. Therefore it is obligatory on all concerned, including policy-makers, law-makers, and social scientists to devise ways and means to decrease the social and physical burden of an aging population in society. The scientists and clinicians must devote greater resources for research to find ways and means to decrease the general and reproductive debility and frailty associated with aging by promoting healthy aging. It is well established that reproductive tissues are among those showing the early manifestations of aging-related decline. Thus reproductive and endocrine-related disorders will continue to be an important area of research in the biology and related medicine of aging.

From all standpoints discussed above, starting with contraception and management of infertility to aging-related endocrine disorders, a clear understanding of reproductive biology is fundamental for solutions to many problems faced by mankind today. This Conference was organized to focus on the latest developments in the area of reproductive biology with a special emphasis on emerging techniques in reproductive medicine. The scientific contributions at this Conference form the basis of this book which deals mainly with the basic aspects of the biology of reproduction with a special focus on fertility control and treatment of infertility–with valuable contributions coming from scientists from India and abroad. The editors would like to thank all the authors for their co-operation and help in making their contributions available for this book.

In addition, the editors are very grateful for the moral support and encouragement provided by Prof P.K. Dave, Director, All India Institute of Medical Sciences, Prof N.K. Ganguli, Director General, Indian Council of Medical Research, Mr A.R. Nanda, Secretary to the Department of Family Welfare, Ministry of Health and Family Welfare, Government of India, Dr Stephan F. von Welck, First Counsellor, Embasy of Germany, New Delhi, Dr M.L. Swarnkar, Director, Jaipur Fertility & Microsurgery Research, Jaipur, Prof Ilpo T. Huhtaniemi, Chief Managing Editor, Molecular and Cellular Endocrinology, Elsevier Science Ltd., Oxford, U.K., Dr Gyanendra Nath, Advisor, Department of Science and Technology, Government of India, Dr B.M. Gandhi, Advisor, Department of Biotechnology, Government of India, Dr V.K.Bahal, Assistant Commissionar (Family Welfare) Ministry of Health and Family Welfare, Government of India and Dr Ravi Rangachari, Deputy Director General, Indian Council of Medical Research. Prof F. Leidenberger, Former Director and Chairman, Governing Body of the Institute for Hormone and Fertility Research, Hamburg University, Hamburg was the originator of the concept and has been a constant source of inspiration to the editors at all stages of the work.

The book would not have seen the light of day but for the generous funding and support made available by the Indian Council of Medical Research, Ministry of Health and Family Welfare, Department of Biotechnology, Council of Scientific and Industrial Research, Government of India, Bundesministerium fur Bildung, Forschung and Technologie (BMBF), Federal Republic of Germany and World Health Organization.

The editors are thankful to Dr Mrs P. Jha, Mr P.K. Dhawan, Dr K. K. Gaur, Mr Deepak Tekchandani and Ms Sonja Schleichert for their help and co-operation in various organizational aspects. Thanks are also due to Mr N.K. Mehra and Mr M.S. Sejwal of Narosa Publishing House, Delhi. Last but not the least, the editors acknowledge the silent contribution of Mrs Gauri Mukhopadhyay behind the stage, who willingly took care of many small needs of the international participants, making the Conference a pleasant event.

EDITORS

Contents

Spiritual View

Follicular Growth, Ovulation and Fertilization: Molecular and Clinical Basis
Anand Kumar and Amal K. Mukhopadhyay (Eds.)
Narosa Publishing House, New Delhi, India, 2001

1

Heredity, Evolution and Population: A spiritual view

Alok Pandey
Command Hospital (Air Force), Agram, Bangalore-560007, India

Introduction

Biological science, in its search for the basis and nature of life has entered a mysterious world teeming with miniature marvels of nature. This minute, quantum basis of life, as we are told, are the genes that hold a secret which unfolds in the blooming of a rose and the marvel of a man. And yet, at this so called first interphase between nonliving and living matter, the riddle of the sphinx meets us, we stand as if at a doorway of some Elusian mystery that leads us to a deeper unknown. The ancient question revisits us as an echo from its depths—What is life? Are genes the lifeforce uses to manifest in matter? This ancient question was raised in modern times by some one no less than the discoverer of this basic unit of life itself. Sir Francis Crick carrying the line of argument of Enrico Fermi, Arrhenius and JBS Haldane believed that the chance association of amino acid coming together in a meaningful polypeptide chain would be exceedingly rare, if possible at all. A loose analogy would be the near impossibility of the event of a million monkeys randomly throwing their fingers at the typewriter to produce a single sonnet of Shakespeare. The feat would be impossible even for intelligent human beings unless of course the rules of the game change and randomness is replaced by a conscious and deliberate action of an intelligent-will that secretly organises the concert of life. Spiritual sciences approaching from another line of experience arrived at the affirmation that there is indeed such an all pervading Creatix Consciousness that bursts into colour and pattern through an interplay of gross and subtle energies. One of the result of this play is what we know as life. But embodied, cellular life as we know today does not exhaust all the possibilities that are latent in the life force just as the visible spectrum does not exhaust all the ranges of light.

The gap between the genetic mechanism and the total phenomenon called man is too wide to be bridged by an equally elusive factor of environment playing with an ambiguous heredity. It is difficult to surmise that genes and heredity dictate the entire range of psychological experiences from Caesar's ambition to a Buddha's enlightment. The gulf between that which explains and that which is explained is not just a quantitative but a qualitative one. The force of heredity is no doubt there but perhaps more as a mechanism than a first or last determinant. Let us then approach the enigma of life from this other standpoint and see if it compliments our material understanding of life.

The One Who Becomes the Many

Two Illusions

Viewed from the standpoint of consciousness, we discover that there is only one fundamental reality that becomes many. The life in a chimpanzee and a fish and a plant, a worm and a man is different only in terms of capacity, force, degree, form and process used. There is so to say, a difference of quality and of quantity. When we focus too much on these differences. We loose the fundamental truth which is the basis of all true understanding. This fundamental truth is that all these qualitative and quantitative differences rest on a basis of oneness. Life is one but manifests differently in different form. This truth has great practical implication for it gives at once a correct orientation and meaning to the entire evolutionary journey of life from dust (where it is concealed) to man. Missing this vital truth, we are life-explorers lost in a forest while counting the individual trees. We count people and discover 'the many' factors that lead to an increase in population but we miss out the one single most important factor without which all our manipulations become a labour of Sisyphus.

Three Faces of One Diety

Life is one but appears to us in three forms of its movement. First, there is the birth or creation of life out of life itself. It extends into another form created by its force. Next, are the thousand processes that balance the force of life in one form in its interchange with other forms. We call this loosely as a biological chain but actually this chain is not merely biological but psychological and spiritual as well. Finally, when the possibility of utility and growth in a particular form is over, life withdraws itself to pass into other forms and substance. That to us is a death. But actually, death is not an opposition but one of the process used by life for its further progression. To use an Upanishadic formula we may say that "nothing dies" or even life is immortal in its nature, only the form perishes. But even in this perishing of form there is a difference. The contact and experience of life remains vibrating in matter and by a contagion spreads in matter all around. It is this unseen commerce of life that is a spiritual basis for 'Morphogenetic resonance" and serves for further growth and evolution in matter.

Inter-Conversion of Energy

The ancient Indian science derived from yogic experience affirmed not only an underlying oneness but because of this oneness it saw the possibility of conversion of one form of energy of life into another. In the material domain, this oneness and the interconversion of energy is well known and its practical utility are enormous. In the domain of living matter, the possibilities become even more immense. It is possible, for instance, to conserve the energy of life from flowing in an unregulated manner and thereby increase health, strength and vigor. Control and channelisation of life can help development of human resources. The process of sublimation of the force of sex into creative mental activity has indeed been admitted by thinkers like Bergson, Schophenhaur and Freud. The ancient Indian theory of Brahmacharya has studied and developed this truth and arrived at the conclusion that the energy for procreation, if rightly conserved and directed can open the doors of a new creation for man.

The Double Helix

For like the DNA, the nature of man is like a double helix. There is below him a pull and burden of other forms of animal and material life. There is also in him an attraction from above for an evolution into a higher form and figure of life. In man, the Lucifer and the angel meet and oppose each other. In man lies the possibility of their reconciliation and a victory of life using both the powers for evolutionary fulfilment.

Overpopulation: The two views

This background was necessary for looking at the problems we face today of which overpopulation is one of them. The ordinary, material view of life occupied with physical facts and figures, isolate the different biosocial and mathematical factors contributing to this problem. To the spiritual view, the reason for the population increase is an imbalance in the creative, preservative and destructive forces. Though the imbalance appears apparently due to an increasing mastery over the causes of destruction and preservation. But there has been also an outburst in creative energy of the race. In fact the growing power of man over the material causes of death is itself one of the results of this creative outbrust. This imbalance and an increase in creative energy has been deliberately introduced primarily because nature has entered in a new mode of activity for a new creation. But the race is unable to hold, contain and utilise the sudden increase in this creative evolutionary energy. The result is its degradation and flow into other channels of which one side effect is an increase in population.

Population Control: The mechanical method

Since the mechanistic view of man does not see this, it simply tries to put a mechanical and physio-chemical control. The result is a temporary success. But the increased evolutionary energy continues to be dissipated and misdirected. The race becomes more and more incapable and if this continues man will get crushed under his own weight. Unable to bear the burden of the increased energy and power into his fagile system, he may well end up destroying himself as a race.

Population Control: Nature's method

But nature is wiser. The degraded life-energy becomes death thus restoring the lost balance! But equally, the overwhelming power may devise ways and means to destroy the power of fertility and conquer the uncontrolled sexual impulse in man. Already we see signs of these. There is a growing desensitization to sex through sheer over exposure and an exhaustion of the impulse through an uncontrolled release. There is simultaneously a blurring of sex distinction and even an effort at bypassing the sexual process of reproduction. All this is no doubt a dangerous process but nature laughs at dangers. It cares little for the price paid even if it be the extermination of an entire species so long as secret intention is fulfiled.

Population Control: The spiritual way

But man can collaborate! Instead of trying to cleverly manipulate Nature, he can try to read her moods and symbol gestures. Instead of directing the evolutionary energy given to him for narrow, hedonistic gains he can turn it towards manifesting the fullest inner potential that man is capable of. He can grow and progress not only outwardly but more importantly inwardly from being merely a vital-mental creature to a spiritual and supramental being.

An Ancient Indian Tale

The Spiritual Sense of Evolution

The ancient Indian parable of the ten incarnations of God precisely points towards this possibility. According to this legend, Vishnu, the Godhead of preservation comes in every evolutionary crisis to rescue the earth. What is interesting in this parable is the form he successively adopts from fish to tortoise to a wild boar and then to a half-animal half human-creature, the lion-man. The subsequent series proceeds from dwarf-man to a wild man and hence to a mentalised moral man. The series does not stop here but carries further till man becomes capable of manifesting in his outer life the kingdom of heaven he ever carries within. We admit the evolution of form, but the real significance is that there is an evolution of consciousness and its powers through an evolution of the form. In fact, it is the evolution of consciousness that through an inner alchemy and subtle process exerts a pressure upon the species to burst its bounds, mutate randomly but successfully till the crawling matter begins to fly and the living matter becomes thinking matter. It is not survival but progress and ascent of life that is the real secret. If survival were the sole point, then there would be no reason to grow beyond the virus who in many ways can still outsmart a man! Man must grow into something more than man if he would really solve the puzzle of life. Or else, he will step aside allowing the expanding force of consciousness to yet shoot beyond man into a more perfect form more capable of manifesting the hidden truth that smiles unvaryingly behind the many moods of nature.

The Unraised Question

The whole question therefore is not whether man can control population or alter heredity but rather whether he can find the key to a conscious evolution! The answer of course is unlikely to be found in the existing machinery of material nature. It has to be first discovered at the appropriate level of consciousness where a splendour of oneness release the billion rays of life that assume a thousand form and figure of itself. We already see hints of such a possibility in the growing stride of science in manipulating the genes and altering heredity. An understanding of the material basis of gene-environment may open yet another door. In the domain of philosophy too this thought of a new race has been put forward by thinkers like Nietze and Teilhard De Chardin. The idea has also been recently catching upon the mind and imagination of the race as is evident in the new creed of science-fiction writers. In our own time, the vision of a new race has been most forcefully put forward and worked out by the seer of Pondicherry, Sri Aurobindo.

The Answer from Sri Aurobindo

Sri Aurobindo came to announce that, "Man is a transitional being; he is not final". "The essential purpose and sign of the growing evolution here is the emergence of consciousness in an apparently inconscient universe, the growth of the consciousness and with it growth of the light and power of the being; the development of the form and its functioning or its fitness to survive, although indispensible, is not the whole meaning or the central motive.... The step from man to superman is the next approaching achievement in the earth's evolution. It is inevitable because it is at once the intention of the inner spirit and the logic of Nature's process." Let us then close with this vision of Sri Aurobindo captured beautifully in the epic-poem Savitri.

"All now seems nature's massed machinery"

...."An endless servitude to material rule
And long determinations rigid chain,
Her firm and changeless habits aping law,
Her empire of unconscious deft device
Annual the claim of man's free human will.

... "But wisdom comes and vision grows within,
Then nature's instrument crowns himself her King's
He feels his witnessing self and conscious power,
A Godhead stands behind the brute machine.

....."If in the meaningless void creation rose
If from a bodiless force matter was born,
If life could climb in the unconscious tree,
If green delight break into emerald leaves
And its laughter of beauty blossom in the flower
If sense could wake in tissue, nerve and cell,
And thought seize the grey matter of the brain
And soul peep from its secrecy through the flesh,
How shall the nameless light not leap on men,
And unknown powers emerge from Nature's sleep?

.....Mind is not all his tireless climb can reach,
.....There is an infinite truth, an absolute power
.....Its greatness shall be felt shaping the world's course.
.....All then shall change, a magic order come

Overtopping this mechanical universe.
A mightier race shall inhabit the mortal's world,
On Nature's luminous tops, on the Spirit's ground,
The superman shall reign as King of life.
Make earth almost the mate and peer of heaven
And lead towards God and truth man's ignorant heart
And lift towards Godhead his mortality."

Gonadotropins

Follicular Growth, Ovulation and Fertilization: Molecular and Clinical Basis
Anand Kumar and Amal K. Mukhopadhyay (Eds.)
Narosa Publishing House, New Delhi, India, 2001

2

New Biology of FSH and LH

S.G. Hillier
Reproductive Medicine Laboratory
Department of Reproductive and Developmental Sciences
University of Edinburgh, 37 Chalmers Street Edinburgh EH3 9EW, Scotland

Introduction

The molecular and cellular mechanisms that underpin follicular growth, and ovulation are fundamental to reproduction. This review briefly summarises our current understanding of how follicle-stimulating hormone (FSH) and luteinising hormone (LH) secreted by the anterior pituitary gland act on target cells in the ovaries to stimulate follicular maturation. The concept is advanced that locally produced steroids and growth/differentiation factors play critical roles in mediating cell-to-cell interactions in the developing preovulatory follicle. The follicular life-cycle recruitment, selection and dominance culminates in LH-induced proteolytic degradation of the follicle wall to permit shedding of the oocyte. Both pro- and anti-inflammatory modulators must be involved in this tightly controlled sequence of tissue injury and repair. It is proposed that altered glucocorticoid metabolism by granulosa cells, induced by LH and pro-inflammatory cytokines, leads to increased accumulation of anti-inflammatory cortisol in the preovulatory follicle, which may promote resolution of the inflammatory process and rapid healing of the ovarian surface.

Follicular Recruitment

During antral follicular growth (recruitment), FSH acts via granulosa cell FSH receptors coupled to cAMP mediated post-receptor signalling to stimulate the formation of factors that locally modulate cell proliferation and differentiation. Locally produced factors that affect FSH-regulated granulosa cell proliferation include growth/differentiation factors such as activin and transforming growth factor-beta that activate serine/threonine kinase mediated post-receptor signalling. Paracrine factors of thecal cell origin capable of influencing FSH action include androgens-i.e. aromatase substrates. There is experimental evidence that expression of the granulosa cell androgen receptor (AR) that mediates paracrine androgen action is developmentally regulated by FSH. Immunocytochemical studies of AR content of non-human primate and rat ovaries reveals presence of AR almost entirely in granulosa cells of preantral, and early-intermediate antral follicles, disappearing in the preovulatory follicle. In-vivo experiments on immature female rats confirm that granulosa cell AR mRNA levels are down-regulated during FSH-induced preovulatory follicular development. A current working hypothesis is that negative regulation of the granulosa cell AR is part of the intraovarian mechanism that determines which follicle(s) is 'selected' to complete its development and ovulate (Hillier et al. 1997).

Selection and Dominance

During follicular 'dominance', LH-stimulated thecal androgen is required as an oestrogen precursor in FSH-stimulated granulosa cells. In-vitro studies reviewed below on isolated granulosa and thecal cells (human, non-human primate and rat) and whole follicles (rat) reveal that FSH stimulates granulosa cells to produce a factor(s) that upregulates LH-stimulated thecal P450c17 mRNA expression and androgen synthesis during preovulatory follicular maturation. Growth/differentiation factors implicated in this positive feedback loop are inhibin and insulin-like growth factors (IGF-I and IGF-II). Although physiological (intrafollicular) concentrations of IGF-I and IGF-II augment LH-stimulated human thecal androgrenproduction *in vitro*, their effects are greatly enhanced by the additional presence of inhibin. Thus paracrine inhibin may have special relevance to the maintenance of follicular dominance in the human menstrual cycle (Hillier 1996).

Ovulation

Research on the paracrine control of folliculogenesis by steroids has invevitably focused on sex steroids-androgens and oestrogens-with less attention being paid to C21 steroids other than progesterone. Although the ovaries do not express all the enzymes necessary for corticosteroid biosynthesis, experimental and clinical data suggest that corticosteroids may also directly influence ovarian function. We have recently produced evidence that ovarian follicles undertake development-related changes in glucocorticoid metabolism (Tetsuka et al. 1997, Hillier & Tetsuka 1998), the pattern of which is consistent with an intraovarian role for cortisol in limiting inflammation mediated tissue damage associated with ovulation. A switch in the expression of mRNAs encoding 11β-hydroxysteroid dehydrogenase (11βHSD) isoforms occurs during granulosa cell luteinisation. Before exposure to LH/HCG, non-luteinised human granulosa cells express mainly 11βHSD2 mRNA and little or no 11βHSD1 mRNA. After exposure to LH/HCG, 11bHSD2 mRNA expression is suppressed while 11β-HSD1 mRNA is enhanced. This developmental shift in potential for glucocorticoid metabolism from inactivation (C11-oxidation) to activation (C11-reduction) is reflected in the ability of luteinising human granulosa cells to undertake predominantly reductive (cortisone to cortisol) metabolism in vitro, and the substantially raised concentrations of cortisol present in follicular fluid collected after the ovulation-inducing LH surge in spontaneous menstrual cycles. We therefore propose that developmental regulation of 11βHSD expression before follicular rupture may be a mechanism to raise the intrafollicular cortisol level in the dominant follicle shortly before it ovulates (Yong et al. 2000).

Resolution

The postulted anti-inflammatory role of intrafollicular glucocorticoid may be necessary for timely resolution of the inflammatory process and healing of the ovarian surface. To test the ability of gonadotrophins and pro-inflammatory cytokines to act directly on granulosa cells to modulate gene 11βHSD expression directly, we optimised a rat granulosa cell culture model in which FSH treatment is first given to induce functional maturation (i.e. LH responsiveness), followed by LH to induce luteinisation. In this system, IL-1β mimicked LH-induced up-regulation of 11βHSD1 mRNA expression (Tetsuka et al. 1999a), consistent with other evidence for the involvement of pro-inflammatory factors in ovulation (Espey 1980, 1994, Adashi 1998, Terranova & Rice 1997). Work on the control of 11βHSD1 expression in glomerular mesangial cells has shown that exposure to pro-inflammatory cytokine increases the expression and reductase activity of 11βHSD1, serving to augment the anti-inflammtory effect of glucocorticoids (Escher et al. 1997). Incresed production of TNFα and IL-1β (Adashi 1998, Terranova & Rice 1997) with up-regulation of PLA2 (Ben-Shlomo et al. 1997), COX-2 (Morris & Richards 1995, Hellberg et al. 1996, Ando et al. 1998) and 11βHSD1 (Tetsuka et al. 1997) are common features of ovulation. Glucocorticoid receptor mRNA expression in rat granulosa cells is not measurably influenced by LH/HCG treatment in vivo (Tetsuka et al., 1999b). Thus local activation of glucocorticoids by 11βHSD could be a major determinant of glucocorticoid action in the ovulatory follicle. It is of interest that an LH-inducible pro-inflammatory component (IL-1β) in the periovulatory cascade has the capacity to set in train a means through which inflammation might be quelled, i.e. through 11βHSD1-mediated formation of anti-inflammatory glucocorticoid. Further work is required to determine if the spatio-temporal expression of genes encoding inflammatory mediators in the wall of the periovulatory follicle is consistent with this postulated sequence of events *in vivo*.

Conclusion

In conclusion, a mechanism is suggested through which the potentially deleterious inflammatory component of follicular rupture is ultimately self-limiting. LH can act directly to up-regulate 11βHSD1 gene expression in granulosa cells *in vitro* and its action is mimicked by the pro-inflammatory mediator IL-1β. This supports a working hypothesis that up-regulation of 11βHSD1 expression during granulosa cell luteinisation (with reciprocal down-regulation of 11βHSD2 gene expression) is a mechanism to augment local production of anti-inflammatory glucocorticoids, serving to minimise inflammatory tissue damage at the time of rupture nd encourage rapid healing of the ovarian surface in anticipation of the next ovulation.

Acknowledgement

This work was supported by a Medical Research Council (UK) Programme Grant G8929853.

References

Adashi EY 1998 The potential role of interleukin-1 in the ovulatory process: an evolving hypothesis. Molecular and Cellular Endocrinology; **140**: 77–81.

Ando M, Kol S, Kokia E, Ruutiainen-Altman K, Sirois J, Rohan RM, Payne DW & Adashi EY 1998 Rat ovarian prostaglandin endoperoxide synthase-I and -2: periovulatory expression of granulosa cell-based interleukin-1-dependent enzymes. *Endocrinology*; **139**: 2501–2508.

Ben-Shlomo I, Kol S, Ando M, Altman KR, Putowski LT, Rohan RM & Adashi EY 1997 ovarian expression, cellular localization, and hormonal regulation of rat secretory phospholipse A2: increased expression by inerleukin-1 and by gonadotropins. *Biology of Reproduction*; **57**: 217–225.

Escher G, Galli I, Vishwanath BS, Frey BM & Frey FJ 1997 Tumor necrosis factor alpha and interleukin 1beta enhance the cortisone/cortisol shuttle. *Journal of Experimental Medicine*; **186**: 189–198.

Espey LL 1980 Ovulation as an inflammatory reaction—a hypothesis. *Biology of Reproduction*; **22**: 73–106.

Espey LL 1994 Current status of the hypothesis that mammalian ovulation is comparable to an inflammatory reaction. *Biology of Reproduction*; **50**: 233–238.

Hellberg P, Larson L, Olofsson J, Brannstrom M & Hedin L 1996 Regulation of the inducible form of prostaglandin endoperoxide synthease in the perfused rat ovary. *Molecular Human Reprodution*; **2**: 111–116.

Hillier SG 1996 Roles of FSH and LH in controlled ovarian hyperstimulation. *Human Reproduction*; **11** (supplement 2): 101–109.

Hillier SG & Tetsuka M, 1998 An anti-inflammatory role for glucocorticoids in the ovaries? *Journal of Reproductive Immunology*; **39**: 21–27.

Hillier SG, Tetsuka, M & Fraser HM 1997 Location and Developmental Regulation of Androgen Receptor in Primate Ovary. *Human Reproduction*; **12**: 107–111.

Morris JK & Richards JS 1995 Luteinization hormone induces prostaglandin endoperoxide synthase-2 and luteinization in vitro by A-kinase and C-kinase pathways. *Endocrinology*; **136**: 1549–1558.

Terranova PF & Rice VM 1997 Review: cytokine involvement in ovarian processes. *American Journal of Reproductive Immunology*; **37**: 50–63.

Tetsuka M, Haines LC, Milne M, Simpson GE & Hillier SG. 1999a Regulation of 11beta-hydroxysteroid dehydrogenase type I gene expression by LJ and interleukin-1beta in cultured rat granulosa cells. *Journal of Endocrinology*; **163**: 417–423.

Tetsuka M, Milne M, Simpson GE & Hillier SG. 1999b Expression of 11beta-hydroxysteroid dehydrogenase, glucocorticoid receptor and mineralocorticoid receptor genes in rat ovary. *Biology of Reproduction*; **60**: 330–335.

Tetsuka M, Thomas FJ, Thomas MJ, Anderson RA, Mason JI & Hillier SG, 1997 Differential expression of messenger ribonucleic acids encoding 11beta-hydroxysteroid dehydrogenase types 1 and 2 in human granulosa cells. *Journal of Clinical Endocrinology and Metabolism*; **82**: 2006–2009.

Yong, PY, thong KJ, Andrew R, Walker BR, Hillier SG. 2000 Development-related increase in cortisol biosynthesis by human granulosa cells. *Journal of Clinical Endocrinology and Metabolism*; **85**: 4728–4733.

Follicular Growth, Ovulation and Fertilization: Molecular and Clinical Basis
Anand Kumar and Amal K. Mukhopadhyay (Eds.)
Narosa Publishing House, New Delhi, India, 2001

3

Neuroendocrinology of Aging in the Reproductive System: Gonadotropin Secretion as an Example

Winfried G. Rossmanith
Department of Obstetrics & Gynecology, Diakonissenkrankenhaus, Karlsruhe, Germany

Introduction

Over the past two decades our understanding of the neuroendocrine processes of aging has been considerably enlarged. It is now generally acknowledged that endocrine function shows age-related changes from the very beginning of development. Thus, neuroendocrine alterations observed during transition states such as the menopause may only coincidentally be related to this period. Endocrine alterations at this time may represent the sum of functional aberrations which were initiated much earlier in life (Meites *et al.*, 1982). Evidence has accumulated to suggest that neuroendocrine aging may relate to changes in the interrelationship between hormonal and neural signals rather than to discrete neuroendocrine events (Timiras, 1983).

The cessation of menstrual cyclicity during the menopause represents a biological hallmark in the chronobiology of female life, since it indicates the end of female reproductive capacity. During this transition period, the progressive decline in sex steroid secretion from the ovaries precipitates profound alterations in the neuroendocrine regulation of the reproductive system. The state of hypergonadotropic hypogonadism during the postmenopausal years represents a prolonged period of endocrine stability during which the sex steroid milieu remains virtually unaltered. Hence, this time provides a unique chance to delineate age-related alterations in the neuroendocrine control of reproductive hormone release. Determination of the changing gonadotropin dynamics during aging in postmenopausal women may thus prove to be useful in facilitating a more profound understanding of the principal mechanisms involved in the aging of endocrine systems.

To address some of the mechanisms presumed to be involved in the process of neuroendocrine aging, we have aimed at determining serum gonadotropin pulsatility in postmenopausal women of different ages. This approach of evaluating age-related processes in humans appears to be justified, since it is founded on the demonstration of tight functional and temporal links between hypothalamic signals and pituitary gonadotropin release (Clarke and Cummins, 1982). Therefore, changes in the serum gonadotropin profiles may reflect the sum of altered neuroendocrine function occurring with progressive age in postmenopausal women. In addition, determination of gonadotropin-releasing hormone (GnRH)-stimulated gonadotropin secretion may discriminate between age-related hypothalamic and pituitary affections. Secretory dynamics determined in this way probably reflect collective age-related alterations in the neuroendocrine mechanisms governing gonadotropin release in postmenopausal women.

Characterization of the LH Pulsatility of Postmenopausal Women

The secretion of gonadotropins is invariably increased in postmenopausal women (Hammond and Ory, 1985; Rossmanith *et al.*, 1990; Rossmanith *et al.*, 1991 a,b). Since significant negative feedback by ovarian sex steroids on the hypothalamo-pituitary system is lacking after the onset of the menopause (Yen, 1999), gonadotropin release may enhance as consequence of an open feedback loop. In the early perimenopausal years, the concentrations of follicle-stimulating hormone (FSH) rise initially, which is then followed by a gradual increase in luteinizing hormone (LH) levels (Judd *et al.*, 1978; Hammond and Ory, 1985). The secretion of gonadotropins has been recognized to be episodic ("pulsatile") in nature (Knobil, 1980; Rossmanith *et al.*, 1990). In fact, these LH pulse characteristics become very prominent in women in the postmenopausal period compared to normally cycling women. Since gonadotropin secretion patterns in postmenopausal women are unaffected by ovarian sex steroid influences (Rossmanith *et al.*, 1991a,b), the pulse attributes of this basic rhythm may reflect the maximal activity of the central pulse generator (Rossmanith *et al.*, 1990; Rossmanith *et al.*, 1994).

In the LH secretory profiles of postmenopausal women, the pulses are set at a frequency similar to those of normally cycling women, albeit with much higher pulse amplitudes (Fig. 1). That this LH pulsatility indeed represents the maximal release rate of the central GnRH-LH pacemaker is demonstrated by the finding that increases and decreases in LH pulse attributes in women during the menstrual cycle are limited by the pulse characteristics of postmenopausal subjects. In particular, the LH pulse frequency represents a threshold which is never exceeded by the changing LH pulse attributes during the menstrual cycle (Rossmanith *et al.*, 1990). Since the LH pulse frequency determined in the gonadotropin profiles of postmenopausal women closely compares with that found for the spontaneous pulsatile GnRH release from human hypothalami *in vitro* (Rasmussen *et al.*, 1989), it is suggested these LH release frequencies indeed reflect an unrestrained pulse rhythm. Even during periods with increasing serum concentrations of sex steroids, as during the menstrual cycle including the midcycle LH surge, the LH pulse attributes approximate those of postmenopausal women. In the presence of residual

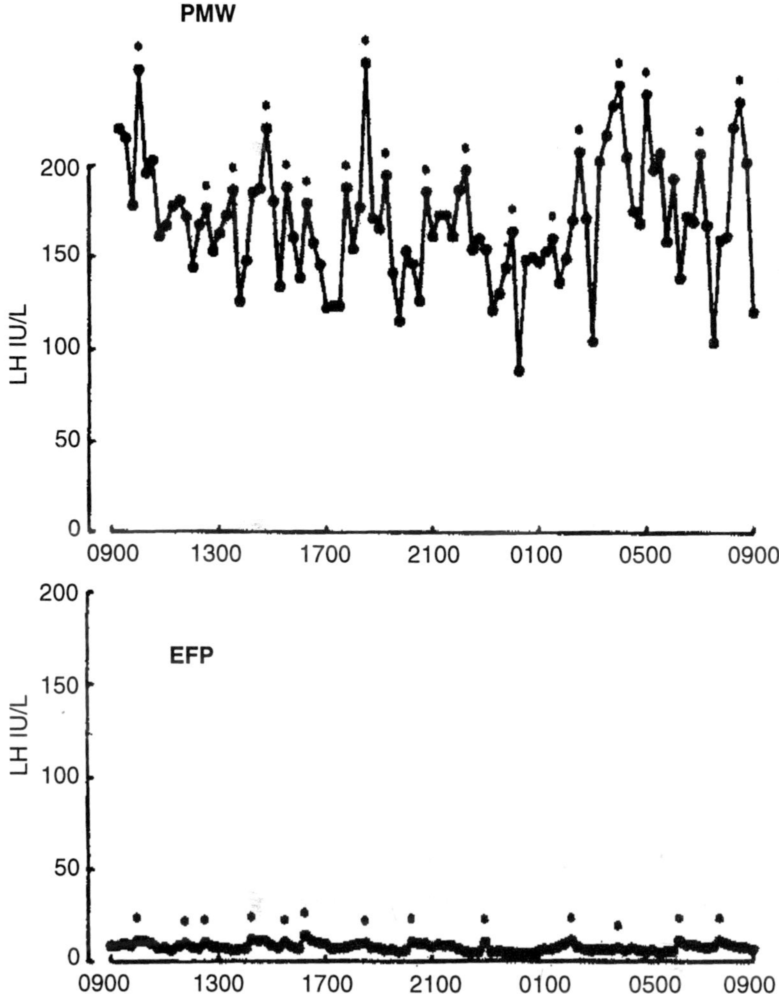

Fig. 1 24-hour LH secretory profiles in a woman during the postmenopause (PMW, top) and in a woman during the early follicular phase of her cycle (EFP, bottom). Asterisks indicate significant pulses.

serum androgen concentrations, this basic pulse rhythm remains unrestrained, since androgen receptor blockade fails to noticeably accelerate the LH pulsatility of post-menopausal women (Rossmanith *et al.*, 1991a). Similarly, changing LH pulsatility during ovarian sex steroid replacement in postmenopausal women (Fig. 2) is also confined to the pulse attributes of the unrestrained LH rhythm during unreplaced conditions (Rossmanith *et al.*, 1994). Collectively, the changing LH pulsatile attributes during high sex steroid exposure, as during the menstrual cycle or during hormone replacement, are limited by the characteristics of the unrestrained pulsatility observed during the unreplaced state in postmenopausal women (Rossmanith, 1994).

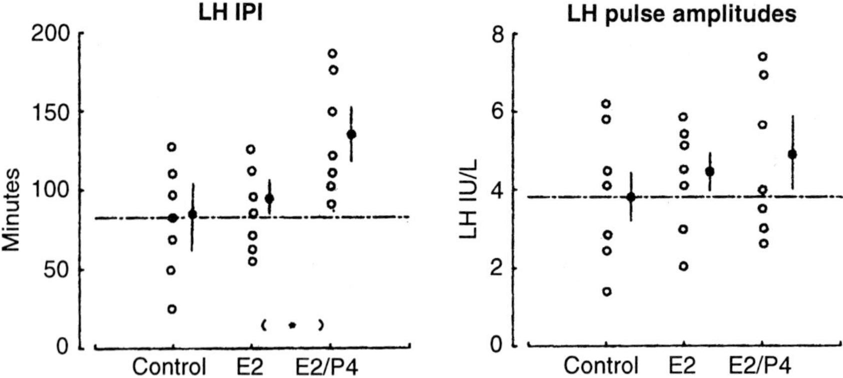

Fig. 2 Mean (±SEM) LH interpulse intervals and pulse amplitudes of seven postmenopausal women before replacement (= control) and during E2 or E2/P4 replacement regimens. The horizontal line iddicates the mean values during control conditions. $^*p < 0.05$ vs values linked with brackets.

Episodic Gonadotropin Secretion during Aging in Postmenopausal Women

In the years during and after the menopause, gonadotropin levels are considerably increased (Judd et al., 1978; Rossmanith et al., 1990; Yen, 1999). While aging progresses in the postmenopause, gonadotropin secretion gradually declines, such that gonadotropin levels during senescence almost approximate those of the premenopausal period (Rossmanith *et al.*, 1991b). Fewer LH and FSH secretory episodes with lower pulse amplitudes are observed in aged women, compared with postmenopausal women during their first decade after natural onset of the menopause (Fig. 3). Since the metabolic clearance rates of gonadotropins remain virtually unaffected by age (Kohler *et al.,* 1968), this attenuated gonadotropin pulsatility in old postmenopausal women may indeed relate to decreased gonadotropin release from the pituitary. As gonadotropin pulse frequency primarily reflects intermittent pituitary activation by episodic GnRH release (Knobil, 1980; Clarke and Cummins, 1982), the finding of a lower gonadotropin pulse frequency in aged postmenopausal women may be interpreted as result of a functional decline in the hypothalamic GnRH pulse generator, although a pituitary site of action cannot be entirely excluded (Rossmanith *et al.*, 1991b; Rossmanith 1992a). This notion is further corroborated by the finding of a progressive loss of GnRH neurons at various hypothalamic sites (Witkins, 1989). Moreover, the GnRH content of the hypothalamus has been reported to decrease with advanced age in humans (Parker and Porter, 1984). Thus, our findings suggest a hypothalamic site as the principal cause of the age-related attenuation in gonadotropin secretion. In fact, prolonged estrogen exposure has been demonstrated to decrease the number and functional integrity of hypothalamic GnRH neurons (Brawer *et al.*, 1980), suggesting a pivotal role for ovarian sex steroids in the hypothalamic processes of aging.

The finding of decreased gonadotropin release in postmenopausal women

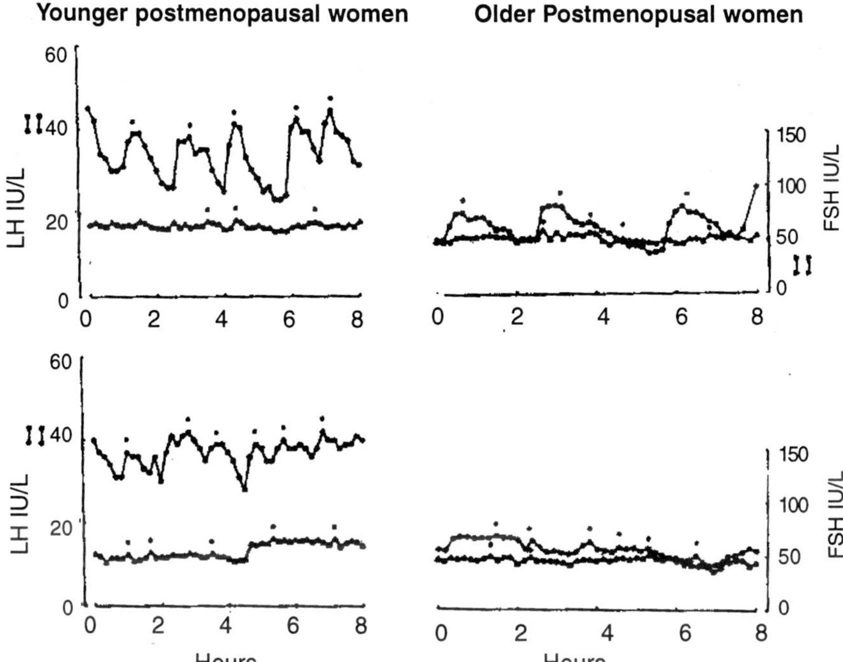

Fig. 3 Representative LH (circles) and FSH (triangles) secretory profiles of two younger (left) and two older postmenopausal women. Asterisks indicate significant pulses (from Rossmanith *et al.*, 1991).

is in keeping with results of some (Judd *et al.*, 1978; Rossmanith *et al.*, 1991b), but not all, previous investigations (Hammond and Ory, 1985; Scaglia *et al.*, 1978). The differences may at least partially be reconciled by the heterogeneity of the studied populations. It is well established that concurrent endocrinological and general illness interferes with gonadotropin secretion in postmenopausal women (Morley *et al.*, 1992). Therefore, results obtained in such affected women need to be carefully excluded from analysis, in order not to confound the interpretation. Moreover, the validity of single hormone determinations performed in early studies may be questioned, in view of the marked differences in nadir and peak hormone concentrations (Rossmanith *et al.*, 1991b).

Is Pituitary Gonadotropin Responsiveness Preserved During Aging in Postmenopausal Women?

The GnRH-stimulated gonadotropin response has been reported to be clearly exaggerated in postmenopausal subjects compared to premenopausal women (Scaglia *et al.*, 1978). We have conducted an investigation in a large cohort of pre- and postmenopausal women to determine the effects, if any, of advancing age on pituitary gonadotropin responsiveness. GnRH was found to stimulate the LH and FSH secretion during different stages of life in women. However, when the percentage gonadotropin increments were calculated to take account of variability in basal gonadotropin concentrations, no differences were found between premenopausal and postmenopausal women even at old age. Thus, the pituitary gonadotroph responsiveness is preserved while aging progresses in women. In particular, it does not vary between postmenopausal women of different ages (*Rossmanith, 1995*). Despite a prolonged period of hypersecretion of gonadotropins in response to low estrogen concentrations in the postmenopause, the pituitary release capacity is apparently not impaired in old women. While earlier studies have claimed a decreased gonadotropin release in response to GnRH stimulation in elderly subjects (Judd *et al.*, 1978), the findings of the current and other investigations (Hanker *et al.*, 1981; Scaglia *et al.*, 1978) are in contrast to this notion. Therefore, pituitary gonadotropin responsiveness is likely to be preserved in old age; this view supports a hypothalamic functional decline as the principal mechanism for the age-related attenuation of gonadotropin secretion in women.

However, it should be remembered that quantitative changes in pituitary hormones may also occur during aging (Wide and Hobson, 1983). The physico-chemical characters of the gonadotropin molecular moieties show an age-related polymorphism. Large LH forms display slower metabolic rates and different biological profiles than the minor ones (Strollo *et al.*, 1981). As deduced from our findings obtained by radioimmunoassay measurements, the gonadotropins may be released in response to GnRH stimulation at fairly constant rates throughout pre-and postmenopausal life. In spite of this, the biological potency of these GnRH-stimulated gonadotropins may vary with advancing of age as a result of different distribution patterns in the gonadotropin moieties, and therefore mask the true characteristics of gonadotropin release.

Can the Feedback Actions of Ovarian Sex Steroids Still be Activated in Aged Postmenopausal Women?

It is well established that ovarian sex steroid replacement decreases both LH and FSH levels in postmenopausal women. Since FSH is more sensitive to the negative feedback effects of estrogen than LH (Yen, 1999), factors other than changes in hypothalamic GnRH release have been assumed to account for the differential suppression of these gonadotropins. Estradiol replacement to postmenopausal women does not completely return the FSH serum levels to normal, indicating that additional regulatory components, such as ovarian inhibin, may play a major role for feedback regulation (Morley et al., 1992). Early studies suggested that postmenopausal women may be more responsive to estrogen negative feedback than premenopausal subjects (Yen, 1999). However, this assumption has been questioned, based on the marked differences in the experimental methodologies (Hammond and Ory, 1985). Whether the negative feedback sensitivity of gonadotropins to sex steroid exposure may be relevant to postmenopausal women of advanced age still remains unresolved.

Accordingly, we have used the anti-estrogen clomiphene citrate to probe the gonadotropin sensitivity to negative feedback in postmenopausal women of different ages (Rossmanith *et al.*, 1994). During the low estrogen milieu of

release of postmenopausal women. Whether derangements in the neurotransmitter activity continue to progress and then account for the profound attenuation of gonadotropin secretion in old postmenopausal women still remains to be elucidated.

Conclusions

Evidence provided by the current and other investigations has accumulated to suggest that biological aging during the postmenopause markedly affects the neuroendocrine system governing gonadotropin release. The studies presented and their interpretation may aid in expanding our knowledge of the neuroendocrine processes involved in aging. The determination of age-related dynamics in gonadotropin secretion of postmenopausal women has proven to be a valid approach in delineating changes as a function of progressive age. As result, major functional derangements, primarily at a hypothalamic rather than a pituitary site, have been delineated as concomitants of aging in women. Moreover, aging may also impair the negative feedback sensitivity to ovarian sex steroids, and derangements in central nurotransmitter parallel this decline. By appraising the presented results in relation to neuroendocrine events in the postmenopausal period, we would speculate that the process of aging relates rather to a hypothalamic decline than to a pituitary hypofunction.

While we have made significant advances, our current understanding of the neuroendocrine processes during aging in humans still remains inconclusive. More detailed insights into the complex neuroendocrine alterations during aging may provide better methodologies for the appropriate prevention or attenuation of deleterious effects by specific interventions (Timiras, 1983).

Acknowledgements

The author is thankful to Professors C. Lauritzen, W.A. Scherbaum and S.S.C. Yen for their support in completion of the studies, to Dr. R. Benz, Dr. A. Handke-Vesely, M. Beuter, S. Schurer and C. Reichelt for their technical assistance, and to Prof. A.B. Grossman for his editorial comments.

References

Brawer, J.R., Schipper, H. F., and Naftolin, F. (1980) Ovary-dependent degeneration in the hypothalamic arcuate nucleus, *Endocrinology;* **107**: 274–279.

Clarke, I.J., and Cummins J.T. (1982) The temporal relationship between gonadotropin-releasing hormone (GnRH) and luteinizing hormone (LH) secretion in ovariectomized ewes. *Endocrinology*; **111**: 1737–1739.

Hanker, J.P., U. Ende and Bohnet, H.G. (1981) Intermittent stimulation with LH-RH in postmenopausal hypergonadotropinism. *Horm. Metab. Res.* **13**: 696–699.

Hammond, C.B., and Ory, S.J. (1985) Endocrine aspects of the menopause. In: Sherman, R.P. (ed) *Clinical Reprodutive Endocrinology*, Livingstone, New York, pp. 185–194.

Judd, J.W.W., Collins, W.P., and Forecast, W.D. (1978). Plasma hormone profiles after the menopause and bilateral oophorectomy. *Postgrad. Med., J.* **54**: 25–30.

Knobil, E. (1980). The neuroendocrine control of the menstrual cycle. *Rec. Prog. Horm. Res.* **36**: 53–88.

Kohler, P.O., Ross, G.T., and O'Dell, W.T. (1968) Metabolic clearance and production rates of human luteinizing hormone in pre- and postmenopausal women. *J. Clin. Invest.* **47**: 38–47.

Lachelin, G.C.L., Leblanc, H., and Yen, S.S.C. (1977). The inhibitory effects of dopamine agonists on LH release in women. *J. Clin. Endocrinol. Metab.* **44**: 728–732.

Meites, J., Huang, H.H., and Simpkins, J.W. (1982) Central nervous system neurotransmitters during the decline of reproductive activity. In: Fioretti P., Martini, L. and Melis, G.B. (eds) *The Menopause: Clinical, Endocrinological and Pathophysiological Aspects*, Academic Press, London, pp. 3–14.

Meites, J. (1988), Alterations in hypothalamic-pituitary function with age. In: Armbrecht, H.J., Coe, R.M., and Wongsurawat, N. (eds) *Endocrine Function and Aging*, Springer, New York, pp. 1–12.

Melis, G.B., Cagnacci, A., Gambacciani, M., Paoletti, A.M., Caffi, T. and Fioretti, P. (1988) Chronic bromocryptine administration restores luteinizing hormone response to naloxone in postmenopausal women. *Neuroendocrinology*; **47**: 159–163.

Melis, G.B., Paoletti, A.M., Gambacciani, M., Mais. V. and Fioretti, P. (1984). Evidence that estrogens inhibit LH secretion through opioids in postmenopausal women using naloxone. *Neuroendocrinology*; **39**: 60–63.

Morley, J.E., Korenman, S.G., and Kaiser, F.E. (1992). The menopause. In: Morley, J.E., and Korenman, S.G. (eds) *Endocrinology and Metabolism in the Elderly*, Blackwell Scientific Publications, Oxford, pp. 322–335.

Parker, C.R., and Porter, J.C. (1984). Luteinizing hormone-releasing hormone and thyrotropin-releasing hormone in the hypothalamus of women: effects of age and reproductive status. *J. Clin. Endocrinol. Metab.* **58**: 488–491.

Rasmussen, D.D., Gambacciani, M., Swartz, W.H., Tueros, V.S. and Yen, S.S.C. (1989). Pulsatile GnRH release from the human mediobasal hypothalamus *in vitro*: opiate receptor mediated suppression. *Neuroendocrinology*; **49**: 150–156.

Reid, R.L., Quigley, M.E., and Yen S.S.C. (1983). The disappearance of opioidergic regulation of gonadotropin secretion in postmenopausal women. *J. Clin. Endocrinol. Metab.* **57**: 1107–1110.

Reymond, M.J., Donda, A., and Lemarchand-Beraud, T. (1989). Neuroendocrine aspects of aging: experimental data. *Horm. Res.* **31**: 32–38.

Rossmanith, W.G., Wirth, U., and Yen S.S.C. (1989). Does prolonged dopaminergic blockade alter the LH pulsatile secretion in normal cycling and postmenopausal women? *Acta Endocrinol.* **121**: 147–152.

Rossmanith, W.G., Liu, C.H., Laughlin G.A., Mortola, J.F., Suh, B.Y., and Yen S.S.C. (1990). Relative changes in LH pulsatility during the menstrual cycle: using data from hypogonadal women as reference point. *Clin. Endocrinol.* **32**: 647–660.

Rossmanith, W.G., Beuter, M., Benz, R. and Lauritzen, C. (1991a). How do androgens affect the episodic gonadotrophin secretion in postmenopausal women? *Maturitas*; **13**: 325–335.

Rossmanith, W.G., Scherbaum, W.A., and Lauritzen C. (1991b) Gonadotropin secretion during aging in postmenopausal women. *Neuroendocrinology*; **54**: 211–218.

Rossmanith, W.G.: (1992a). Circulating luteinizing hormone (LH) patterns during female reproductive life. In: Genazzani, A.R., and Petraglia, F. (eds), Progress in Gynecological Endocrinology, Parthenon Publishing, Carnforth, pp. 225–240.

Rossmanith, W.G. (1994) Endogenous opioid regulation of luteinizing hormone (LH) secretion in women. In: Negri, M., Lotti, G., and Grossman, A. (eds) *Clinical Perspectives of Opioid Peptide Production*, Wiley and Sons, London, pp. 134–159.

Rossmanith, W.G., Reichelt, C., and Scherbaum, W.A. (1994). Neuroendocrinology of aging in humans: Attenuated sensitivity to sex steroid feedback in elderly postmenopausal women. *Neuroendocrinology*, **59**: 355–362.

Rossmanith, W.G. (1995) Gonadotropin secretion during aging in women. *Experimental Gerontology*, **30**: 369–381.

Scaglia, H., Medina, M., Pinto-Ferreira, A.L., Vasques, G., Gual, C. and Perez-Palacios, G. (1978). Pituitary LH and FSH secretion and responsiveness in women of old age. *Acta Endocrinol.* **81**: 673–679.

Simpkins, J.W., and Millard, W.J. (1987), Influence of age on neurotransmitter function. In: Sacktor, B. (ed) *Endocrinology of Aging*. Endocrinology and Metabolism Clinics of North America, Saunders, Philadelphia, pp. 893–917.

Strollo, F., Harlin, J., and Hernandez-Montes, H. (1981), Qualitative and quantitative differences in the isoelectro-focusing profile of biologically active lutropin in the blood of normally menstruating and post-menopausal women. *Acta. Endocrinol.* **97**: 166–175.

Timiras, P.S. (1983) Neuroendocrinology of aging. In: Meites, J. (ed) Neuroendocrinology of Aging, Plenum Press, New York, pp. 5–29.

Wide, L., and Hobson, B.M. (1983). Qualitative difference in follicle-stimulating hormone activity in the pituitary of young women compared to that of men and elderly women. *J. Clin. Endocrinol. Metab.* **56**, 371–375.

Witkins, J.W. (1989), Morphology of luteinizing hormone-releasing hormone neurons as a function of age and hormonal condition in the male rat. *Neuroendocrinology*; **49**: 344–348.

Yen, S.S.C., (1999) Neuroendocrinology of reproduction. In: Yen, S.S.C., Jaffe, R.B. and Barbieri, R.L., (eds) *Reproductive Endocrinology*, Saunders, Philadelphia, pp. 30–81.

Follicular Growth, Ovulation and Fertilization: Molecular and Clinical Basis
Anand Kumar and Amal K. Mukhopadhyay (Eds.)
Narosa Publishing House, New Delhi, India, 2001

4

Neuroendocrine Regulation of Follicle-Stimulating Hormone[1]

Vasantha Padmanabhan
Department of Pediatrics and the Reproductive Sciences Program
University of Michigan, Ann Arbor, Michigan 48109, USA

Introduction

Follicle-stimulating hormone (FSH) is a key reproductive hormone involved in the regulation of follicular development. Dissociations in luteinizing hormone (LH) and FSH release accompany many anovulatory disorders and menopausal transition. In spite of its pivotal role, our understanding of the factors involved in the control of FSH is incomplete. The long circulating half-life of FSH, the unavailability of assays to detect the variant isoforms of FSH and cross reactivity with gonadotropin α subunit and LH have all made it rather difficult to assess secretory patterns of FSH from peripheral measurements. Nonetheless several advances have been made in understanding the regulation of FSH. In general FSH appears to be secreted in two modes, a basal or constitutive mode and an episodic mode.

Basal Mode of FSH Release

Support for the contention that substantial portion of circulating FSH originates from constitutive secretion comes from the following findings. FSH is continued to be released for prolonged times from (1) hypothalamic disconnected sheep (Clarke *et al.*, 1983, Padmanabhan *et al.*, 1997a); (2) hypophysectomized rats (Depaolo, 1991) and sheep (Padmanabhan *et al.*, 1997a) bearing pituitary transplants under the kidney capsule, and (3) long-term pituitary cultures (Sheridan *et al.*, 1979). This component of FSH release appears to be dictated by the availability of translatable FSHβ mRNA (Farnworth, 1995).

Episodic Mode of FSH Release

In addition to the basal mode, an episodic mode of FSH release has also been

[1]Supported by USPHS grants HD23812, HD34731, HD35620 and NSF IBN 9725943.

predicted from peripheral measurements. Because pulses of FSH measured at the periphery are not as discrete as that of LH, investigators had to rely heavily on statistical approaches to define them. Monitoring FSH patterns at the secretory site appears to be the only resource by which one can establish the true nature of FSH secretion. Hypophyseal portal collection technique pioneered by Clarke and Cummins (1982) for monitoring hypophysiotropic hormones provides us a tool to define FSH secretory dynamics. Because hypophyseal portal vessels are lesioned at the surface of the pituitary, hypophyseal portal blood provides a resource for measuring not only hypothalamic releasing and release-inhibiting peptides but also of secretory products originating from the pituitary.

Comparison of patterns of LH in the hypophyseal portal and peripheral circulation (Midgley *et al.*, 1997) have revealed that: (1) each gonadotropin-releasing hormone (GnRH) pulse is associated with a LH pulse both at the peripheral and hypophyseal portal level, (2) LH concentrations are several-fold higher in hypophyseal portal than peripheral circulation, (3) patterns of LH are more discrete in hypophyseal portal than in the periphery (Fig. 1, top right panels). Utilizing this approach it was recently found that FSH is indeed

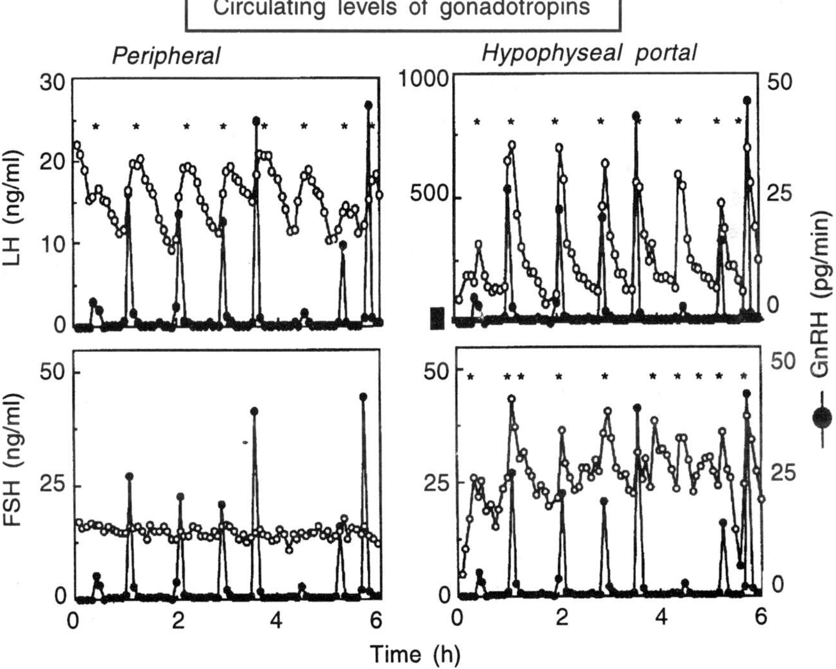

Fig. 1 Patterns of GnRH, hypophyseal portal and jugular LH and FSH from an ovariectomized sheep. Asterisks indicate pulses identified by a pulse detection algorithm. Note: (1) the discreteness and magnitude of LH in hypophyseal portal blood, (2) the one to one relationship of both hypophyseal portal and peripheral LH with GnRH, (3) dynamic pattern of FSH in hypophyseal portal and not peripheral secrtion and (4) the existence of GnRH-associated and non-GnRH-associated pulses of FSH. Differences between peripheral and hypophyseal portal levels of LH and FSH may relate to influences of plasma half lives on recirculating levels.

secreted in a dual mode a basal or constitutive mode and an episodic mode (Padmanabhan *et al.*, 1997b). This affirms what has been surmised all along from peripheral measurements. Furthermore, these studies have unequivocally established that each pulse of GnRH, in addition to its association with LH, is associated with a pulse of FSH and that additional non-GnRH-associated pulses of FSH exist (Fig. 1, bottom panels). How are these non-GnRH-associated pulses of FSH triggered?

Neuroendocrine vs Paracrine Regulation of FSH Production/Secretion

Believers of the one releasing hormone (GnRH) hypothesis have explained the paradoxical differences in release patterns of LH and FSH as due to: (1) an alteration in GnRH pulse frequency with slower frequencies favoring FSH release (Wildt *et al.*, 1981), (2) differential sensitivity of subpopulation of *gonadotrophs* that secrete LH and FSH (Denef *et al.*, 1978) and consequent threshold differences in response to GnRH (amplitude modulation) with low levels favoring FSH secretion (Wise *et al.*, 1979) and (3) different response times of LH and FSH to GnRH. If GnRH is the only regulator of episodic mode of FSH release, one should be able to obliterate FSH pulsatility by blocking GnRH action. This however is not the case. Episodic FSH secretion has been shown to persist even after elimination of endogenous GnRH activity/action (Culler and Negro-Vilar, 1987, Pau *et al.*, 1991, Padmanabhan *et al.*, 1997a) thus supporting the existence of additional FSH-regulatory mechanisms.

Well-documented differences in metabolic clearance rates of LH and FSH (reviewed in Ulloa-Aguirre *et al.*, 1995), differential effects of gonadal steroids and peptides (McNeilly 1988) offer explanations for changes in threshold levels of circulating LH and FSH but do not explain the asynchronous presence and initiation of FSH pulses, especially in gonadectomized animals (Culler and Negro-Vilar, 1987; Pau *et al.*, 1991, Padmanabhan *et al.*, 1997b). Two other explanations could potentially explain the existence of the non-GnRH associated pulses of FSH (1) temporal changes in the local pituitary milieu of inhibins, activins and follistatins and/or (2) the existence of a separate neural trigger.

Inhibins, activins and follistatins, are all produced at the pituitary level (DePaolo *et al.*, 1991, Mather *et al.*, 1992). In defining the local feedback loop, one needs to recognize the functional overlap that exists among them. Inhibins and activins have functionally opposite roles (Ying, 1988); inhibins suppress FSH and activins stimulate FSH production (Ying, 1988). Inhibin can bind activin receptors (Mathews and Vale, 1991) and antagonize activin's action. Follistatin is a binding protein and neutralizer of activin's action (Robertson, 1992). Relative abundance of these three regulators is likely, therefore, to set the local regulatory tone for FSH (Fig. 2). While increases in activin/activin receptors or decreases in inhibin/follistatin would increase FSH production, the converse would inhibit FSH production/secretion.

To account for the episodic mode of FSH secretion, the regulatory tone

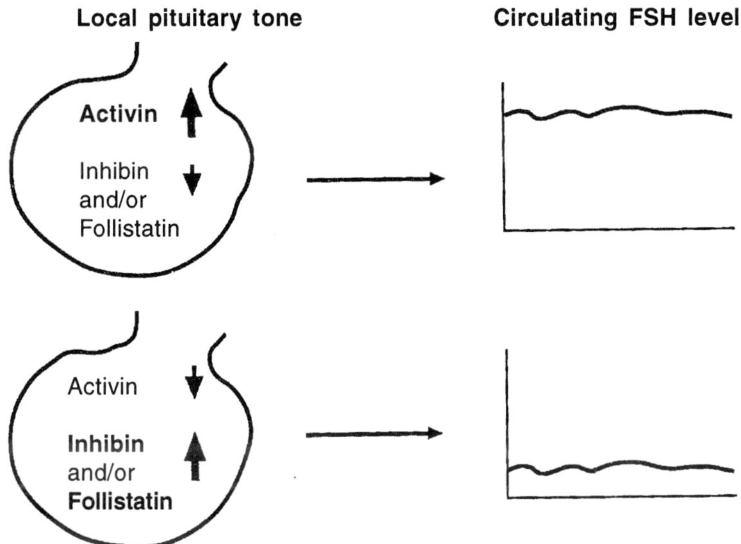

Fig. 2 Schematic representation of the local pituitary regulatory loop. Model predicts that increases in pituitary activin or decreases in inhibin and follistatin would increase and decreases in activin or increases in follistatin/inhibin would decrease FSH secretion.

must undergo frequent changes. Temporal changes in individual components of the FSH regulatory loop in concert with estradiol may result in "pulse" like changes in regulatory tone that lead to an episodic mode of FSH secretion. An inherent pituitary rhythm or an external trigger would be however required for enforcing changes in local pituitary tone. Absence of a pulsatile pattern of FSH release from pituitary tissue *in vitro* (Turgeon and Waring, 1982) and hypothalamic pituitary-disconnected sheep (Padmanabhan *et al.*, 1997a) do not support the existence of an inherent pituitary-driven mechanism.

More recent studies have shown that the local regulatory loop is under neuroendocrine control. A reciprocal relationship appears to exist in the regulation of FSHβ and follistatin mRNA by GnRH. High frequency GnRH pulses increase follistatin mRNA and protein expression (Besecke *et al.*, 1996) but decrease FSH (Wildt *et al.*, 1981). Low frequency GnRH stimulates only FSHβ and not follistatin mRNA expression. More recent studies have shown that GnRH may mediate its effect by changing local activin availability (Besecke et al., 1996). In concert, these findings suggest that the changes in local pituitary milieu may be a means by which changes in GnRH could ultimately lead to differential secretion patterns of LH and FSH. What other mechanisms could explain the non-GnRH associated pulses of FSH?

Evidence Supporting the Existence of a Separate FSH-Releasing Factor

Evidence that a specific site of control for FSH release may exist stems from studies which showed selective release of FSH following ablation (Lumpkin and McCann, 1984), deafferentation (Lamperti and Hill, 1987) of the dorsal anterior hypothalamic area or electrochemical stimulation of hypothalamic regions apart from those which regulate LH secretion (Chappel and Barraclough, 1976). Subsequently, Lumpkin *et al.* (1989) demonstrated that radiofrequency lesions of the dorsal anterior hypothalamic area abolished FSH and not LH pulsatility thus providing the strongest evidence in support of the existence of specific site of control for FSH. Selective suppression of FSH pulsatility also occurs following lesion of the caudal and mid-median eminence (Marubayashi *et al.*, 1999; Mizunuma *et al.*, 1983). McCann's group succeeded in achieving partial separation of FSH and LH-releasing activities from extracts of porcine and ovine hypothalami (Lumpkin *et al.*, 1987). Despite the supportive neuroanatomical and physiological evidence, a physiologically relevant FSH-RF is yet to be isolated and characterized. Independent of this line of research, identification of variant forms of GnRH has opened up new possibilities. Recently, it was postulated that lamprey GnRH III might be the counterpart of mammalian FSH-RF (Yu *et al.*, 1997). More rigorous testing needs to be done before establishing that lamprey GnRH-III is indeed the long sought out FSH-RF.

FSH Heterogeneity and Its Functional Relevance

In addition to the regulation discussed thus far, there appears to be yet another level of control that involves the potency of FSH delivered. FSH consists of a family of related isoforms, which can be separated on the basis of size, charge, biologic to immunologic potency relationships and plasma clearance. Considering the various FSH-induced functions and the potential for functional differences in the target site actions of the various FSH isoforms, determination of the exact nature of circulating FSH isoform mix is an important consideration in measuring FSH in circulation. At least three criteria have to be met for establishing the biologic significance of FSH heterogeneity namely: (1) it should exist, (2) it should be regulated, and (3) the various FSH isoforms should vary in potency or have differing functions.

The first criteria namely existence of FSH heterogeneity is uncontested. A mix of FSH isoforms has been identified at the level of the pituitary, peripheral circulation and urine in several species (reviewed in Ulloa-Aguirre *et al.*, 1995). Considerable biologic evidence documenting neuroendocrine regulation of FSH heterogeneity (Fig. 3) also exists in support of criteria 2. For example, qualitative differences in pituitary FSH content that correlate with age, sex, and stage of the estrous cycle have been reported in several species (reviewed in Ulloa-Aguirre *et al.*, 1995). In general, increases in estradiol and GnRH in gonad-intact models are associated with increases in less acidic FSH isoforms and circulating bioactive FSH (reviewed in Beitins and Padmanabhan, 1991; Ulloa-Aguirre *et al.*, 1995). For example, administration of GnRH to prepubertal boys (Phillips and Wide, 1994), and women (Zambrano *et al.*, 1995) increases release of less acidic FSH isoforms. Conversely, treatment of women with GnRH antagonist results in release of highly basic and

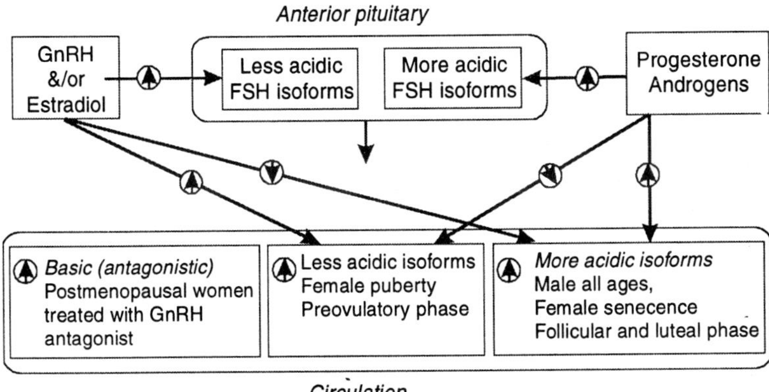

Fig. 3 Schematic representation of the neuroendocrine regulation of FSH and the nature of FSH isoforms during different physiologic states.

antagonistic forms of FSH (Dahl *et al.*, 1988). In these studies, because estradiol can act at the hypothalamic level and alter GnRH secretion, it is difficult to separate the independent effects of estradiol and GnRH in altering FSH heterogeneity.

Studies utilizing *ovariectomized*, nutritionally growth-retarded (hypogonadotropic) sheep have shown that, in the absence of estradiol, pulsatile administration of GnRH does not alter the distribution profile of FSH or bioactive FSH secretion (Hassing *et al.*, 1993). In contrast, estradiol administration increases the release of less acidic FSH isoforms in these animals (Padmanabhan *et al.*, 1997c) even after blockade of GnRH action (Padmanabhan *et al.*, unpublished). More recent studies characterizing FSH distribution near the site of secretion have shown that estradiol selectively increases the *secretion* of less acidic FSH isoforms (Lee *et al.*, 1998). Furthermore, estradiol alters pituitary glycosyl-transferase activity levels, enzymes involved in sialylation/ sulphation of gonadotropins (Damian-Matsumura *et al.*, 1999; Dharmesh and Baenziger, 1993) thus having the potential to contribute to glycosylation differences. Together, these findings support a role for estradiol in mediating FSH heterogeneity. Progesterone, on the other hand, negates the effects of estradiol and increases the presence of more acidic FSH isoforms. For instance, during the luteal phase of the human menstrual cycle (Padmanabhan *et al.*, 1988) and the prepartum period in cattle (Crowe *et al.*, 1998), when levels of progesterone and estradiol are both high, the predominant circulating form of FSH is acidic. Similarly, high androgenic states have also been found to be associated with increased release of acidic FSH isoform (Beitins and Padmanabhan, 1991; Ulloa-Aguirre *et al.*, 1995).

To meet the third criteria one needs to establish that changes in FSH heterogeneity are biologically meaningful. In general, more acidic FSH isoforms predominate in males, during senescence in both males and females, and during the early follicular and luteal phases of the estrous/menstrual cycles (Beitins and Padmanabhan, 1991; Ulloa-Aguirre *et al.*, 1995) (Fig. 3). In contrast, increased release of less acidic FSH isoforms are evident at the onset

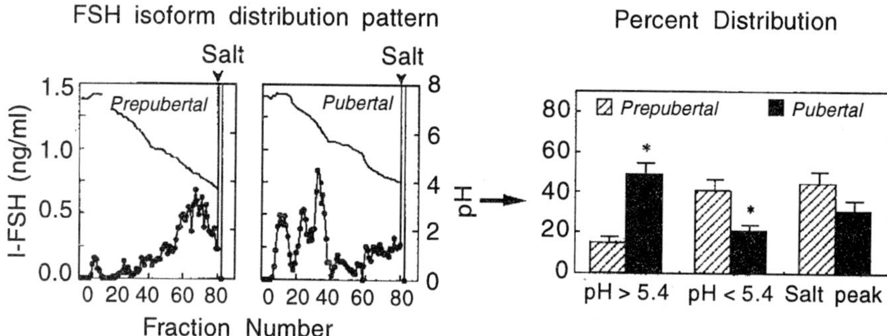

Fig. 4 Changes in FSH heterogeneity during experimental induction of puberty in female lambs. Distribution pattern of circulating FSH from a representative prepubertal and pubertal lamb are shown on the left. Percent distribution of circulating FSH isoforms eluting at various PH ranges are shown on the right.

of puberty and the preovulatory period. Experimental induction of puberty in female lambs increases release of circulating bioactive FSH and less acidic isoforms of FSH (Padmanabhan *et al.*, 1992) (Fig. 4). Similarly, increased release of less acidic circulating FSH isoforms are associated with the preovulaory period in humans (Padmanabhan *et al.*, 1988; Wide and Bakos, 1993; Zambrano *et al.*, 1995). Considering that pubertal onset and preovulatory period are associated with increased gonadal activity, the less acidic FSH isoforms prevailing at that time have the ability to provide a potent and acute signal to the ovary.

Are the observed changes in FSH heterogeneity during various physiologic states of sufficient magnitude to alter the net potency and/or function of the hormone? Less acidic isoforms of human recombinant FSH were found to be more potent than those in the acidic range in inducing mouse follicular development *in vitro* (Vitt *et al.*, 1998). Estimation of the biopotency of FSH isoforms on the basis of FSH mass has also predicted a 5–8-fold difference in radioreceptor and biopotency of human pituitary FSH (Burgon *et al.*, 1993). Considering that a subtle increase in FSH is sufficient to facilitate ovarian responses (Ben-Rafael *et al.*, 1995), the magnitude of changes such as those seen during midcycle (Padmanabhan et al., 1988) and puberty (Padmanabhan *et al.*, 1992) appear to be capable of yielding meaningful biological consequences.

Considering the various functions FSH mediate, the involvement of several second messenger systems, and the potential for receptor cross talk, the different FSH isoforms may also encode different functions. There is precedence for this with other glycoproteins. Asialo human chorionic gonadotropins have been shown to have higher thyrotropin-like activity in human thyroid follicles (Yamazaki *et al.*, 1995). Certain LH isoform(s) have also been shown to possess renotrophic activity (Nomura *et al.*, 1988). Studies with deglycosylated and native mixes of ovine FSH have shown isoform specific cAMP and estradiol responses (Beitins and Padmanabhan, 1991). Overall, while additional work remains to be done to establish the biologic relevance of FSH heterogeneity, data accumulated thus far support the premise that endocrine-induced changes in FSH heterogeneity are of biologic importance.

Methodological Considerations in Measuring FSH

The adequacy of assays and standards in current use have been reviewed in depth recently (Rose et al., 2000). Briefly, several classes of assays are available to a practicing endocrinologist. These include: (1) immunoassays that are designed to recognize an epitope in the FSH molecule, (2) radioreceptor assays that assess the binding ability of the hormone to the receptor, and (3) bioassays which measure an end response *in vitro* or *in vivo* (reviewed in Ulloa-Aguirre *et al.*, 1995). While immunoassays using different antisera can differentially recognize the various FSH isoforms, substances present in the test material can interfere or modulate the measurement of FSH in bioassays (Beitins and Padmanabhan, 1991; Chappel, 1995). Furthermore, FSH standards that are in use in the various assays do not have the full repertoire of FSH isoforms existing in biologic fluids and may hence add a source of error (Rose et al., 2000). Depending on the end point measured and the source of tissue utilized conclusions drawn from bioassays may vary. The newer FSH bioassays that utilize cells transfected with recombinant FSH receptors and cAMP as the end point ignore the involvement of other second messenger systems and may not distinguish between agonistic and antagonistic forms of FSH (Ulloa-Aguirre et al., 1995). Therefore when comparing FSH measures, it is important to take into consideration the nature of the assay and the standard being utilized.

Summary

Considering the importance of FSH in regulating folliculogenesis, multiple regulatory mechanisms appear to be in place to ensure its production and release (Fig. 5). In addition to GnRH, gonadal steroids and FSH regulatory proteins such as activins, inhibins and follistatins are shown to have major FSH regulatory roles. More recent studies suggest these regulators to be produced at the pituitary level thus having the ability to exert local effects on FSH secretion. Neuroanatomical, biochemical and physiological evidence provide support for the existence of separate FSH releasing factor(s), although none have been isolated so far. From a qualitative perspective, while many issues still remain to be addressed, it is clear that a mix of circulating gonadotropin isoforms reaches target tissues to influence a variety of biologic end points. FSH isoforms with longer half-lives may facilitate progression of maturational events by providing a long-acting stimulus. Conversely, FSH isoforms with shorter half-life that are secreted intermittently may provide an acute yet potent stimulus. Relative proportions of the various gonadotropin isoforms within the circulation should be taken into consideration in interpreting pathophysiology.

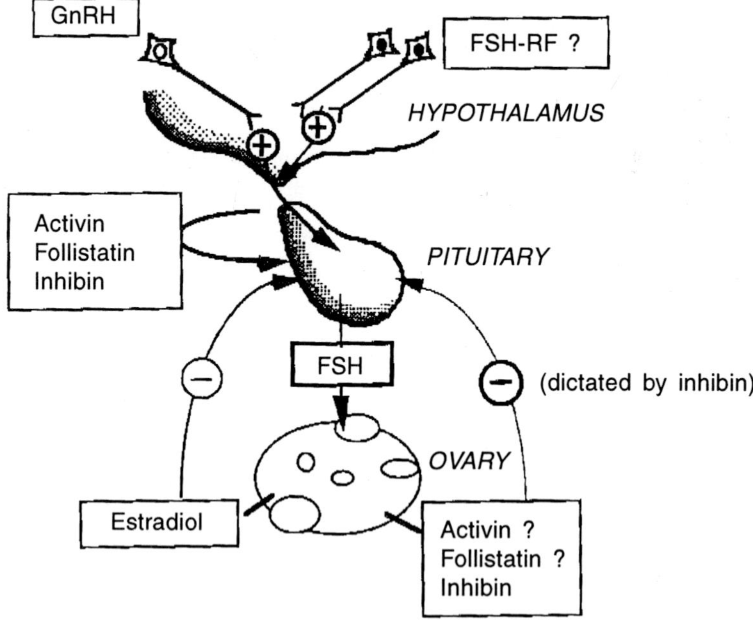

Fig. 5 Shown are the hypothalamic, ovarian and pituitary regulation of FSH prodution/ secretion. At the hypothalamic level, in addition to GnRH, there is evidence supporting the existence of a separate FSH-releasing factor. At the pituitary level, locally-produced activin, inhibin and follistatin appear to dictate the threshold of FSH release. At the ovarian level estradiol and inhibin are two major negative feedback regulators of FSH (activin and follistatin are not believed to have an endocrine role).

References

Beitins, IZ and Padmanabhan, V. (1991) Bioactivity of gonadotropins. *Endocrinology and Metabolism Clinics of North America*; **20**: 85–120.

Ben-Rafael Z, Levy T, Schoemaker J (1995) Pharmacokinetics of follicle-stimulating hormone: clinical significance. *Fertility and Sterility*; **63**: 689–700.

Besecke LM, Guendner MJ, Schneyer A. Bauer-Dantoin AC, Jameson JL and Weiss J (1996) Gonadotropin-releasing hormone regulates follicle-stimulating hormone-beta gene expression through an activin/follistatin autocrine or paracrine loop. *Endocrinology*; **137**: 3667–3673.

Burgon PG, Robertson DM, Stanton PG, and Hearn MTW (1993). Immunological activities of highly purified isoforms of human FSH correlate with *in vitro* bioactivities *Journal of Endocrinology* **139**: 511–518.

Chappel SC (1995). Heterogeneity of follicle stimulating hormone: control and physiological function *Human Reproduction update* **1**: 479–487.

Chappel SC and Barraclough CA (1976) Hypothalamic regulation of pituitary FSH secretion. *Endocrinology* **98**: 927–935.

Clarke IJ and Cummins JT (1982). The temporal relationship between gonadotrophin-releasing hormone (GnRH) and luteinizing hormone (LH) secretion in ovariectomized ewes *Endocrinology* **111**: 1737–1739.

Clarke IJ, Cummins JT and de Kretser DM (1983). Pituitary gland function after disconnection from hypothalamic influences in the sheep. *Neuroendocrinology* **36**: 376–384.

Crowe MA, Padmanabhan V, Mihm M, Beitins, IZ, and Roche JF (1998). Resumption of follicular waves in beef cows is not associated with peri-parturient changes in FSH heterogeneity despite major changes in steroid and gonadotropin concentrations. *Biology of Reproduction* **58**: 1445–1450.

Culler MD and Negro-Vilar A (1987) Pulsatile follicle-stimulating hormone secretion is independent of luteinizing hormone-relasing hormone (LHRH): pulsatile replacement of LHRH bioactivity in LHRH-immunoneutralized rats. *Endocrinology* **120**: 2011–2021.

Dahl KD, Bicsak TA, and Hsueh AJW (1988) Naturally occurring antihormones: secretion of FSH antagonists by women treated with a GnRH analog. *Science;* **239**: 72–74.

Damián-Matsumura P, Zaga V, Maldonado A, Sánchez-Hernández C, Timossi C and Ulloa-Aguirre A (1999). Oestrogens regulate pituitary alpha2,3-sialyltransferase messenger ribonucleic acid levels in the female rat. *Journal of Molecular Endocrinology* **23**:153–165.

Dharmesh SM and Baenziger JU (1993). Estrogen modulates expression of the glycosyltransferases that synthesize sulfated oligosaccharides on lutropin. *Proceedings of the National Academy of Sciences USA* **90**: 11127–11131.

Denef C, Hautekeete E and Dewals R (1978). Monolayer cultures of gonadotrophs separated by velocity sedimentation: heterogeneity in response to luteinizing hormone-releasing hormone. *Endocrinology* **103**: 736–747.

DePaolo LV (1991). Hypersecretion of follicle-stimulating hormone (FSH) after ovariectomy of hypophysectomized, pituitary-grafted rats: implications for local regulatory control of FSH. *Endocrinology* **128**: 1731–1740.

DePaolo LV, Bicsak TA, Erickson GF, Shimasaki S and Ling N (1991). Follistatin and activin: A potential intrinsic regulatory system within diverse tissues. *Proceedings of the Society for Experimental Biology and Medicine* **198**: 500–512.

Farnworth PG (1995). Gonadotropin secretion revisited. How many ways can FSH leave a gonadotroph *Journal of Endocrinology;* **145**: 387–395.

Hassing JM, Kletter GB, I'Anson H, Woods RI, Beitins IZ, Foster DL, and Padmanabhan V (1993). Pulsatile administration of gonadotropin-releasing hormone does not alter the follicle-stimulating hormone (FSH) isoform distribution pattern of pituitary or ciculating FSH in nutritionally growth-restricted ovariectomized lambs *Endocrinology;* **132**: 1527–1536.

Lamperti A and Hill L (1987). The effects of anterior hypothalamic deafferentation on FSH-releasing acivity in the intact female hamster. *Experimental Brain Research*; **68**:189–194.

Lee JS, Manning JM, Foster DL, and Padmanabhan V (1998). Estrogen increases the secretion of less-acidic FSH isoforms in ovariectomized prepubertal and peripubertal lambs. *Program and Abstracts of the 80th Annual Meeting of the Endocrine Society*, New Orleans, (Abstract#P2–474), p. 348.

Lumpkin MD and McCann SM (1984). Effect of destruction of the dorsal anterior hypothalamus on follicle-stimulating hormone secretion in the rat. *Endocrinology* 115: 2473–2480.

Lumpkin MD, Moltz JH, Yu WH, Samson WK and McCann SM (1987). Purification of FSH-releasing factor: its dissimilarity from LHRH of mammalian, avian, and piscian origin. *Brain Research Bulletin* **18**: 175–178.

Lumpkin MD, McDonald JK, Samson WK and McCann SM (1989). Destruction of the dorsal anterior hypothalamic region suppresses pulsatile release of follicle stimulating hormone but not luteinizing hormone. *Neuroendocrinology* **50**: 229–235.

Marubayashi U, Yu WH, McCann SM (1999). Median eminence lesions reveal separate hypothalamic control of pulsatile follicle-stimulatng hormone and luteinizing hormone release. *Proceedings of the Society for Experimental Biology & Medicine* **220**: 139–146.

Mather JP, Woodruff TK and Krummen LA (1992). Paracrine regulation of reproductive function by inhibin and activin. *Proceedings of the Society for Experimental Biology and Medicine*; **201**: 1–15.

Mathews LS and Vale WW (1991). Expression cloning of an activin receptor, a predicted transmembrane serine kinase. *Cell* **65**: 973–982.

McNeilly As (1988). The control of FSH secrtion Acta Endocrinologia Suppl (Copenh); **288**: 31–40.

Midgley AR Jr., McFadden K, Ghazzi M, Karsch FJ, Brown MB, Mauger DT, and Padmanabhan V (1997). Nonclassical secretory dynamics of LH revealed by hypothalamo-hypophyseal portal sampling of sheep *Endocrine* **6**: 133–143.

Mizunuma H, Samson WK, Lumpkin MD and McCann SM (1983) Evidence for an FSH-releasing factor in the posterior portion of the rat median eminence *Life Sciences*; **33**: 2003–2009.

Nomura K, Tsunasawa S, Ohmura K, Sakiyama F, and Shizume K (1988) Renotropic activity in ovine luteinizing hormone isoform(s) Endocrinology **123**: 700–772.

Padmanabhan V, Lang LL, Sonstein J, Kelch RP, and Beitins IZ (1988) Modulation of serum follicle-stimulating hormone bioactivity and isoform distribution by estrogenic steroids in normal women and in gonadal dysgenesis *Journal of Clinical Endocrinology and Metabolism* **67**: 465–473.

Padmanabhan V, Mieher CD, Borondy M, I'Anson H, Wood RI, Landefeld TD, Foster DL, and Beitins IZ (1992) Circulating bioactive follicle-stimulating hormone and less acidic follicle-stimulating hormone isoforms increase during experimental induction of puberty in the female lamb *Endrocrinology* **131**: 213–220.

Padmanabhan V, Van Cleeff, J, McLeod MK, Nett TM and Karsch FJ (1997a) Is there a GnRH-independent hypothalamic control of FSH secretion? *Program and Abstracts of the Annual Meeting of the Society for Neuroscience*, St. Louis. p. 588 (Abstract).

Padmanabhan V, McFadden K, Mauger DT, Karsch FJ and Midgley AR Jr. (1997b) Neuroendocrine control of FSH secretion: I. Direct evidence for separate episodic and basal components of FSH secretion *Endocrinology* **138**: 424–432.

Padmanabhan V, I'Anson H, Foster DL, and Beitins IZ (1997c) Estradiol increases the release of less acidic FSH isoforms in nutritionally growth-retarded lambs. *Program and Abstracts of the 79th Annual Meeting of the Endocrine Society* Abstract p3–336.

Pau KY-F, Gleissman PM, Oyama T and Spies HG (1991). Disruption of GnRH pulses by anti-GnRH serum and Phentolamine obliterates pulsatile LH but not FSH secretion in ovariectomized rabbits *Neuroendocrinology* **53**: 382–381.

Phillips DJ and Wide L (1994). Serum gonadotropin isoforms become more basic after an exogenous challenge of gonadotropin-releasing hormone in children undergoing pubertal development *Journal of Clinical Endocrinology and Metabolism* **79**: 814–819.

Robertson DM (1992). Follistatin/Activin-binding protein *Trends Endocrinol Metab* **3**: 65–68.

Rose MP, Gaines-Das RE and Balen AH (2000) Definition and measurement of follicle-stimulating hormone *Endocrine Reviews* **21**: 5–22.

Sheridan R, Loras B, Surardt L, Ectors F and Pasteels JL (1979). Autonomous secretion of follicle-stimulating hormone by long term organ cultures of rat pituitaries *Endocrinology* **104**: 198–204.

Turgeon JL and Waring DW (1982). Differential changes in the rate and pattern of follicle-stimulating hormone secretion from pituitaries of cyclic rats superfused in vitro *Endocrinology* **11**: 66–73.

Ulloa-Aguirre A, Midgley Jr. AR, Beitins IZ and Padmanabhan V (1995). Follicle Stimulating Isohormones: Biological characterization and physiological relevance *Endocrine Reviews* **16**: 765–787.

Vitt UA, Kloosterbocr HJ, Rose UM, Kiesel PS, Bete A, and Nayudu PL (1998). Isoforms of human recombinant follicle-stimulating hormone: Comparison of effects on murine follicle development *in vitro: Biology of Reproduction* **59**: 854–861.

Wildt L, Hausler A, Marshall G, Hutchison JS, Plant TM, Belchetz PE and Knobil E (1981). Frequency and amplitude of gonadotropin-releasing hormone stimulation and gonadotropin secretion in the rhesus monkey *Endocrinology* **109**: 376–385.

Wise PM, Rance N, Barr GD and Barraclough CA (1979). Further evidence that luteinizing hormone-releasing hormone also is follicle-stimulating hormone-releasing hormone *Endocrinology* **104**: 940–947.

Yamazaki K, Sato K, Shizume K, Kanaji Y, Ito Y, Obara T, Nakagava T, Koizumi T, and Nishimura R (1995). Potent thyrotropic activity of human chorionic gonadotropin variants in terms of ^{125}I incorporation and de novo synthesized thyroid hormone release in human thyroid follicles *Journal of Clinical Endocrinology and Metabolism* **80**: 473–479.

Ying S (1988) Inhibins, activins and follistatins: gonadal proteins modulating the secretion of follicle-stimulating hormone *Endocrine Reviews* **9**: 267–293.

Yu WH, Karanth S, Walczewska A, Sower SA and McCann SM (1997) A hypothalamic folicle-stimualting hormone-releasing decapeptide in the ram *Proceedings of the National Academy of Sciences of the United States of America* **94**: 9499–9503.

Zambrano E, Olivares A, Mendez JP, Guerrero L, Diaz-Cueto L, Veldhuis JD, and Ulloa-Aguirre, A (1995). Dynamics of basal and gonadotropin-releasing hormone-releasable serum follicle stimulating hormone charge isoform distribution throughout the human menstrual cycle *Journal of Clinical Endocrinology and Metabolism* **80**: 1647–1656.

Follicular Growth, Ovulation and Fertilization: Molecular and Clinical Basis
Anand Kumar and Amal K. Mukhopadhyay (Eds.)
Narosa Publishing House, New Delhi, India, 2001

5

A Common Genetic Variant of Luteinizing Hormone: Physiological and Functional Consequences

Pulak R. Manna, Lata Joshi and Ilpo T. Huhtaniemi
Department of Physiology, Institute of Biomedicine, University of Turku
FIN 20520 Turku, Finland

Introduction

The gonadotrophic hormones, luteinizing hormone (LH) and follicle-stimulating hormone (FSH), are composed of a dimer of noncovalently linked α- and β-subunits. The α-subunit is common to all glycoprotein hormones, including thyroid-stimulating hormone (TSH) and chorionic gonadotropin (CG), whereas β-subunit confers the hormone-specific biological activity (Pierce and Parsons, 1981). The biopotency and circulating clearance rate of glycoprotein hormones are dependent on specific composition of their carbohydrate moieties which in any given situation display considerable degree of microheterogeneity. The function of LH is initiated by its binding to the LH receptor (R), a member of the family of G-protein coupled seven-transmembrane domain receptors that is primarily expressed in testicular Leydig cells, and in ovarian theca, granulosa and luteal cells (McFarland *et al.*, 1990; Camp *et al.*, 1991; Segaloff and Ascoli, 1993). On the other hand, FSH action is crucial for normal spermatogenesis in male and folliculogenesis in females (Simoni *et al.*, 1997; Tapanainen *et al.*, 1997). Interaction of these hormones with their receptors results in activation of adenylate cyclase, phospholipase C dependent inositol phosphates, and modulation of intracellular Ca^{2+} signaling pathways (Gudermann *et al.*, 1992; Segaloff and Ascoli, 1993; Zhang *et al.*, 1997).

The human LHβ gene is approximately 1.5 kb in size and located in chromosome 19 p13.32. The amino acid sequences of LH- and hCG-β-subunits show about 85% homology, and exhibit almost similar biological properties when dimerized with α-subunit. It has also been reported that carbohydrate moieties of α- and not β-subunit are involved in activation of the LHR as well as G protein-coupled signal transduction system (Sairam, 1989). A terminal sulfate residue is present in carbohydrate moieties attached to the LHβ gene and is important for hormone binding, while terminal sialic acid chains (hCGβ and FSHβ genes) affects the *in vivo* stability (Pierce and Parsons, 1981; Jia *et al.*, 1991). The LH and TSH synthesized in the pituitary possess oligosaccharide with Sulphate (SO_4) and N-acetylgalactoseamine (GalNAc) at termini, whereas hCG and FSH bear oligosaccharides terminating at sialic acid and galactose (Gal) (Furuhashi *et al.*, 1995). The SO_4 group of LH shortens the circulating half-life of this hormone (Baenziger *et al.*, 1992).

Multiple gonadal abnormalities have so far been documented in studies on subjects with mutations in the LHβ gene and/or the cognate receptor (Haavisto *et al.*, 1995; Kraaji *et al.*, 1995; Suganuma *et al.*, 1995; Jameson, 1996; Suganuma *et al.*, 1996). In fact, only one inactivating mutation of the LHβ has so far been reported (Weiss *et al.*, 1992), but there is a number of both activating (gain-of-function) and inactivating (loss-of-function) mutations in the LHR gene. In the case of LH and FSH, all mutations have been demonstrated in their β-subunits, whereas no mutation has been found in the common α-subunit gene. The mutations of the present study concern with a common genetic variant, which has phenotypic expression in the form of mild reproductive malfunctions (Haavisto *et al.*, 1995; Rajkhowa *et al.*, 1995; Suganuma *et al.*, 1995; Suganuma *et al.*, 1996).

Based on the above, and to characterize in more details the functional effects of variant (V)-LH *in vitro* and *in vivo*, nucleotide sequences of the wild type (WT)-LHβ gene were altered by site-directed mutagenesis at two positions, Trp^8 (TGG) to Arg (**C** GG) and Ile^{15} (ATC) to Thr (A**C**C) and recombinant (rec) forms of WT- and V-LH were produced. The present work was started with the objectives in purifying rec LHs produced in Chinese hamster ovary (CHO) cells, followed by assessing their functional properties. These observations provide evidence that amino acid substitutions in the LHβ gene affect signal transduction pathways beyond the receptor level. The V-LH associated with increased bioactivity *in vitro* and short half-life *in vivo*, may predispose its carrier to disorders in reproductive function.

Materials and Methods

Plasmids, Transfections, and Production and Purification of rec LHs

The mutations in exon 2 of the V-LHβ gene, at codon 8 Trp (TGG) to Arg (CGG) and 15 Ile (ATC) to Thr (ACC), were constructed by site directed mutagenesis and confirmed by sequencing with fluorescent dye termination reaction (Prism Ready Reaction Dye Termination Cycle Sequencing Kit) using automated sequencer (Perkin-Elmer, Foster City, CA). The full-length LHβ gene (1511 bp), a *Bgl* II-*Bam* HI fragment, was inserted into the eukaryotic expression vector PM^2, downstream of the Harvey murine sarcoma virus long terminal repeat (DeFeo *et al.*, 1981; Matzuk *et al.*, 1987; Suganuma *et al.*, 1996).

Chinese hamster ovary (CHO) cells were maintained in Dulbecco's Modified Eagle's Medium (DMEM)-Ham's F-12 medium (1:1, life Technologies, Grand Island, NY) supplemented with 10% heat-inactivated fetal calf serum (FCS) containing 0.1 g/L gentamycin (biological Industries, KB Haemek, Israel) and

2.5 mg/L fungizone (Life Technologies) at 37°C in humidified atmosphere (95% O_2/5% CO_2). These cells were transfected at 65–75% confluency by lipofectamine (Gibco-BRL, MD), using 2:1 ratio of the α-gene (DeFeo *et al.*, 1981; Matzuk *et al.*, 1987; Matzuk and Boime, 1988; Suganuma *et al.*, 1996) and either WT- or V-LHβ gene. Clonal lines were selected in the presence of 800 μg/L of geneticin (G418, Sigma Chemical Co., MO), expanded and screened for LH levels.

The clonal lines of choice, WT-32 and V-16, were subsequently cultured in the presence of 300 mg/L of G418, in a siliconized membrane fitted with flat culture cassette in a Technomouse bioreactor (Integra Biosciences Inc., Woburn, MA). Synthesized proteins were collected in serum free CHO SF-Speciality Media (Grand Centrla Avenue, Lavallette, USA), supplemented with 0.5 μg/L insulin, 5 mg/L transferrin and 1 μmol/L of 2, 3 dehydro-2 deoxy N-acetyl neuraminic acid (Sigma). The production of human rec LHs was monitored by a solid phase, two-site immunofluorometric assay technique (IFMA, Delfia LH spec, Turku, Finland).

The media containing WT- and V-LHs were first concentrated through Pellicon™-2 ultrafiltration (Millipore Corporation, Bedford, MA) system followed by different purification steps. The first step of purification was carried out with red sepharose CL-6B (Amersham-Pharmacia-Biotech, Uppsala, Sweden) affinity chromatography. Upon loading, bound proteins were eluted with increasing concentrations of KCl (0.5, 1.0 and 3.0 mol/L), and fractions (750 μl) were collected, pooled, and processed further.

Immunoaffinity chromatography was performed next using cyanogen bromide (CNBr)-activated Sepharose 4 Fast Flow matrix (Amersham-Pharmacia). The pooled fractions of rec LH were equilibrated with buffer A (10 mmol/L sodium phosphate with 0.6% saline, pH 7.2), loaded onto the column and allowed to bind for 4–6 h at 4°C. Bound LHs were eluted with buffer B (buffer A containing 3.0 mol/L sodium iodide). The immunoreactive peak fractions (500 μ*l*) were pooled, dialyzed extensively against 10 mmol/L PBS, pH7.2, lyophilized and stored at – 70°C.

The bound fraction of immunoaffinity was redissolved in Tris-HCl buffer (20 mmol/L), desalted and subjected to FPLC Mono Q anion exchange (HR 5/5) chromatography (Amersham-Pharmacia). The fractions (250 μl) were eluted with 20 mmol/L Tris-HCl, pH 6.5, containing 1.0 mol/L NaCl. The immunoactive peak fractions were pooled, lyophilized and stored at – 70 °C for further studies, and their purity was assessed by sodium dodecyl sulphate-polyacrylamide gel electrophoresis (SDS-PAGE).

Chromatofocusing

The distribution pattern of rec LH isoforms was analyzed with HR 5/20 column, using FPLC. The column used was pre-packed with polybuffer-exchanger 94 (PBE-94, Amersham-Pharmacia). Rec LHs were loaded, equilibrated with 25 mmol/L Tris-HCl, pH 8.0, and eluted with elution buffer (80% PB 74 and 20% PB 96; Pharmacia) as demonstrated previously (Hakola *et al.*, 1997b). The flow rate was adjusted to 1.0 ml/min and pH of each fraction (300 μ*l*) was determined after diluting with 10 mmol/L PBS containing 0.1% bovine serum albumin (Sigma), and LH levels were subsequently measured (Delfia LH spec).

Effects of WT- and V-LHs on Progesterone (P) and cAMP Production of Immortalized Murine Leydig Cells

The mLTC-1 mouse Leydig tumor cells (Rebois, 1982) were maintained in Waymouth's growth medium, as described previously (Manna *et al.*, 1999). Prior to stimulation, cells were washed, trypsinized and subcultured (~70,000 cells/well) in 24-well plates. Following 24 h, cells were stimulated for 6 h (P) and 2 h (cAMP) with increasing concentrations of rec LHs (0-1000 µg/L) and hCG (CR-127, NIDDK). The media were collected from different treatment groups, and measured for P and cAMP levels with specific RIAs (Harper and Brooker, 1975; Vuorento *et al.*, 1989).

Receptor Binding Activity of rec LHs

hCG was radio-iodinated with Na[^{125}I]-iodide (IMS 300, Amersham-Pharmacia), using a solid phase lactoperoxidase method (Karonen *et al.*, 1975). The specific activity of the labeled hormone preparations was found to be 20-32 Ci/g. Binding studies were carried out under optimized conditions (Zhang et al., 1997), using intact mLTC-1- cells. Briefly, similar aliquots of cell suspensions were incubated with fixed amount of [^{125}I]iodo-hCG (1.5×10^5 cpm/tube) in the absence (total) or presence (non-specific) of increasing amounts of unlabeled hCG (Pregnyl; Organon), or WT- or V-LH (0-1000 µg/L). The incubations were terminated by placing the tubes on ice, and followed by 3-ml ice-cold mixture of polyethylene glycol (PEG 6000, 20%) in 10 mmol/L sodium phosphate buffer, pH 7.2. Specific binding was determined by subtracting non-specific binding from total binding, as measured by ν-spectrometer (1260 Multigamma II, AGG Wallac, Turku, Finland).

Efficacy of rec WT- and V-LHs in Ovulation Induction

Mature Sprague-Dawley female rats, weighing 180–230 g, were housed under controlled lighting schedule (14 h light and 10 h darkness), fed commercial diet and water *ad libitum*. The animals were maintained in accordance with guidelines of the Turku University Ethical Committee for the Use and Care of Experimental Animals. The rats primed with gonadotropin (PMSG, 10 IU/ 100 g body weight; Sigma) were treated with increasing amounts of rec LHs (0–100 IU/rat). Following 24 h, rats were anesthetized, and ovulation was assessed by counting number of oocytes from oviducts of different groups.

Data analysis

The data presented are the mean ± SEM, whenever appropriate. The statistically significant differences were analyzed using the Statview program fitted with the Macintosh system, following ANOVA (Fisher's protected least significant differences test) as noted. $p < 0.05$ was considered statistically significant.

Results

Purification and Characterization of rec WT- and V-LHs

The results presented in Table 1 demonstrate that the recovery of these WT- and V-LH by red sepharose CL-6B chromatography corresponded to 62 and 69%, respectively, while the degree of purification were found to be 110-fold in both cases. Approximately 43% rec LHs was recovered by immunoaffinity chromatography. Although the yield was low, the specific activity of purified preparations increased to 8832 and 7095 IU/mg in WT-and V-LHs, respectively. The bound materials obtained upon immunoaffinity chromatography were subjected to FPLC Mono Q anion exchange chromatography. The purity of the Mono Q P-II fraction was assessed in both cases by SDS-PAGE, which demonstrated a clear single band of approximately 31 kD in size (data not shown). The degree of purification at this step was found to be 180-fold over starting materials, while the yields of the purified proteins were 0.615 and 0.577 mg in WT- and V-LHs respectively.

Table 1 Summary of purification of the rec wild type (WT)- and variant (V)-LHs

Steps of purification	*Immunoactivity (units)*	*Protein (mg)*	*Specific Activity (units/mg)*	*Yield (%)*	*Fold of purification*
Medium	**WT**-31360	490	64	100	1
	V-29700	540	55		
Ultraconcentration	**WT**-29304	7.47	3923	93	61.3
	V-26160	8.2	3190	88	58
Red-Sepharose	**WT**-19244	3.1	6208	61.4	97
	V-20552	3.46	5940	69	108
Immunoaffinity	**WT**-19894	1.46	8832	41	138
	V-13380	1.88	7095	45	129
Mono Q	**WT**-6927	0.615	11265	22	176
(FPLC)	**V**-5879	0.577	10188	19.8	185

The WT- and V-LHs were purified employing these steps. Fold of purification was calculated using specific activities of these LHs, considering media as 1.

The biological activity of the rec LH preparations was evaluated by determining P and cAMP stimulation in comparison to hCG (CR-127), utilizing mLTC-1 cells. The data presented in Fig. 1 (A and B) demonstrate a dose dependent increase of P(6h) and cAMP (2h) synthesis with increasing concentrations of hormones (0–1000 μg/L). Increased response in both cases was significant at ≥ 1 μg/L, half-maximal stimulation occurred at about 12 μg/L, and maximum increase at doses ≥ 30 μg/L of these hormones. The sensitivities of the responses to V-LH were about 4- and 7-fold higher than to WT-LH in P and cAMP production, respectively.

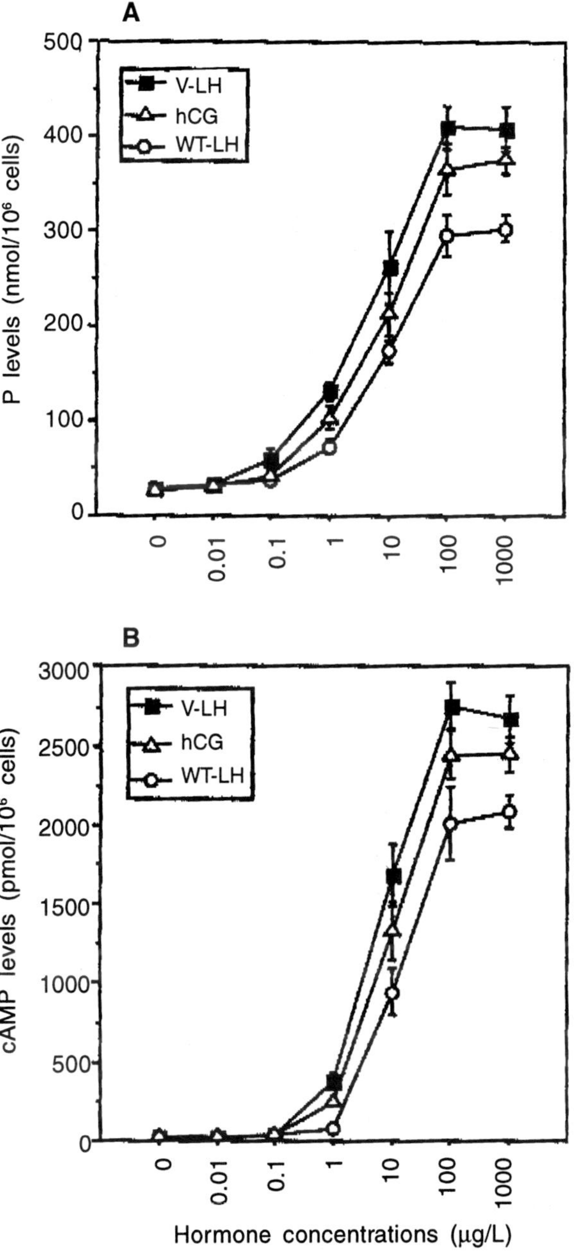

Fig. 1 Effects of WT- and V-LHs on P and cAMP production in mLTC-1 cells. Cells were stimulated for 6 h (P; *Panel A*) and 2 h(cAMP; *Panel B*) with increasing concentrations of rec LHs and hCG (0–1000 μg/L), and measured for P and cAMP synthesis by RIAs. Data presented are the mean ± SEM of four independent experiments.

In spite of the increased bioactivity of V-LH, distribution pattern of its isoforms did not differ significantly with WT-LH, where pI was found to be between 6.6 and 7.8 in both cases (Fig. 2).

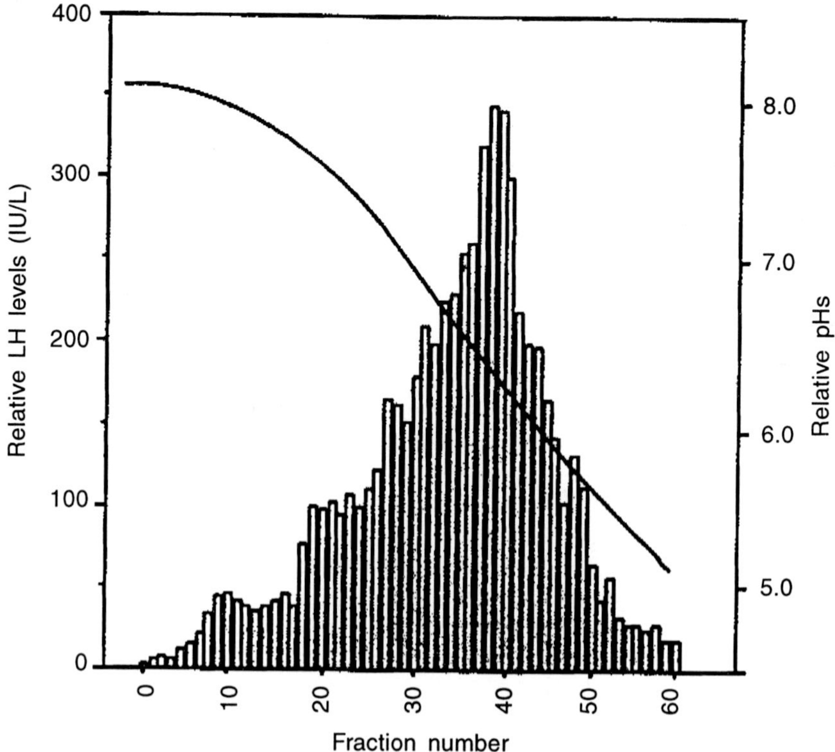

Fig. 2 Elution pattern of purified rec V-LH in chromatofocusing using PBE-94 column with a pH ranges between 5 and 8. The relative LH levels in each fraction were determined with Delfia LH spec (bar diagram) together with corresponding pH values (line drawing), are illustrated.

Receptor Binding Properties of rec LHs

The receptor binding studies of these LHs were carried out with [^{125}I]iodo-hCG using intact mLTC-1 cells. As illustrated in Fig. 3, displacement of [^{125}I]iodo-hCG increased gradually with increasing amounts of unlabeled hormones, where concentrations of rec LHs and hCG occurred for 50% displacement were found to be 48 $\pm$ 9.6 μg/L. In fact, no significant differences were observed with WT- and V-LHs in inhibiting [^{125}I]iodo-hCG binding, suggesting that the mutational changes did not affect significantly the affinity of LH to its cognate receptor.

Effects of rec WT- and V-LHs in Ovulation Induction

To corroborate the receptor binding studies, we next assessed in vivo effects of rec LHs in ovulation induction. Adult female primed rats, injected with either WT- or V-LHs (0-100 IU/rat), did not differ significantly in their ovulation rates, and at higher doses, all animals ovulated with similar number of oocytes (Fig. 4). These results demonstrate that although conformational changes in the V-LHβ gene augment P and cAMP synthesis, they have no effect on capacity of the hormone to induce ovulation.

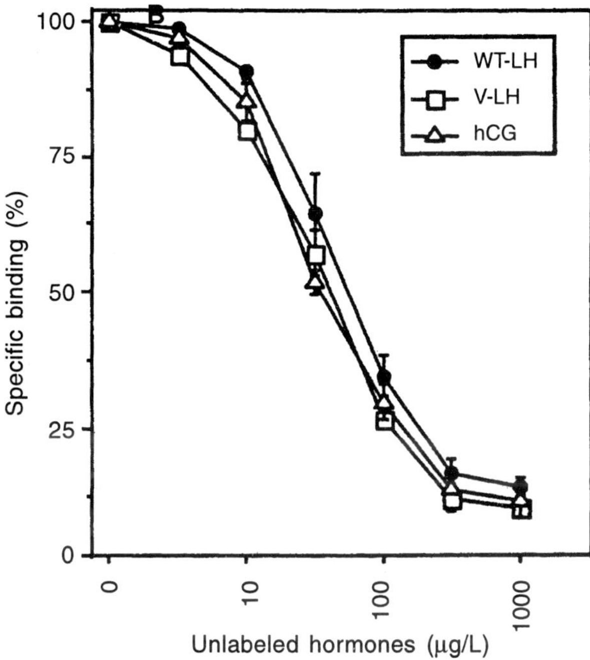

Fig. 3 [^{125}I]iodo-hCG binding to the intact mLTC-1 cell membrane. Inhibition of [^{125}I]iodo-hCG studies were carried out with increasing amounts of unlabeled hormones (0–1000 µg/L), as described under *Materials and Methods*. The results are the mean ± SEM of three experiments.

Discussion

The gonadotropic hormones (LH and FSH), together with TSH, have many structural and chemical similarities, constituting the major family of adenohypophyseal glycoprotein hormones. They possess an α-subunit that is common to all glycoprotein hormones, including hCG, and hormone specific β-subunit (Pierce and Parsons 1981, Fiddes and Talmadge, 1984; Gharib *et al.*, 1990; Catt and Dufau, 1991). It is well established that gonadal development and maturation associated with reproductive function are intricately regulated by actions of LH and FSH. It has been unequivocally reported that structural alterations in gonadotropins or their receptors can be involved in abnormalities of gonadal function (Aittomaki *et al.*, 1995; Haavisto *et al.*, 1995; Kraaji *et al.*, 1995; Suganuma *et al.*, 1995; Jameson, 1996; Suganuma *et al.*, 1996; Simoni *et al.*, 1997; Liao *et al.*, 1998; Elter *et al.*, 1999; Tapanainen *et al.*, 1999). Recently, conformational changes in the LHβ gene (Trp^{8}Arg and Ile^{15}Thr) were found to be associated with mild reproductive disorders in women (Suganuma *et al.*, 1995; Suganuma *et al.*, 1996; Liao *et al.*, 1998; Elter *et al.*, 1999; Tapanainen *et al.*, 1999) The present investigation was aimed at gaining further insight information about the molecular effects of mutations in the LHβ gene, employing recombinant LH preparations. The functional properties of the recombinant LH forms clearly demonstrate that mutational changes modulate the signal transduction pathways of LH.

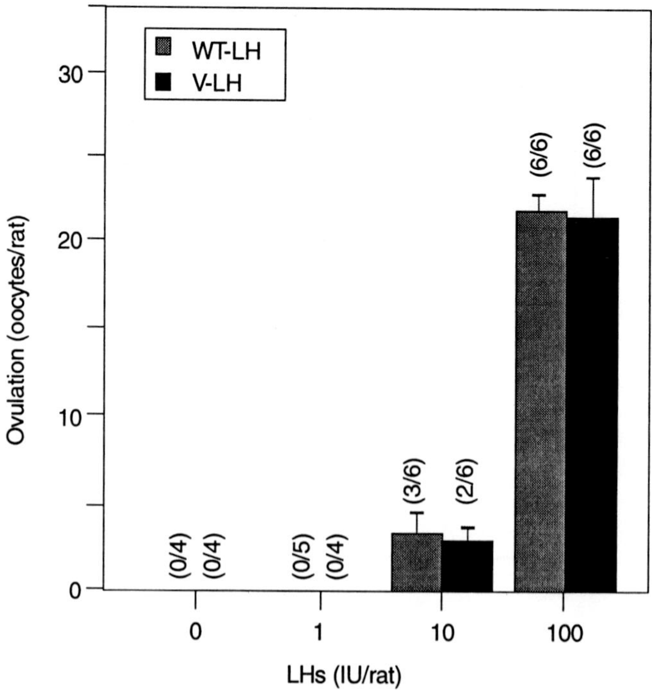

Fig. 4 Effect of purified rec LHs on ovulation induction. Adult female PMSG primed rats were administered with increasing doses of rec WT- and V-LHs (1–100 IU/rat). Number of rats ovulated among total is indicated in parenthesis. Number of occytes recovered from the oviducts are counted and presented. Data are the mean ± SD.

The physiological relevance of the effects of LHβ mutations was studied using rec forms of WT- and V-LH. This procedure has been shown to allow normal assembly of α-with β-subunits and secretion, as regards hCG and LH production in CHO cells (Matzuk *et al.*, 1989). Purification of the LH preparations was carried out by red-sepharose CL-6B affinity, immunoaffinity and FPLC Mono Q anion exchange chromatography. These techniques were found to be very effective in purifying glycoprotein hormones, as they have recently been successfully utilized in purifying rat rec LH and FSH in CHO cells (Hakola *et al.*, 1997a; 1997b). Mono Q anion exchange chromatography was used to achieve the highest purity. Lyophilization of fractions eluted on Mono Q (P-II), followed by SDS-PAGE, demonstrated a discrete sharp band of 31 kDa in both cases (not illustrated), which is in accordance with the rat rec LH (33 kDa) and pit LH (32 kDa) (Hakola *et al.*, 1997b). The slight differences in molecular weight are apparently due to differences in glycosylation and/or terminal N-acetyl-neuraminic acids and sulfates in the β-subunit (Kaetzel *et al.*, 1985; Green and Baenziger, 1998).

The purification of the rec LHs provided opportunity to characterize their functional effects *in vitro*. Presently, the activating function of V-LH in signal transduction could be due to the alterations (Trp^8Arg and $Ile^{15}Thr$) of hydrophobicity to hydrophilicity in the NH_2-terminus. Studies also demonstrate that these mutations have been detected in women with reproductive abnormalities, including menstrual disorders and polycystic ovary syndrome (PCOS) (Kremer *et al.*, 1995; Suganuma *et al.*, 1995; Tapanainen *et al.*, 1997; Liao *et al.*, 1998; Tulppala *et al.*, 1998;). Our results show that these LHs bind specifically to the LH receptor with similar affinity. However, we observed that V-LH was a more potent stimulator of P and cAMP accumulation than WT-LH in mLTC-1 cells. In accordance, the increased bioactivity of V-LH was also observed in IP_3 production, following expression of hLH receptor in HEK 293 cells (data not shown). It is interesting to note that the alteration of Ile^{15} to Thr results in an extra oligosaccharide attachment site at Asn^{13} in the LHβ that did not affect hormonal binding but modulated signal transduction. On the other hand, importance of Asn^{13} in hCGβ and bovine LHβ has been shown to be associated with hormonal binding and signal transduction (Kaetzel et al., 1989; Matzuk *et al.*, 1989).

Previous reports form this and other laboratories provide evidence that the bio-immunoassay ratio of LH in homozygotes for V-LHβ is higher than that WT-LH in controls (Haavisto *et al.*, 1990, 1995; Patterson *et al.*, 1992; Furui *et al.*, 1994; Rajknowa *et al.*, 1995; Suganuma *et al.*, 1996). As regards the clearance rate in circulation, V-LH has shorter half-life (6–9 min) compared to WT-LH (12–20 min) or hCG (50–70 min) (data not shown). In accordance, studies also show that human LH injected in ewes cleared with half-life of 23 min, while ovine LH cleared between 5–25 min (Ascoli *et al.*, 1975; Kalyan and Bhal, 1983). Importantly, the mutational changes altered the LHβ structure closer to that of hCGβ, whereas the circulating clearance rate of hCGβ has been shown to be very long (50–140 min) in rats (Kalyan and Bhal, 1983). It has been demonstrated that recombinant glycoproteins produced in CHO cells possess sulfated carbohydrates (Skzudlinski *et al.*, 1993). It is evident that V-LH has an extra carbohydrate side chain with terminally sulfated oligasaccharides that may be responsible for its faster turn-over from circulation than WT-LH (Baenziger *et al.*, 1992).

Mutational changes in glycoprotein hormones so far demonstrated are mostly concerned with loss-of-function instead of gain-of-function. Subsequently, mutation in the FSHβ (Val^{61}) gene is associated with primary amenorrhea, while in the case of TSHβ it caused familial hypothyroidism leading to infertility (Dacou *et al.*, 1990; Mattews *et al.*, 1993). Occurrence of a point mutation ($Gln^{54}Arg$) in the LHβ gene has been shown on patients affected with hypogonadism (Weiss *et al.*, 1992). In contrast, alterations in LHβ gene in the present study clearly showed activation of function of different signaling pathways. However, this increased in vitro bioactivity is at least partly compensated for by shorter circulatory half-time.

Collectively, the conformational changes caused by mutations in the LHβ gene significantly affected sensitivity of the different signalling pathways of LH action, without altering affinity of receptor-ligand interaction. On the basis of collective *in vitro* and *in vivo* data, it is tempting to propose that the increased bioactivity of V-LH with shorter half-life may create an abnormal environment to gonadal function, associated with the multiple reproductive disorders detected in cases of the V-LHβ allele, including infertility.

Acknowledgements

This work was supported by grants from the Academy of Finland, the Sigrid Jusélius Foundation and the Foundation for the Finnish Cancer Societies. We would like to thank S. Lindgren (Department of Cell Biology, Univ. of Turku) and M. Vehniäinen (Department of Biochemistry, Univ. of Turku) for their help in relation to large-scale production of rec WT- and V-LHs and with ultraconcentration system, respectively. Our special thanks to H. Peuravuori (Department of Pathology, Univ. of Turku) for this active cooperation in relation to FPLC system.

References

Aittomaki, K., Lucena, L.J., Pakarinen, P. *et al.*, (1995) Mutation in the follicle-stimulating hormone receptor gene causes hereditary hypergonadotropic ovarian failure. *Cell,* **82**: 959–968.

Ascoli, M., Liddle, R.A., and Puett, D. (1975) The metabolism of luteinizing hormone. Plasma clearance, urinary excretion, and tissue uptake. *Mol Cell. Endocrinol.*, **3**: 21–36.

Baenziger, J.U., Kumar, S., Brodbeck, R.M., *et al.*, (1992) Circulatory half-life but not interaction with the lutropin/chorionic gonadotropin receptor is modulated by sulfation of bovine lutropin oligosaccharides, *Proc. Natl. Acad. Sci. USA,* **89**: 334–338.

Camp, T.A., Rahal, J.O. and Mayo, K.E. (1991) Cellular localization and hormonal regulation of follicle-stimulating hormone and luteinizing hormone receptor messenger RNAs in the rat ovary. *Mol. Endocrinol.*, **5**: 1405–1417.

Catt, K.J. and Dufau, M.L. (1991) *Gonadotropic hormones: Biosynthesis, secretion, receptors, and actions. In: Reproductive endocrinology*, Edn 3, (Yen, S.S.C., and Jaffe, R.B.), W.B. Saunders C., Philadelphia, PA, pp. 105–155.

Dacou, V.C., Feltquate, D.M., Drakopoulou, M. *et al.*, (1990) Familial hypohyroidism caused by a nonsense mutation in the thyroid-stimulating homone β-subunit gene. *Am. J. Hum. Genet.*, **46**: 988–993.

DeFeo, D., Gonda, M.A., Young. H.A. *et al.*, (1981) Analysis of two divergent rat genomic clones homologous to the transforming gene of Harvey sarcoma virus. *Proc. Natl. Acad. Sci. USA,* **78**: 3328–3332.

Elter, K., Erel, T., Cine, N. *et al.*, (1999) Role of the mutations Trp^8Arg and $Ile^{15}Thr$ of the human luteinizing hormone β-subunit in women with polycystic ovary syndrome. *Fertil. Steril.,* **71**: 425–430.

Fiddes, J.C. and Talmadge, K. (1984) Structure, expression and evolution of the genes for the human glycoprotein hormones (review). *Recent Prog. Horm. Res.*, **40**: 43–78.

Furuhashi, M., Suzuki, S., Tomoda, Y. and Suganuma, N. (1995) Role of the Pro-Leu-Arg motif in glycosylation of human gonadotropin alpha-subunit. *Endocrinology,* **136**: 2270–2275.

Furui, K., Suganuma, N., Tsukahara, S. *et al.*, (1994) Identification of two point mutations in the gene encoding luteinizing hormone (LH) β-subunit, associated with immunologically anomalous LH variants. *J. Clin. Endocrinol. Metab.*, **78**: 107–113.

Gharib, S.D., Wierman, M.E., Shupnik, M.A. and Chin, W.W. (1990) Molecular biology of the pituitary gonadotropins. *Endocr. Rev.,* **11**: 177–199.

Green, E.D. and Baenziger, J.U. (1998) Aspergine-linked oligosaccharides on lutropin, follitropin and thyrotropin. *J. Biol. Chem.,* **263**: 36–44.

Gudermann, T., Birnbaumer, M. and Birnbaumer, L. (1992) Evidence for dual coupling of the murine luteinizing hormone receptor to adenylate cyclase and phosphoinositide breakdown and Ca^{2+} mobilization, *J. Biol. Chem.*, **267**: 4479–4488.

Haavisto, A-M., Dunkel, L., Pettersson, K. and Huhtaniemi, I. (1990) LH measurements by *in vitro* bioassay and highly sensitive immunofluorometric assay improve the distinction between boys with constitutional delay of puberty and hypogonadotropic hypogonadism. *Pediatr. Res.*, **27**: 211–214.

Haavisto, A.M., Pettersson, K., Bergendhal, M. *et al.*, (1995) Occurrence and biological properties of a common genetic variant of luteinizing hormone, *J. Clin. Endocrinol. Metab.*, **80**: 1257–1263.

Hakola, K., Van der Boogaart, P., Mulders, J. *et al.*, (1997a) Recombinant rat follicle-stimulating hormone, production by Chinese hamster ovary cells, purification and functional characterization. *Mol. Cell. Endocrinol.*, **127**: 59–69.

Hakola, K., Van der Boogaart, P., Mulders, J. *et al.*, (1997b) Recombinant rat luteinizing hormone; production by Chinese hamster ovary cells, purification and functional characterization. *Mol. Cell. Endorcinol.*, **128**: 47–56.

Harper, J. and Brooker, G. (1975) Femtomole sensitive radioimmunoassay for cyclic-AMP and cyclic-GMP after 2′-O-acetylation by acetic anhydride in aqueous solution. *J. Cycl. Nuleotide Res.*, **4**: 207–218.

Jameson, J.L. (1996) Inherited disorders of the gonadotropin hormones. *Mol. Cell Endocrinol.*, **125**: 143–149.

Jia, X-C Bo, M., Tanaka, T. *et al.*, (1991) Expression of human luteinizing hormone (LH) receptor: interaction with LH and chorionic gonadotropin from human but not equine, rat and ovine species. *Mol. Endocrinol.* **5**: 759–768.

Kaetzel, D.M., Browne, J.K., Wondisford, F. *et al.*, (1985) Expression of biologically active luteinizing hormone in chinese hamster ovary cells. *Proc. Natl. Acad. Sci. USA,* **82**: 7280–7283.

Kaetzel, D.M., Vergin, J.B., Clay, C.M. and Nilson, J.H. (1989) Disruption of N-liked glycosylation of bovine luteinizing hormone β-subunit by site-directed mutagenesis dramatically increases its intracellular stability but does not affect biological activity of the secreted heterodimer. *Mol. Endocrinol.*, **3**: 1765–1774.

Kalyan, N.K. and Bahl. O.P. (1983) Role of carbohydrate in human chorionic gonadotropin: effect of deglycosylation on the subunit interaction and its *in vitro* and *in vivo* biological properties. *J. Biol. Chem.* **258**: 67–74.

Karonen, S., Mörsky, P., Siren, M, and Söderling, U. (1975) An enzymatic solid-phase method for trace iodination of proteins and peptides with ^{125}I-iodine, *Anal. Biochem.*, **67**: 1–10.

Kraaij, R., Post, M., Kremer, H. *et al.*, (1995) A missense mutation in the second transmembrane segment of the luteinizing hormone receptor causes familial male-limited precocious puberty. *J. Clin. Endocrinol. Metab.*, **80**: 3168–3172.

Kremer, H., Kraaij, R., Toled, S. P. *et al.*, (1995) Male pseudohermaphroditism due to a homozygous missense mutation of the luteinizing hormone receptor gene. *Nat. Genet.,* **9**: 160–164.

Liao, W-X., Roy, A.C., Chan, C. *et al.*, (1998) A new molecular variant of luteinizing hormone associated with female infertility. *Fertil. Steril.,* **69**: 102–106.

Manna, P.R., Pakarinen, P., El-Hefnawy, T. and Huhtaniemi, I.T., (1999) Functional assessment of the calcium messenger system in cultured mouse Leydig tumor cells: Regulation of human chorionic gonadotropin-induced expression of the steroidogenic acute regulatory protein. *Endocrinology,* **140**: 1739–1751.

Mattews, C.H., Borgato, S., Beck-Peccoz, P. *et al.*, (1993) Primary amenorrhoea and infertility due to a mutation in the β-subunit of follicle-stimulating hormone. *Nat. Genet.,* **5**: 83–86.

Matzuk, M.M. and Boime, I. (1988) The role of aspergine-linked oligosaccharides of the α-subunit in the secretion and assembly of human chorionic gonadotropin. *J. Cell Biol.,* **106**: 1049–1059.

Matzuk, M.M., Keene, J.L. and Boime, I. (1989) Site specificity of the chorionic gonadotropin N-linked oligosaccharides in signal transduction. *J. Biol. Chem.,* **264**: 2409–2414.

Matzuk, M.M., Krieger, M., Corless, C.L. and Boime, I. (1987) Effects of preventing O-glycosylation on the secretion of human chorionic gonadotropin in Chinese hamster ovary cells. *Proc. Natl. Acad. Sci. USA,* **84**: 6354–6358.

Matzuk, M.M., Spangler, M.M., Camel, M. *et al.*, (1989) Mutagenesis and chimeric genes define determinants in the β-subunits of human chorionic gonadotropin and lutropin for secretion and assembly. *J. Cell Biol.,* **109**: 1429–1438.

McFarland, K.C., Sprengel, R., Philips, H.S. *et al.*, (1990) Lutropin/choriogonadotropin receptor: an unusual member of the G protein-coupled receptor family. *Science,* **245**: 494–499.

Pattersson, K., Ding, Y-Q. and Huhtaniemi, I. (1992) An immunologically anomalous luteinizing hormone variant in a healthy woman. *J. Clin. Endrocrinol. Metab.*, **74**: 161–171.

Pierce, J.G. and Parsons, T.F. (1981) glycoprotein hormones: structure and function. *Annu. Rev. Biochem.*, **50**: 465–495.

Rajkhowa, M., Talbot, J.A., Jones, P.W. *et al.*, (1995) Prevalence of an immunological LHβ-subunit variant in a UK population of healthy women and women with polycystic ovary syndrome. *Clin. Endocrinol. (Oxf),* **43**: 297–303.

Rebois, R.V. (1982) Establishment of gonadotropin-responsive murine Leydig tumor cell line. *J. Cell Biol.* **94**: 70–76.

Sairam, M.R. (1989) Role of carbohydrates in glycoprotein hormone signal transduction. *FASEB. J.*, **3**: 1915–1926.

Segaloff, D.L., and Ascoli, M. (1993) the lutropin/choriogonadotropin receptor... 4 years later. *Endocr. Rev.*, **14**: 324–347.

Simoni, M., Gromoll, J., Hoppner, W. and Nieschlag, E. (1997) The follicle-stimulating homone receptor: biochemistry, molecular biology, physiology, and pathophysiology. *Endocr. Rev.* **18**: 739–773.

Skzudlinski, M.W., Thotakura, N.R., Bucci, I. *et al.,* (1993) Purification and characterization of recombinant human thyrotropin (TSH) isoforms produced by Chinese hamster ovary cells: the role of sialylation and sulfation in TSH bioactivity. *Endocrinology,* **108**: 1490–1503.

Suganuma, N., Furui, K., Furuhashi, M. *et al.,* (1995) Screening of the mutations in the luteinizing hormone β-subunit in patients with menstrual disorders. *Fertil. Steril.,* **63**: 989–995.

Suganuma, N., Furui, K., Kikkawa, F. *et al.,* (1996) Effects of the mutations ($Trp^8 \rightarrow$ Arg and $Ile^{15} \rightarrow$ Thr) in human luteinizing hormone (LH) β-subunit on LH bioactivity *in vitro* and *in vivo. Endocrinology*; **137**: 831–838.

Tapanainen, J.S., Aittomaki, K., Min, J. *et al.*, (1997) Men homozygous for an activating mutation of the follicle-stimulating hormone (FSH) receptor gene present variable suppression of spermatogenesis and fertility. *Nat. Genet.,* **15**: 205–206.

Tapanainen, J.S., Koivunen, R., Fauser, B.M.J. *et al.*, (1997) Frequency of a varient form of LH in women with PCOS, *Hum. Reprod.* **12**: 85–86.

Tapanainen, J.S., Koivunen, R., Fauser, B.C.J.M. *et al.,* (1999) A new contributing factor to polycystic ovary syndrome: The genetic variant form of luteinizing hormone. *J. Clin. Endocrinol. Metab.,* **84**: 1711–1715.

Tulppala, M., Huhtaniemi, I. and Ylikorkala, O. (1998) Genetic varient of luteinizing hormone occurring frequently in women with a history of recurrent miscarriage and with relative overweight: no relation to luteal function and miscarriage rate. *Hum Reprod.,* **13**: 2699–2702.

Vuorento, T., Lahti, A., Hovatta, O. and Huhtaniemi, I.T., (1989) Daily measurements of salivary progesterone reveal a high rate of anovulation in healthy students. *Scand. J. Clin. Lab. Invest.* **49**: 395–401.

Weiss, J., Axelrod, L., Whitecomb, R.W. *et al.*, (1992) Hypogonadism caused by a single amino acid substitution in the β-subunit of luteinizing hormone. *N. Engl. J. Med.*, **326**: 179–183.

Zhang, F-P., Rannikko, A., Manna, P.R. *et al.*, (1997) Cloning and functional expression of the luteinizing hormone receptor complementary deoxyribonucleic acid from the marmoset monkey testis: Absence of sequences encoding exon 10 in other species. *Endocrinology*; **138**: 2481–2490.

Follicular Growth, Ovulation and Fertilization: Molecular and Clinical Basis
Anand Kumar and Amal K. Mukhopadhyay (Eds.)
Narosa Publishing House, New Delhi, India, 2001

6

Humanized Anti-hCG Antibodies for Safe and Wide Time Window Emergency Contraception

G.P. Talwar[1], Sonal Kathuria[1,2] and Rainer Fischer[2]
[1]Talwar Research Foundation, New Delhi-11068, India
[2]RWTH, Aachen-52074

Introduction

There is a continuing demand for methods that can prevent unwanted pregnancies caused by unprotected sex, incest or rape. Yuzpe's regime of steroids must be taken within 48 to 72 hours of the event, or it is ineffective. This regime has the known side effects of vomiting and nausea. A number of alternative methods are under development for emergency contraception, which include the anti progestin RU 486, insertion of an intra-uterine device or use of high doses of progestins. We propose the passive use of antibodies against the human chorionic gonadotropin (hCG) as a wide time window safe method for avoiding pregnancy.

Rationale

Human chorionic gonadotrophin (hCG) is an early signal of conception. It is synthesized by the embryo at the pre-implantation stage. *In vitro* fertilized eggs make and secrete hCG at the morula-blastocyst stage, as reported by Fischel, Edwards and Evans in a historical paper [1]. HCG has a crucial role in implantation. Marmoset embryos exposed to anti β-hCG antibodies (reactive with both human and marmoset CG) fail to implant [2]. The intervention is exercised by antibodies against the hormone, as exposure of the embryo to gamma globulins devoid of such antibodies has no effect on implantation.

The involvement of hCG in implantation is also indicated in humans. Sexually active women immunized with a vaccine that causes the generation of antibodies against hCG do not become pregnant [3]. The luteal phase of these women is not lengthened [4], which would be the case if the intervention occurred at the post-implantation stage. Thus anti-hCG antibodies intercept implantation and thereby prevent the onset of the pregnancy. Support to this contention is also provided by the observation that the protective threshold of antibodies for preventing pregnancy is 35 ng/ml of hCG bioneutralization capacity [4]. This level of hCG prevails before the attachment of the embryo to the endometrium, whereas hCG rises steeply beyond this figure post-implantation [5].

Embryonic implantation in humans is believed to occur on the 8th to 9th day post ovulation-fertilization. This period is much longer than the time limit for use of Yuzpe's regime. Besides implantation, hCG also exercises a crucial role in rescue of the ovarian corpus luteum for synthesis and secretion of progesterone which is vital for sustaining the gestation during the first seven weeks of pregnancy. Antibodies against hCG would be effective in aborting pregnancy if their administration is delayed beyond the 9th day of unprotected sex. In such circumstances, the antibodies would act as menstrual regulator.

Evidence for the Contra-Gestation Action of Anti-hCG Antibodies

Baboon chorionic gonadotropin is cross-reactive with human chorionic gonadotropin. Therefore, experiments to determine the action of anti hCG antibodies on pregnancy were performed in this sub human primate. Females were mated with fertile males and the onset of pregnancy was gauged by appearance of chorionic gonadotropin (CG) activity in serum as assayed by Leydig cell response and by non-decline of progesterone at the tail end of the luteal phase of the cycle. At this stage, immunoglobulins extracted from sera of monkeys immunized with an hCG vaccine were administered to the pregnant baboons on two consecutive days. Following this treatment, serum CG activity declined below the detection limit and was accompanied by a fall in serum progesterone. Vaginal bleeding ensued with termination of pregnancy. Complete termination of pregnancy was confirmed by sex steroid profiles and sex swelling in the following cycle (Fig. 1(A)). A parallel experiment performed with immunoglobulins extracted from the serum of non vaccinated monkeys exerted no effect and pregnancy continued with suppression of cyclicity and sexual skin swelling. Termination of pregnancy by anti-hCG antibodies had no residual effect on the future fertility of female baboons, Menstrual cycling post treatment was normal and when mated in a subsequent cycle, the baboon conceived and carried to term the progeny [6] (Fig. 1(B)).

The progeny born to females whose pregnancy was terminated earlier by passive administration of antibodies was normal in all respects. The developmental features of offspring were similar to the offspring of the untreated animals in the colony [7]. On reaching adulthood they cycle, mate and reproduce normally [8].

Safety of Circulating Anti-hCG Antibodies in Humans

The safety of anti-hCG antibodies and the lack of any adverse side effects due to these antibodies in women is amply documented in three series of Phase I

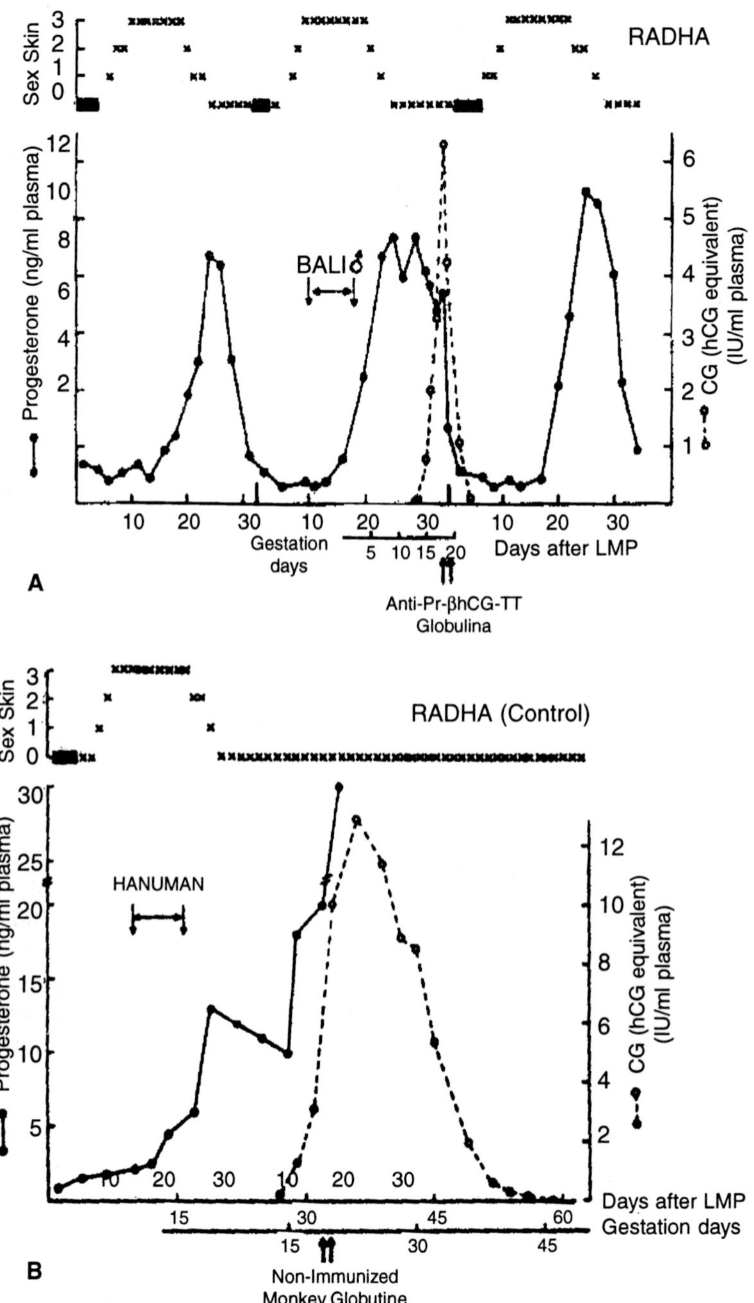

Fig 1 (A) Termination of pregnancy in a baboon by administration of anti-hCG antibodies induced in monkeys by the vaccine β-hCG-TT. (B) Reestablishment of fertility in the same baboon in a subsequent cycle and failure of normal monkey globulins to terminate pregnancy. (Reproduced from Talwar *et. al* [18].

clinical trials conducted on the hCG vaccines that we have developed [9, 10, 11] as well as on an anti-hCG vaccine developed by Stevens [12]. The trails were conducted on our vaccines in five centres in India and in Finland, Sweden, Brazil and Chile under the supervision of expert clinical scientists. The trials of the Steven's vaccine was carried out in Australia. Thorough clinical, laboratory and immunopathological investigations showed in every trial the safety of the anti-hCG vaccines. Other studies have ruled out any auto-reactivity of the generated anti-hCG antibodies [13] and lack of immunopathology as determined on autopsy of immunized primates [14]. No adverse immune complex deposits were seen in kidneys, choroid plexus and pituitary of monkeys immunized with the hCG vaccine and repeatedly challenged with hCG [15].

Further support for the safety of injected anti-hCG antibodies is provided by the Phase II clinical trials carried out on the HSD-HCG vaccine [4]. Amongst the 148 women of reproductive age who were sexually active, 110 generated antibodies above protective threshold of 50ng/ml. Approximately 30 per cent of these women had very high antibody titres, several fold higher than the protective threshold, which were maintained at high levels by periodic booster injections over a long time period (> $2^1/_2$ years). None of these women experienced any side-effects on body functions, other than the fact that they did not become pregnant. Their menstrual cycles remained regular and of normal duration. They ovulated normally, and no change in blood chemistry, hematology or organ functions was detectable. Fig. 2, is an illustrative example of antibodies produced by 4 women, in whom where the titre remained above 800 ng/ml for about 30 months without any side-effect.

Regain of Fertility and Normality of Progeny Born

Women protected from becoming pregnant at and above 50 ng/ml antibody titre readily conceived when antibody titres declined below 35 ng/ml. Twenty two such pregnancies were recorded [4]. In most cases these pregnancies were terminated by vacuum aspiration. However, four female subjects decided to maintain the pregnancy. Their offspring have been carefully followed up by a pediatrician and psychologist over the last few years. Their physical development and cognitive capabilities are normal in relation to their siblings [16].

These long term observations clearly demonstrate that anti hCG antibodies are well tolerated by women and have no deleterious effect on body functions other than that of prevention of pregnancy during the period that the antibodies are present at and above 50 ng/ml titres.

Bio-effective Monoclonal Antibodies

For large scale therapeutic interventions, it is necessary to have pure antibodies of high titres, which can be produced in unlimited amounts and whose characteristics are consistent. Hybridoma technology permits the achievement of this objective. We developed a mouse monoclonal antibody (MoAb PIPP) reactive with hCG, with a high affinity ($K_a = 3 \times 10^{10} M^{-1}$) and which has high

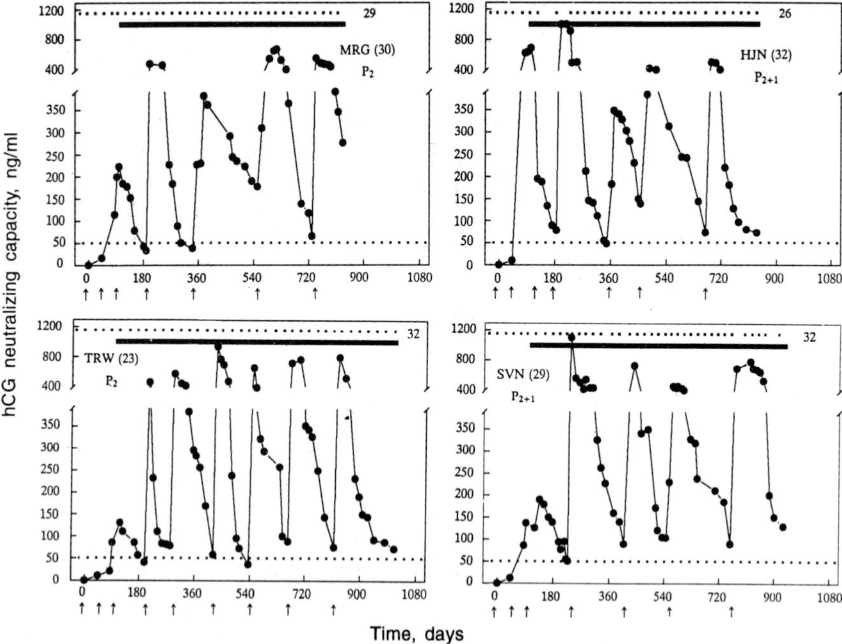

Fig. 2 Kinetics of anti-hCG response in four female volunteers after immunization with the HSD vaccine. All subjects (MRG,HJN,TRW,SVN; 30,32,23 and 29 years old respectively) were of proven fertility with two previous surviving children (P_2); HJN and SVN had also undergone elective termination of an unwanted pregnancy (P_{2+1}). Arrows denote injections with the vaccine at a gonadotropin dose of 300 μg. Rectangles near the top abscissa indicate menstrual events. Periods denoted by solid bars represent cycles in which the woman was exposed to the risk of pregnancy but did not become pregnant (Reproduced from Talwar *et. al* [4]).

specificity (< 5% cross reaction with LH and totally non-reactive with any other pituitary hormones) [17]. This MoAb neutralizes hCG bioactivity in both *in vitro* and *in vivo*, and is in principle an appropriate candidate for passive immunization to intercept an unwanted pregnancy. Although mouse monoclonals have been approved for use in terminal cancer cases the mouse monoclonals will however, be not permitted for use in healthy humans as contraceptives. Therefore, for therapeutic intervention in women, it was considered appropriate to humanize this MoAb.

Engineering and Humanization of MoAb PIPP

The part of antibody involved in recognition and binding with the antigen is contained in the variable portion of heavy and light chains. Accordingly the portion of genes coding for these regions in MoAb PIPP were identified and cloned. Hybrid cells actively synthesize the Moab. Accordingly RNA from these cells was extracted and reverse transcribed employing specific primers for variable fragments of mouse immunoglobulins. The gene fragments were amplified by PCR and cloned. Fragments were sequenced to confirm that these accorded to the sequences of immunoglobulins in the data bank. The DNA sequence of the three complementarity determining regions (CDRs) was determined in both heavy and light chains along with the sequences of frame work residues.

A single chain variable fragment was assembled incorporating the variable regions of both light and heavy chains coupled by a peptide linker. It was necessary to verify that this assembled gene can be expressed in a prokaryotic system to make a recombinant antibody specific for hCG. The recombinant ScFv was expressed in bacteria and the recombinant product demonstrated high affinity binding to hCG as determined by Surface Plasmon Resonance.

After engineering a murine recombinant anti-hCG antibody, analogues were sought in a human phage and combinatorial library. Repeated panning did not isolate a product of sufficient affinity. New libraries are being constructed to overcome this technical hurdle.

Meanwhile a chimeric antibody, in which the variable regions are mouse derived and the constant regions human in both light and heavy chains, has been constructed and transiently expressed in tobacco leaves. Agrobacterium mediated co-infiltration of two vectors encoding light and heavy chains was necessary to obtain antibody activity. The yield was high, even at this stage of transient expression. Plants make faithfully the active chimeric antibody of requisite properties which is bio-effective *in vitro* and *in vivo* systems.

Concluding Comments

Antibodies of high affinity and specificity against the human chorionic gonadotropin (hCG) can be used for intercepting the onset of pregnancy following unprotected sex. The antibodies prevent implantation, which is considered to take place between 6 and 9 days after fertilization. This time window is substantially longer than that of Yuzpe's regime. The antibodies would also be effective in aborting early pregnancy if given any time upto the expected day of menstruation, as hCG has a central role in sustaining ovarian corpus luteum functions during this period.

The use of specific antibodies against hCG would be totally safe and devoid of any side effects for preventing an unwanted pregnancy. By virture of their intrinsic properties high affinity, high specificity antibodies react only with the target antigen and thus no other molecule or systemic function is interfered with, in contracts to pharmacological drugs which though chosen for the desired efficacy, have invariably some minor side reactions.

Previous experience with the hCG vaccines generating antibodies against this hormone has clearly demonstrated the safety of anti-hCG antibodies. No long term residual side effects on future fertility of women or progeny born to previously immunized women or primates are observed. Phase II clinical trials with the hCG vaccine have indicated the effective dose levels of antibodies for intervention to be 50 ng/ml.

Humanized anti-hCG antibodies are now being engineered. Chimeric (mouse-human) anti hCG antibodies of high affinity and specificity have been expressed at high yield in plants. This technology offers the possibility to make a therapeutic recombinant antibody for prevention of pregnancy.

Acknowledgements

This work was supported by a research grant (98024#88) of the Rockefeller Foundation and the Talwar Research Foundation. S.K. was supported by DLR scholarship from the International Buro of the German Research Council (BMBF).

References

1. Fishel S.B., Edwards R.G., Evans C.J. Human chorionic gonaddotropin secreted by pre-implantation embryos cultured *in vitro. Science* 1984; **223**: 816–818.
2. Hearn J.P., Gidley–Baird A.A., Hodges J.K., Summers P.M., Wilbey G.E. Embryonic signals during the pre-implantation period in primates. *J. Reprod. Fertil (Suppl.) 1988*; **36**: 49–58.
3. Talwar G.P., Singh Om, Gupta S.K., Hasnain S.E., Pal R. *et al.* HSD-hCG vaccine prevents pregnancy in women, feasibility study of a reversible safe contraception vaccine. *Am. J. Reprod. Immunol.* 1997; **37**:153–160.
4. Talwar G.P., Singh Om, Pal R., Chatterjee N., Sahai P. *et al.* A vaccine that prevents pregnancy in women, *Proc. Natl. Acad. Sci. USA* 1994; **91**: 8532–8536.
5. Lenton E.A. Pituitary and ovarian hormones in implantation and early pregnancy. In: Implantation Biological and Clinical Aspects. Eds: H. Chapman, G. Grudzinskas, T. Chard. London. *Springer verlag 1988*; 17–29.
6. Tandon A, Das C, Jailkhani B.L., Talwar, G.P., Efficacy of antibodies generated by Pr-β-hCG-TT to terminate pregancy in baboons; its reversibility and rescue by medroxy progesterone acetae, *Contraception 1981*; **24**: 83–95.
7. Talwar G.P., Das C., Tandon A., Sharma M.G., Salahuddin M., Dubey S.K. Immunization against hCG : Efficacy and teratological studies in baboons. In:T.C. Anand Kumar Ed: Non-human Primate Models for Study of Human Reproduction. *Basel, S. Karger 1980*, 190–201.
8. Talwar G.P., Gupta S.K., Tandon A.K. Immunologic interruption of pregnancy, **In**: Norbert Gleicher, Reproductive Immunology. New York, *Alan R-Liss, 1981*, pp. 451–459.
9. Talwar G.P. *et al.* 14 papers in CONTRACEPTION 1976; **13**: 129–268 and 7 papers in CONTRACEPTION 1978; **18**: 19–104.
10. Nash H., Talwar G.P., Segal S., Luukkainen T., Johansson EDB, Vasquez J., Coutinho E., Sundaram K. Observations on the antigenicity and clinical effects of a candidate anti pregnancy vaccine: β subunit of human chorionic gonadotropin linked to tetanus toxoid. Fertil Steril 1980; **34**: 328–335.
11. Talwar G.P., Hingorani V., Kumar S., Roy S., Bannerjee A., Shahani S.M., Krishna U., Dhall K., Sawhney H., Sharma N.C., Om Singh, Gaur A., Rao L.V., Arunan K. Phase I clinical trials with three formulations of anti-human chorionic vaccine. Contraception 1990; **41**: 301–316.
12. Jones W.R., Bradley J., Judd S.J., Denholm E.H., ing RMY, Muelkr U.W., Powell J., Griffin P.D., Stevens V.C., Phase-I clinical trials of a World Health Organiation birth control vaccine. Lancet 1988; 1: 1295–98.

13. (a) Rose N.R., Burek C.L., Smith J.P., Safety evaluation of hCG vaccine in primates: auto antibody production, In: Talwar G.P. ed., Contraceptive Reseach for Today and the Nineties. New York. Springer Verlag 1988; 231–240.
    (b) Nath I, Whittingham S., Lambert P.H., Talwar G.P., Screening for autoantibodies in human subjects immunized with Pr-β-hCG-TT. Contraception 1976; **13**: 225–230.
14. Nath I, Gupta P.D., Bhuyan, U.N., Talwar, G.P. Autopsy report on rhesus monkeys immunized with Pr-β-hCG-TT vaccine. Contraceptoin 1976; **13**: 213–224.
15. Gupta P.D., Nath I, Talwar G.P., Immunofluroscence and electronic microscopic studies on kidney, choroids plexus and pituitary in rhesus monekeys immunized with Pr-β-hCG-TT vaccine. Contraception 1978; **18**: 91–104.
16. Mehrban Singh, Das, S.K., Suri, S., Om Singh, Talwar, G.P., Regain of fertility and normality on progency born during below protective threshold antibody antibody titers in women immunized with the HSD-hCG vaccine. Am J. Reprod. Immunol, 1998; **39**: 395–398.
17. Gupta S.K., Talwar, G.P., Development of hybridomas secreting anti-human chorionic gonadotropin antibodies. Ind. J. Exp. Biol. 1980; **18**: 1361–65.

Ovarian Functions

Follicular Growth, Ovulation and Fertilization: Molecular and Clinical Basis
Anand Kumar and Amal K. Mukhopadhyay (Eds.)
Narosa Publishing House, New Delhi, India, 2001

7

The Steroidogenic Acute Regulatory (StAR) Protein

Douglas M. Stocco, Lance P. Walsh, Adam J. Reinhart and Xing Jia Wang
Department of Cell Biology and Biochemistry
Texas Tech University Health Sciences Center, Lubbock, Texas 79430

Introduction

Steroid hormones are a group of small molecules derived from cholesterol that are synthesized in specialized cells in the adrenal, ovary, testis, placenta and brain. They have very diverse physiological functions in the body however, during the early steps, the steroid hormones are synthesized utilizing common pathways. Steroid biosynthesis begins with the conversion of cholesterol to the first steroid synthesized, pregnenolone, a reaction catalyzed by the cytochrome P450 side chain cleavage enzyme (P450*scc*) that is located on the matrix side of the inner mitochondrial membrane (Farkash *et al.*, 1986). Once pregnenolone is formed it is converted into a variety of steroids in the microsomal compartment, the final product being dependent upon the specific steroidogenic cell type. The activity of the P450scc enzyme in converting cholesterol to pregnenolone was long considered as the regulated and rate-limiting step in steroidogenesis, however, it later became clear that the *true* rate limiting step was the delivery of the substrate cholesterol to the inner mitochondrial membrane and to the P450scc (Simpson *et al.*, 1979; Crivello and Jefcoate, 1980; Privalle *et al.*, 1983).

Early observations also indicated that steroid biosynthesis had an absolute requirement for the synthesis of new proteins as indicated by Ferguson (Ferguson, 1962; Ferguson, 1963) and Garren (Garren *et al.*, 1965). The requirement for *de novo* protein synthesis in steroid production has been verified many times, and further studies were able to localize the step at which the putative protein(s) acted, namely, the delivery of the substrate from the outer to the inner mitochondrial membrane (Privalle *et al.*, 1983; Ohno *et al.*, 1983). Additional studies indicated that the acute production of steroids was dependent upon a hormone stimulated, rapidly synthesized cycloheximide sensitive, highly labile protein whose function was to mediate the transfer of cholesterol from the outer to the inner mitochondrial membrane and the P450scc enzyme. An intense search for this protein ensued and several candidate proteins have arisen from these efforts. Their candidacies have been previously reviewed (Stocco and Clark, 1996), and will not be repeated in detail here. Rather, this review will focus on a discussion of only one candidate protein, the Steroidogenic Acute Regulatory (StAR) protein. In addition, given space constraints, this review will only focus on those topics which were covered in the lecture presented at the AIIMS, namely some background information on StAR, what is currently known concerning the regulation of the StAR gene, the role of the arachidonic acid signalling pathway in steroidogenesis and StAR expression and lastly, the potential role of StAR in the inhibition of steroid hormone biosynthesis as a result of exposure to a number of environmental endocrine disruptors.

The Steroidogenic Acute Regulatory (StAR) Protein

StAR was initially described in ACTH-treated rat and mouse adrenocortical cells and in luteinizing hormone (LH)-treated rat corpus luteum cells and mouse Leydig cells as a 30 kDa phosphoprotein, (Kreuger and Orme-Johnson, 1983; Pon and Orme-Johnson, 1986; Pon *et al.*, 1986). Later, identical proteins were characterized in hormone stimulated MA-10 mouse Leydig tumor cells (Stocco and Kilgore, 1988; Stocco and Sodeman, 1991; Stocco and Chen, 1991; Stocco, 1992; Stocco and Ascoli, 1993). These proteins were found to be localized to the mitochondria and consisted of several forms of a newly synthesized 30 kDa protein. A 37 kDa precursor form of these proteins containing an N-terminal mitochondrial targeting sequence was also detected (Epstein and Orme-Johnson, 1991; Stocco and Sodeman, 1991). In 1994 the 37 kDa precursor form of this mitochondrial protein was cloned in our laboratory from mouse MA-10 Leydig tumor cells and when compared with other sequences in the data base both the nucleic acid sequence and protein sequence indicated it represented a novel protein (Clark *at al.*, 1994). A critical observation demonstrated that expression of the cDNA-derived protein in MA-10 cells in the absence of hormone stimulation or in COS-I monkey kidney tumor cells resulted in a several fold increase in the conversion of cholesterol to pregnenolone (Stocco and Clark, 1996; Clark *et al.*, 1996; Sugawara *et al.*, 1995a; Lin *et al.*, 1995), indicating a direct role for the 37 kDa protein in hormone-regulated steroid production. These observations were quickly followed by the finding that mutations in the StAR gene cause the potentially lethal disease, lipoid congenital adrenal hyperplasia (lipoid CAH), in which afflicted individuals are unable to synthesize adequate amounts of steroids for survival (Lin *et al.*, 1995). This condition essentially constituted a human knockout and offered further proof for the indispensable role of StAR in regulated steroid hormone biosynthesis. Recently, a mouse model of lipoid CAH was developed using targeted disruption of the StAR gene, and it was readily apparent that this animal model closely mimicked the human condition and once again demonstrated the essential role of StAR in steroid biosynthesis (Caron *et al.*, 1997a).

Regulation of the StAR Gene

The trophic hormone induced increase in steroid production in steroidogenic cells is accompanied by a rapid increase in StAR mRNA levels (Clark *et al.*, 1995; Caron *et al.*, 1997b). Since this is generally considered to be a cAMP mediated event, attention focused on the role of cAMP in the regulation of the StAR gene (Caron *et al.*, 1997b; Sugawara *et al.*, 1995b; Sugawara *et al.*, 1996; Sugawara *et al.*, 1997; Sandhoff *et al.*, 1998; LaVoie *et al.*, 1999; Rust *et al.*, 1998). Like several other cAMP-regulated genes involved in steroid biosynthesis, the promoter region of the StAR gene lacks a recognizable cAMP response element (CRE). Thus, cAMP stimulation is likely to be mediated by transcription factors that either are not members of the cAMP response element binding protein (CREB) family or by CREB proteins that recognize non-CRE consensus elements. Other studies have shown that the cAMP responsive site is retained within a 245 nucleotide region relative to the transcription start site (Caron *et al.*, 1997b), therefore, this region has been the focus of studies to identify promoter elements and their cognate-binding proteins that could mediate the cAMP response.

One of the first transcription factors considered to be potentially important in the regulation of the StAR gene was steroidogenic factor I (SF-1). SF-1 was first identified in adrenal cortical cells, and has been shown to be instrumental in regulating the cytochrome P450 steroid hydroxylase genes (Ikeda *et al.*, 1993). SF-1 also has an essential role in the development of steroidogenic and other cells types as shown in studies with SF-1 knockout mice (Shinoda *et al.*, 1995; Luo *et al.*, 1995). Of special interest to the regulation of the StAR gene, SF-1 knockout mice do not express StAR mRNA, indicating that SF-1 is required for proper StAR gene expression (Caron *et al.*, 1997a). Several SF-1 consensus binding sites have been found in the StAR promoter. In rat, five SF-1 sites have been reported (Sandhoff *et al.*, 1998), and one additional site has been identified in human and mouse (Caron *et al.*, 1997b; Sugawara *et al.*, 1997). Two of these sites, which are located at positions-97 and -42, are highly conserved in several species, whereas the -132 site may only be present in mouse and rat. Most importantly, SF-1 has been demonstrated to transactivate the StAR promoter in transient transfection assays in several cell types (Caron *et al.*, 1997b; Sugawara *et al.*, 1996; Sugawara *et al.*, 1997; Sandhoff *et al.*, 1998). It is also possible that SF-1 may play some role in the developmental regulation of the StAR gene as StAR mRNA is not detected in the urogenital ridge of SF-1 null mice (Caron *et al.*, 1997a).

Following initial studies with SF-1, it quickly became apparent that other elements were likely involved in StAR's tissue specific and time specific expression. As such the search for additional transacting factors that may be involved in the regulation of the StAR gene has begun. The CCAAT/Enhancer Binding Proteins (C/EBPs), are a family of basic region/leucine zipper transcription factors implicated as regulators of differentiation and function of multiple cell types (Johnson and Williams, 1994). Previous studies have demonstrated that two family members, C/EBPα and C/EBPβ are expressed in steroidogenic cells, including Leydig cells and ovarian granulosa cells (Nalbant *et al.*, 1998; Sirois and Richards, 1993). Two putative C/EBP binding sites in the StAR promoter have recently been identified in our laboratory (Reinhart *et al.*, 1999). We have determined that the StAR promoter is transactivated by C/EBPβ during transient transfection assays and that SF-1transactivation of the StAR promoter is dependent upon the presence of functional C/EBP binding sites, suggesting that SF-1 and C/EBPβ may form a complex on this promoter. In yet another recent and most interesting study, it has been demonstrated that the Sterol Regulatory Element Binding Protein (SREBP) may also be involved in the regulation of the StAR gene as it was reported that SREBP-la was capable of transactivating the StAR promoter (Christenson *et al.*, 1998). While no consensus binding sites for SREBP-la have yet been found in the StAR promoter, this may prove to be a good example of the convergence of the regulation of cholesterol metabolism and steroidogenesis.

Since it is becoming abundantly clear that regulation of the StAR gene is a complex process, the biggest challenge in this area will be to characterize those elements that are involved in this regulation and that serve to provide tissue specific and temporal specific expression of the gene.

Role of Arachidonic Acid in Steroidogenesis and StAR Expression

Arachidonic acid is produced in cells mainly through the activation of the enzyme phospholipase A_2 (PLA_2), which catalyzes its release from phospholipids. Further arachidonic acid metabolism through one of three enzyme pathways, lipoxygenase, cyclooxygenase or cytochrome P450-dependent epoxygenase, produces various metabolites (Needleman *et al.*, 1986; Willis and Smith, 1994). Arachidonic acid and its metabolites have been demonstrated to act as mediators or regulators in the cardiovascular system, the immune system, the central nervous system, the reproductive system and during the course of inflammation (Vane and Botting, 1994). It has been well documented that arachidonic acid or one of its metabolites plays a critical role in the regulation of steroidogenesis and these regulatory effects have been studied for more than two decades.

It has been reported that arachidonic acid can function as a stimulatory mediator in luteinzing hormone-releasing hormone-stimulated progesterone production in the ovary (Wang and Leung, 1989). It also enhanced human chorionic gonadotrophin (hCG)-or forskolin-induced steroid production in goldfish preovulatory ovarian follicles (Van der Kraak and Chang, 1990). Arachidonic acid also induced a dose dependent increase of progesterone and estradiol in cultured granulosa cells, theca cells or corpus luteum cells isolated from goat ovary (Band *et al.*, 1986). In testicular Leydig cells, arachidonic acid regulated both basal and hormone-stimulated steroid production as demonstrated by the induction of dose-dependent increases of testosterone formation in rats and goldfish (Romanelli *et al.*, 1996; Wade and Van der Kraak, 1993). This observation was strengthened when it was found that inhibition of arachidonic acid release from phospholipids inhibited luteinizing hormone (LH)-stimulated testosterone production in rat Leydig cells (Cooke *et al.*, 1991). Also, a dose-and time-dependent biphasic effect of arachidonic acid on LH-or dibutyryl cyclic AMP (dbcAMP)-induced steroid production was reported in rat Leydig cells. In adrenal cells, arachidonic acid stimulation induced a dose-dependent increase of pregnenolone production (Yamazaki *et al.*, 1996). Also, Mikami and colleagues (Mikami *et al.*, 1990) demonstrated that blocking arachidonic acid metabolism in adrenal cells inhibited adrenal corticotrophic hormone (ACTH)-induced pregnenolone production, but this inhibition could be reversed by prior addition of arachidonic acid metabolites to the cells. In rat glomerulosa cells it was observed that angiotensin II (AII) stimulation released arachidonic acid from diacylglycerol (DAG), and that inhibition of this release with a DAG lipase inhibitor inhibited aldosterone production (Natarajan *et al.*, 1990). From observations such as these, it was suggested that the mechanisms for ACTH and AII stimulated steroidogenesis may involve an increase of intracellular arachidonic acid release (Solano *et al.*, 1987).

Although regulatory effects of arachidonic acid and its metabolites on trophic hormone-induced steroid production have been clearly demonstrated, the mechanism(s) for these effects is not clear. It was reported that arachidonic acid and its metabolites appeared to regulate the rate-limiting step in steroidogenesis, namely, cholesterol transfer from the outer to inner mitochondrial membrane where the P450scc resides. This hypothesis was strengthened when it was demonstrated that arachidonic acid and its metabolites had no effect on the activities of P450scc and 3β-hydroxysteroid dehydrogenase (3β-HSD) (Lopez-Ruiz *et al.*, 1992; Mele *et al.*, 1997). Since StAR protein has been demonstrated to play a critical role in mediating cholesterol transfer to the inner mitochondrial membrane in response to hormone stimulation (Clark *et al.*, 1994; Lin *et al.*, 1995; Stocco, 1997; Wang *et al.*, 1998), we reasoned that StAR may be involved in the arachidonic acid pathway as well. In a recent study we showed that arachidonic acid release is one of the requirements needed to regulate steroidogenesis and appears to do so through participation in the regulation of StAR protein expression (Wang *et al.*, 1999a).

It is generally accepted that cAMP transduces the signal from trophic hormones, such as LH, follicle-stimulating hormone (FSH), hCG, and ACTH, to the nucleus to induce StAR gene expression followed by translation of the StAR mRNA and subsequently increased steroidogenesis. However, LH-or dbcAMP-induced StAR protein and progesterone production in MA-10 cells were shown to be inhibited by the PLA_2 inhibitor quinacrine which blocked the release of arachidonic acid from phospholipids (Wang *et al.*, 1999a). The inhibition in steroid production was dose-dependent, as was that of StAR protein expression. Progesterone production and StAR protein expression were almost completely inhibited by 20μM quinacrine and, importantly, the inhibitory effect was reversed by the addition of arachidonic acid, with both StAR protein and progesterone being restored as arachidonic acid concentration in the cells increased. Additional studies have recently demonstrated that arachidonic acid regulates StAR protein expression at the level of transcription (Wang *et al.*, 1999b). While dibutyryl cyclic AMP stimulated StAR mRNA synthesis, inhibition of arachidonic acid release from phospholipids by the PLA_2 inhibitor, quinacrine, inhibited dbcAMP-induced StAR mRNA. This inhibition in transcription of the StAR mRNA was also reversed by co-incubation of the cells with exogenous arachidonic acid. As a further demonstration of this transcriptional level regulation, StAR promoter activity was tested using the dual-luciferase reporter assay system, and it was clearly seen that dbcAMP-induced promoter activity was significantly inhibited by quinacrine. This inhibition was also reversed by addition of exogenous arachidonic acid (Wang *et al.*, 1999b). These results clearly indicate that release of arachidonic acid plays an important role in LH or cAMP-induced StAR gene expression.

Effects of Xenobiotics on Steroidogenesis and StAR Expression

A large number of environmental pollutants have been shown to disrupt male reproductive function. While it is known that many of these toxicants can inhibit steroid hormone biosynthesis and reduce serum testosterone levels, the mechanism of steroidogenic inhibition is poorly understood. Until recently, the rate-limiting step in steroidogenesis has largely been overlooked by researchers as a potential target for steroidogenic inhibition by environmental pollutants. In the past, investigators typically used radioactively labeled steroid hormone precursors to study the effects of xenobiotics on steroidogenic enzymes located in the pathway downstream of cholesterol transfer, thus overlooking what may potentially be an important site of steroidogenic blockade. Several observations led us to hypothesize that environmental toxicants may block steroidogenesis via the disruption of StAR protein expression. First, in contrast to the steroidogenic enzymes which have long half-lives and are chronically regulated, StAR protein is not an enzyme, is acutely regulated and its active form is highly labile and must be continuously synthesized for steroidogenesis to occur. Second, StAR protein mediates the rate-limiting step in steroidogenesis, and as a result, steroidogenesis is very sensitive to disruption in StAR protein expression. Finally, recent studies have shown that the environmental pollutants, α, δ, and γ-hexachlorocyclohexane (Zisterer *et al.*, 1996), and the antifungal drugs, econazole and miconazole (Zisterer *et al.*, 1997), inhibit steroidogenesis at the level of cholesterol transfer, thus implicating StAR protein as a potential target for these compounds. Recent results obtained in our laboratory (in preparation) indicates that the organochlorine insecticide lindane (γ-hexachlorocyclohexane) and its α and δ isomers, the herbicide Roundup, the organophosphate insecticide Dimethoate, and the imidazole antifungals econazole and miconazole, inhibit steroidogenesis either in part or entirely through the inhibition of StAR protein expression. Although these chemicals possess different chemical structures and physical characteristics, they all have been demonstrated to interrupt steroidogenesis at the level of StAR protein expression. Therefore, these results support our hypothesis that at least some environmental toxicants can block steroidogenesis via the disruption of StAR protein expression and that this protein should be included in the list of factors to be studied when investigating the effects of environmental endocrine disruptors on reproductive function.

Acknowledgments

Authors wish to acknowledge the support of NIH grant HD 17481 during the course of these studies. LPW was supported by NH grant T32 HD07271 and by the Lubbock ARCS Chapter.

References

Band, V., Kharbanda, S.M., Murugesan, K. *et al.*, (1986) Prostacyclin and steroidogenesis in goat ovarian cell types *in vitro*. Prostaglandins, **31**, 509–525.

Caron, K.M., Soo, S.C., Wetsel, W. *et al.*, (1997a) Targeted disruption of the mouse gene encoding steroidogenic acute regulatory protein provides novel insights into congenital lipoid adrenal hyperplasia. *Proc. Natl. Acad. Sci.,* **94**, 11540–11545.

Caron, K.M., Ikeda, Y., Soo, S.C. *et al.*, (1997b) Characterization of the promoter region of the mouse gene encoding the steroidogenic acute regulatory (StAR) protein. *Molec. Endocrinol.*, **11**, 136–147.

Christenson, L.K., McAllister, J.M., Martin, K.O., *et al.*, (1998) Oxysterol regulation of steroidogenic acute regulatory protein gene expression: structural specificity and transcriptional and postranslational actions. *J. Biol. Chem.*, **273**, 30729–30735.

Clark, B.J., Wells, J., King, S.R. *et al.* (1994) The purification, cloning, and expression of a novel LH-induced mitochondrial protein in MA-10 mouse Leydig tumor cells: characterization of the steroidogenic acute regulatory protein (StAR). *J. Biol. Chem.,* **269**, 28314–28322.

Clark, B.J., Soo, S.C., Caron, K.M. *et al.* (1995) Hormonal and developmental regulation of the steroidogenic acute regulatory (StAR) protein. *Molec. Endocrinol.*, **9,** 1346–1355.

Cooke, B.A., Dirami, G., Chaudry, L. *et al.*, (1991) Release of arachidonic acid and the effects of corticosteroids on steroidogenesis in rat testis Leydig cells. *J. Steroid. Biochem. Mol. Biol.* **40**, 465–471.

Crivello, J.F. and Jefcoate, C.R. (1980) Intracellular movement of cholesterol in rat adrenal cells. *J. Biol. Chem.*, **255**, 8144–8151.

Epstein, L.F. and Orme-Johnson, N.R. (1991) Regulation of steroid hormone biosynthesis: Identification of precursors of a phosphoprotein targeted to the mitochondrion in stimulated rat adrenal cortex cells. *J. Biol. Chem.,* **266,** 19739–19745.

Farkash, Y., Timberg, R. and Orly, J. (1986) Preparation of antiserum to rat cytohrome P-450 cholesterol side chain cleavage, and its use for ultrastructural localization of the immunoreactive enzyme by protein A-gold technique. *Endocrinology*, **118**, 1353–1365.

Ferguson, J.J. (1962) Puromycin and adrenal responsiveness to adrenocorticotropic hormone. *Biochem. Biophys. Acta*, **57**, 616–617.

Ferguson, J.J. (1963) Protein synthesis and adrenocorticotropin responsiveness. *J. Biol. Chem.*, **238**, 2754–2759.

Garren, L.D., Ney R.L. and Davis, W.W. (1965) Studies on the role of protein synthesis in the regulation of corticosterone production by ACTH *in vivo. Proc. Natl. Acad. Sci. USA*, **53**, 1443–1450.

Ikeda, Y., Lala, D.S., Luo, X. *et al.*, (1993) Characterization of the mouse FTZ-Fl gene, which encodes a key regulator of steroid hydroxylase gene expression. *Mol. Endocrino.*, **7**, 852–860.

Johnson, P.F. and Williams S.C. (1994) CCAAT/Enhancer binding (C/EBP) proteins. In *Liver Gene Expression* (ed. F. Tronch and M. Yaniv), pp. 231–258, Landes Company.

Krueger, R.J. and Orme-Johnson, N.R. (1983) Acute adrenocorticotropic hormone stimualtion of adrenal corticosteroidogenesis. *J. Biol. Chem.*, **258**, 10159–10167.

LaVoie, H., Garmey, J.C. and Veldhuis, J.D. (1999) Mechanisms of insulin-like growth factor-1 augmentation of follicle-stimulating hormone-induced porcine steroidogenic acute regulatory protein promoter activity in granulosa cells. *Endocrinology,* **140**, 146–153.

Lin, T. (1985) Mechanism of action of gonadotropin-releasing hormone stimulated Leydig cell steroidogenesis. III. The role of arachidonic acid and calcium/phospholipid dependent protein kinase. *Life Sci.,* **36**, 1255–1264.

Lin, D., Sugawara, T., Strauss III, J.F. *et al.*, (1995) Role of steroidogenic acute regulatory protein in adrenal and gonadal steroidogenesis. *Science*, **267**, 1828–1831.

Lopez-Ruiz, M.P., Choi, M.S.K., Rose, M.P. *et al.*, (1992) Direct effect of arachidonic acid on protein kinase C and LH-stimualted steroidogenesis in rat Leydig cells; Evidence for tonic inhibitory control of steroidogenesis by protein kinase C. *Endocrinology*, **130**, 1122–1130.

Luo, X., Ikeda, Y. and Parker, K.L. (1995) A cell-specific nuclear receptor is essential for adrenal and gonadal development and sexual differentiation. *Cell*, **77**, 481–490.

Mele, P.G., Dada, L.A., Paz, C. *et al.*, (1997) Involvement of arachidonic acid and lipoxygenase pathway in mediating luteinizing hormone-induced testosterone synthesis in rat Leydig cells. *Endocr. Res.,* **23**, 15–26.

Mikami, K., Omura, M., Tamura, Y. *et al.*, (1990) Possible site of action of 5-hydroperoxyeicosatetraenoic acid derived from arachidonic acid in ACTH-stimulated steroidogenesis in rat adrenal glands. *J. Endocrinol.*, **125**, 89–96.

Nalbant, D., Williams, S.C., Stocco, D.M. *et al.*, (1998) Luteinizing hormone-dependent gene regulation in Leydig cells may be mediated by CCAAT/enhancer-binding protein-beta. *Endocrinology,* **139**, 272–279.

Natarajan, R., Dunn, W.D., Stern, N. *et al.*, (1990) Key role of diacylglycerol-mediated 12-lipoxygenase product formation in angiotensin II-induced aldosterone synthesis. *Mol. Cell. Endocrinol.*, **72**, 73–80.

Needleman, P., Turk, J., Jakschik, B.A., *et al.*, (1986) Arachidonic acid metabolism, *Ann, Rev. Biochem.*, **55**, 69–102.

Ohno, Y., Yanagibashi, K., Yonezawa, Y. *et al.* (1983) A possible role of "steroidogenic factor" in the corticoidogenic response to ACTH; effect of ACTH, cycloheximide and aminoglutethimide on the content of cholesterol in the outer and inner mitochondrial membrane of rat adrenal cortex. *Endocrinology (Jpn)*, **30**, 335–338.

Pon, L.A. and Orme-Johnson, N.R. (1986) Acute stimulation of seroidogenesis in corpus luteum and adrenal cortex by peptide hormones: rapid induction of a similar protein in both tissues. *J. Biol. Chem.*, **261**, 6594–6599.

Pon, L.A., Epstein, L.F. and Orme-Johnson, N.R. (1986) Acute cAMP stimulation in Leydig cells: rapid accumulation of a protein similar to that detected in adrenal cortex and corpus luteum. *Endocr. Res.,* **12**, 429–446.

Privalle, C.T., Crivello, J. and Jefcoate, C.R. (1983) Regulation of intramitochondrial cholesterol transfer to side-chain cleavage cytochrome P450scc in rat adrenal gland. *Proc. Natl. Acad. Sci. USA*, **80**, 702–706.

Reinhart, A.J., Williams, S.C., Clark, B.J. *et al.*, (1999) SF-1 (Steroidogenic Factor 1) and C/EBPβ (CAAT Enhancer Binding Protein β) cooperate to regulate the murine StAR (Steroidogenic Acute Regulatory Protein) promoter, *Molec. Endocrinol.* **13,** 729–741.

Romanelli, F., Valenca, M., Conte, D. *et al.* (1995) Arachidonic acid and its metabolites effects on testosterone production by rat Leydig cells. *J. Endocrinol. Invest.*, **18**, 186–193.

Rust, W. Stedronsky, K., Tillmann, G. *et al.*, (1998) The role of SF-1/Ad4BP in the control of the bovine gene for the steroidogenic acute regulatory (StAR) protein. *J. Mol. Endocrinol.*, **21**, 189–200.

Sandhoff, T.W., Hales, D.B., Hales, K.H., *et al.* (1998) Transcriptional regulation of the rat steroidogenic acute regulatory protein gene by steroidogenic factor 1. *Endocrinology*, **139**, 4820–4831.

Shinoda, K., Lei, H., Yoshii, H. *et al.*, (195) Developmental defects of the ventromedial hypothalmamic nucleus and pituitary gonadotroph in the Ftz-F1 disrupted mice. *Dev. Dyn.*, **204**, 22–29.

Simpson, E.R., McCarthy, J.L. and Peterson, J.A. (1979) Evidence that the cycloheximide-sensitive site of ACTH action is in the mitochondrion. *J. Biol. Chem.*, **253**, 3135–3139.

Sirois, J. and Richards, J.S., (1993) Transcriptional regulation of the rat prostaglandin endoperoxide synthase 2 gene in granulosa cells. Evidence for the role of a cis-acting C/EBP beta promoter element. *J. Biol. Chem.*, **268**, 21931–21938.

Solano, A.R. Dada, L.A., Luz Sardanons, M. *et al.* (1987) Leukotrienes as common intermediates in the cyclic AMP dependent and independent pathways in adrenal steroidogenesis. *J. Steroid Biochem.*, **27**, 745–751.

Stocco, D.M. and Kilgore, M.W. (1988) Induction of mitochondrial proteins in MA-10 Leydig tumour cells with human choriogonadotropin. *Biochem. J.*, **249**, 95–103.

Stocco, D.M. and Sodeman, T.C. (1991) The 30-kDa mitochondrial proteins induced by hormone stimulation in MA-10 mouse Leydig tumor cells are processed from larger precursors. *J. Biol. Chem.*, **266**, 19731–19738.

Stocco, D.M. and Chen, W. (1991) Presence of identical mitochondrial proteins in unstimulated constitutive steroid-producing R2C rat Leydig tumor and stimulated nonconstitutive steroid-producing MA-10 mouse Leydig tumor cells. *Endocrinology*, **128**, 1918–1926.

Stocco, D.M. (1992) Further evidence that the mitochondrial proteins induced by hormone stimulation in MA-10 mouse Leydig tumor cells are involved in the acute regulation of steroidogenesis. *J. Steroid Biochem. Mol. Biol.* **43**, 319–333.

Stocco, D.M. and Ascoli, M. (1993) The use of genetic manipulation of MA-10 Leydig tumor cells to demonstrate the role of mitochondrial proteins in the acute regulation of steroidogenesis. *Endocrinology*, **132**, 959–967.

Stocco, D.M. and Clark, B.J. (1996) Regulation of the acute production of steroids in steroidogenic cells. *Endocr. Rev.* **17**, 221–244.

Stocco, D.M. (1997) The steroidogenic acute regulatory (StAR) protein two years later: an update. *Endocrine*, **6**, 99–109.

Sugawara, T., Holt, J.A., Driscoll, D. *et al.*, (1995a) Human steroidogenic acute regulatory protein: functional activity in COS-1 cells, tissue-specific expression and mapping of the gene to 8p11.2 and a pseudogene to chromosome 13. *Proc. Natl. Acad. Sci. USA*, **92**, 4778–4782.

Sugawara, T., Lin, D., Holt, J.A. *et al.*, (1995b) Structure of the human steroidogenic acue regulatory protein (StAR) gene: StAR stimulates mitochondrial cholesterol 27-hydroxylase activity. *Biochemistry*, **34**, 12506–12512.

Sugawara, T., Holt, J.A., Kiriakidou, M. *et al.*, (1996) Steroidogenic factor 1-dependent promoter activity of the human steroidogenic acute regulatory protein (StAR) gene. *Biochemistry*, **35**, 9052–9059.

Sugawara, T., Kiriakidou, M., McAllister, J.M., *et al.* (1997) Multiple steroidogenic factor 1 binding elements in the human steroidogenic acute regulatory protein gene

5'-flanking region are required for maximal promoter activity and cyclic AMP responsiveness. *Biochemistry*, **36**, 7249–7255.

Van der Kraak G. and Chang, J.P. (1990) Arachidonic acid stimulates steroidogenesis in goldfish preovulatory ovarian follicles. *Gen. Comp. Endocrinol.*, **77**, 221–228.

Vane, J.R. and Botting, R.M. (1994) Biological properties of cyclooxygenase products. In "Handbook of Immunopharmacology: Lipid Mediators" (edited by Cunningham, F.M.) pp. 62–97. Academic Press, San Diego, CA.

Wade, M.G. and Van der Kraak, G. (1993) Arachidonic acid and prostaglandin E2 stimulate testosterone production by goldfish testis *in vitro. Gen. Comp. Endocrinol,* **90**, 109–118.

Wang, J. and Leung, P.C. (1989) Arachidonic acid as a stimulatory mediator of luteinizing hormone-releasing hormone action in the rat ovary. *Endocrinology*, **124**, 1973–1979.

Wang, X., Liu Z., Eimerl, S. *et al.* (1998) Effect of truncated forms of the steroidogenic acute regulatory protein on intramitochodrial cholesterol transfer. *Endocrinology*, **139**, 3903–3912.

Wang, X., Walsh, L.P. and Stocco, D.M. (1999a) The role of arachidonic acid on LH-stimulated steroidogenesis and steroidogenic acute regulatory protein accumulation in MA-10 mouse Leydig tumor cells. *Endocrine*, **10**, 7–12.

Wang, X., Reinhart, A.J., Walsh, L.P. *et al.* (1999b) Arachidonic acid regulation of Steroidogenic Acute Regulatory (StAR) protein gene expression and steroid hormone production. FASEB J., **13**, A556.

Willis, A.L. and Smith, D.L. (1994) Metabolism of arachidonic acid. In "Handbook of Immunopharmacology: Lipid Mediators" (edited by Cunningham, F.M.,) pp. 2–32. Academic Press, San Diego, CA.

Yamazaki, T., Higuchi, K., Kominami, S. *et al.* (1996) 15-lipoxygenase metabolite(s) of arachidonic acid mediates adrenal corticotropin action in bovine adrenal steroidogenesis. *Endocrinology*, **137**, 2670–2675.

Zisterer, D.M., Moynagh, P.N. and Williams, D.C. (1996) Hexachlorocyclohexanes inhibit steroidogenesis in Y1 cells. Absence of correlation with bonding to the peripheral-type benzodiazepine receptor. *Biochem. Pharmacol.*, **51**, 1303–1308.

Zisterer, D.M., and Williams, D.C. (1997) Calmidazolium and other imidazole compounds affect steroidogenesis in Y1 cells: lack of involvement of the peripheral-type benzodiazepine receptor. *J. Ster. Biochem. Mol. Biol.,* **60**, 189–195.

Follicular Growth, Ovulation and Fertilization: Molecular and Clinical Basis
Anand Kumar and Amal K. Mukhopadhyay (Eds.)
Narosa Publishing House, New Delhi, India, 2001

8

Ovarian Hyperstimulation in Bonnet Monkeys Using Gonadotrophins

P.B. Seshagiri, K.K. Acharya, D. Jayaprakash, K.S. Satish and G. Shetty
Department of Molecular Reproduction, Development and Genetics
Indian Institute of Science, Bangalore 560012, India

Introduction

A remarkable breakthrough in the area of primate reproduction is the production of genetically identical rhesus monkeys by either nuclear transfer (Meng *et al.*, 1997) or embryo splitting (Chan *et al.*, 2000), although the birth of the first test tube-rhesus monkey was reported much earlier (Bavister *et al.*, 1984). Despite these achievements, the regulation of primate preimplantation embryo development and implantation is still one of the least studied aspects of mammalian reproduction. While the necessity and advantages of non-human primate models, such as rhesus and bonnet monkeys, for the study of human reproduction are well recognized (Murthy *et al.*, 1979; Wolf *et al.*, 1990; Moudgal *et al.*, 1991; Bavister and Boatman, 1993), they have been of limited use, unlike rodents, in embryology research due to the lack of adequate supply of preimplantation embryos. There is a need to increase in vivo embryo production per donor to facilitate basic studies on the nature and regulation of early development and implantation, which cannot be examined in women for ethical and practical reasons. Induction of superovulation using gonadotrophins, followed by nonsurgical recovery of multiple embryos, could be one of the strategies to enhance the rate of embryo recovery from primates. This however, has not been pursued. The purpose of this report is to describe data on our attempts to produce multiple embryos in bonnet monkeys, following ovarian hyperstimulation using gonadotrophins.

Gonadotrophin Dose and Follicular Development

In the past, stimulation of growth and maturation of multiple follicles using various formulations of gonadotrophins in primates have been extensively studied and most of them however, involved retrieving oocytes by follicular aspiration and fertilizing them in vitro (Wolf *et al.*, 1990; McCarthy *et al.*, 1991; Bavister and Boatman, 1993; Zelinski-Wooten *et al.*, 1994). While this approach can produce a reasonable number of cleavage-stage embryos by culturing them in vitro, their developmental competence is not always comparable to those produced in vivo. In this regard, a nonsurgical uterine flushing technique has earlier been standardized to obtain in vivo developed preimplantation embryos from naturally-ovulating, naturally-bred rhesus and bonnet monkeys (Goodeaux *et al.*, 1990; Seshagiri *et al.*, 1993; Jayaprakash *et al.*, 1997). These embryos successfully develop through the peri-attachment stages (Seshagiri and Hearn, 1993), facilitating studies on embryonic endocrine secretions (Seshagiri *et al.*, 1994).

Multiple follicular development and ovulation, as indicated by the presence of many corpora lutea (CLs) have been induced in rhesus monkeys using crude preparations of gonadotrophins (Simpson and van Wagenen, 1958; 1962) and also by the use of recombinant gonadotrophins (Simon *et al.*, 1988; Weston *et al.*, 1996). Similarly, it has been achieved in baboons using a 11-day regimen of hMG and hCG (McCarthy *et al.*, 1991). However, serious efforts have not been made in any species to recover multiple embryos. Moreover, there are virtually no studies on the induction of multiple ovulation in bonnet monkeys. By "a staircase method" of administration of hMG and hCG, induction of single ovulation has been achieved in this species by Ovadia *et al.*, (1971). In view of this, we attempted to induce multiple ovulation using pure preparations of gonadotrophins in bonnet monkeys with timed mating. Such studies describing methods for use in the bonnet monkeys have not been undertaken so far.

Two protocols for hormone stimulation were employed (Fig. 1). The first one (n = 4) involved two daily (9.45 h in the morning and 17.00 h in the evening) i.m. injections of 12.5 IU each of hFSH on days 3–8 of menstrual cycle. In the morning of day 9, 30 IU of hMG was injected followed by an ovulatory dose (1000 IU) of hCG in the evening on the same day (Fig. 1). Of the 4 monkeys subjected to protocol-I, the monkey #70 produced an unfertilized oocyte (UFO) and others did not yield any ovulation product (Table 1). Only two females (#46, #71) showed raising levels of circulating estradiol (peak: 1.4–2 ng/ml) and in the remaining two monkeys (#70, #72), undetectable or low levels of estradiol were seen (data not shown). Laparotomic observations revealed that ovaries had neither the expected increase in their sizes nor multiple follicles/CLs in monkeys #46 and #72 (Fig. 2A,B; Table 1). Though, follicular steroidogenic activity was evident in monkey #46, no antral follicles could be detected in ovaries, suggesting that a normal and complete development of ovulable follicles did not occur even though the animal showed elevated levels of serum estradiol. Moreover, we observed that the hormone profile of the monkey #70, that yielded an ovulation product, was comparable to that of a natural (unstimulated) cycle, thereby indicating that the recovery of ovulation product from this monkey might have not been due to exogenous gonadotrophin administration.

These findings indicate that the protocol-I fails to induce multiple follicular growth and hence multiple ovulation. This is attributable to inadequate amounts of gonadotrophin administration and/or to the refractoriness of animals to low doses of gonadotrophins. Hence, we designed a second protocol (II; Fig. 1), a more aggressive stimulation regimen to achieve a large number of mature follicles with a subsequent increased likelihood of producing multiple embryos. In this protocol (n = 4), similar to the one attempted in rhesus monkeys to induce multiple follicular development (Wolf *et al.*, 1990), the amount of hFSH was not only doubled but the administration began earlier and hMG was administered for a longer period. Briefly, two daily injections of 25 IU each of hFSH were given on days 1-6 of the menstrual cycle, followed by two daily injections of 25 IU each of hMG on days 7-9 and a single injection of 1000 IU of hCG in the morning of day 10 (Fig. 1).

Protocol I

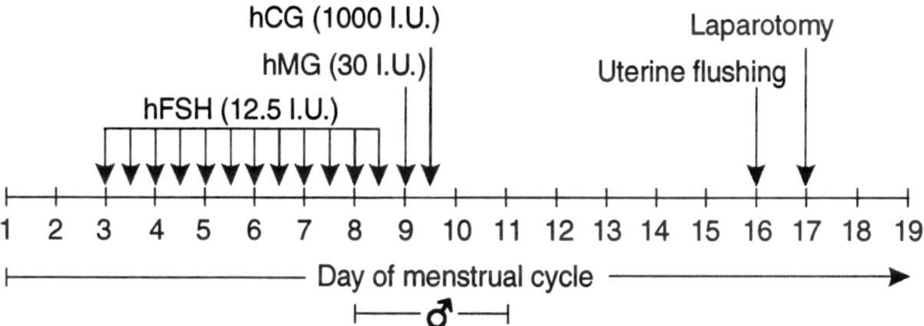

Protocol II

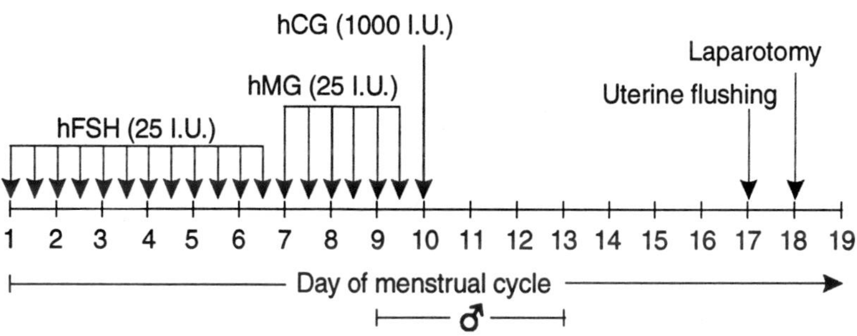

Fig. 1 Hormone stimulation protocol employed in bonnet monkeys. Abbrevations used: hFSH, human follicle stimulating hormone; hCG, human chorionic gonadotrophin; hMG, human menopausal gonadotrophin.

All females responded very well to this protocol, characterized by enlarged ovaries, multiple follicular growth (Table 1) and massive production of estradiol (peak value: 4-7 ng/ml; Fig. 3). Following hMG administration, progesterone levels increased and remained high (>7 ng/ml) until day 18–20 of the cycle (Fig. 3). However, multiple ovulation products were not recovered from any of the monkeys and only a degenerated UFO was obtained from the monkey #66, while others did not yield any product (Table 1). Moreover, there was a

CL each in monkeys #17, #66 and #73 (Fig. 2C, arrow head) but none of the females had more than one CL (Table 1). These observations indicate that high amount of hFSH and/or longer exposure to hMG results in premature luteinization of developing follicles, leading to impaired ovulation.

Table 1 Status on ovarian hyperstimulation using gonadotrophins in bonnet monkeys

Protocol	*Animal No.*	*Parameters monitored*					
		Size of ovary (cm)		*No. of follicles*[a]	*No. of luteinized follicles*[b]	*No. of corpora lutea*	*Ovulation Products*
		Right	*Left*				
I	# 46	0.6 × 1.0	0.6 × 1.0	Nil	Nil	Nil	Nil
	# 70	ND	ND	ND	ND	ND	UFO
	# 71	ND	ND	ND	ND	ND	Nil
	# 72	0.5 × 0.5	0.6 × 1.0	Nil	Nil	Nil	Nil
II	# 17	0.5 × 1.0	1.0 × 2.5	2	Nil	1	Nil
	# 65	2.0 × 3.0	1.5 × 2.5	3	21	Nil	Nil
	# 66	2.5 × 3.0	2.5 × 3.5	2	17	1	UFO
	# 73	1.5 × 3.0	1.5 × 2.0	1	19	1	Nil

[a]Most follicles had a diameter of about 0.5 × 1.0 cm; however, two follicles in #65 measured 0.2 × 0.3 cm.

[b]Most luteinized follicles had a diameter of about 0.5 × 1.0 cm; however those in #65 measured 0.2 × 0.3 cm. Moreover, many luteinized follicles also had hemorrhagic spots (all in #65 and 5 of 19 in #66).

Abbreviations used. ND: not determined; UFO: unfertilized Oocytes.

Ovarian Hyperstimulation and Premature Luteinization

Interestingly, animals in protocol-II had enlarged ovaries, contained 1–3 multiple follicles and often prematurely luteinized follicles with hemorrhagic spots in large numbers (17–21; Table 1; Fig. 2 E, F; arrow heads). These structures, resembling unruptured luteinized follicles (ULFs), had a brownish red tinge in contrast to the whitish or yellow, pearl-like shining appearance of normal or atretic follicles. The number of hemorrhagic spots in luteinized follicles varied i.e., all in monkey #65 and 5 of 19 in #66. These hemorrhagic spots nevertheless, were not stigmata of CL. Most luteinized follicles had a diameter of about 0.5 × 1 cm but, those in monkey #65 measured 0.2 × 0.3 cm. Their histological analysis revealed predominance of luteal cells (Fig. 2G-I). Bioassay of follicular cells, biopsied from ULF structures, revealed their ability to produce massive amounts of progesterone, in response to hCG stimulation, in a dose-dependent manner (Fig. 4). It is important to note that circulating progesterone was elevated (1–3 ng/ml) in all the 4 monkeys before hCG administration (Fig 3). This is strongly indicative of premature luteinization of follicles. This event coupled with the very high amount of serum estradiol might have prevented ovulation in developing antral follicles, resulting in the formation of ULFs.

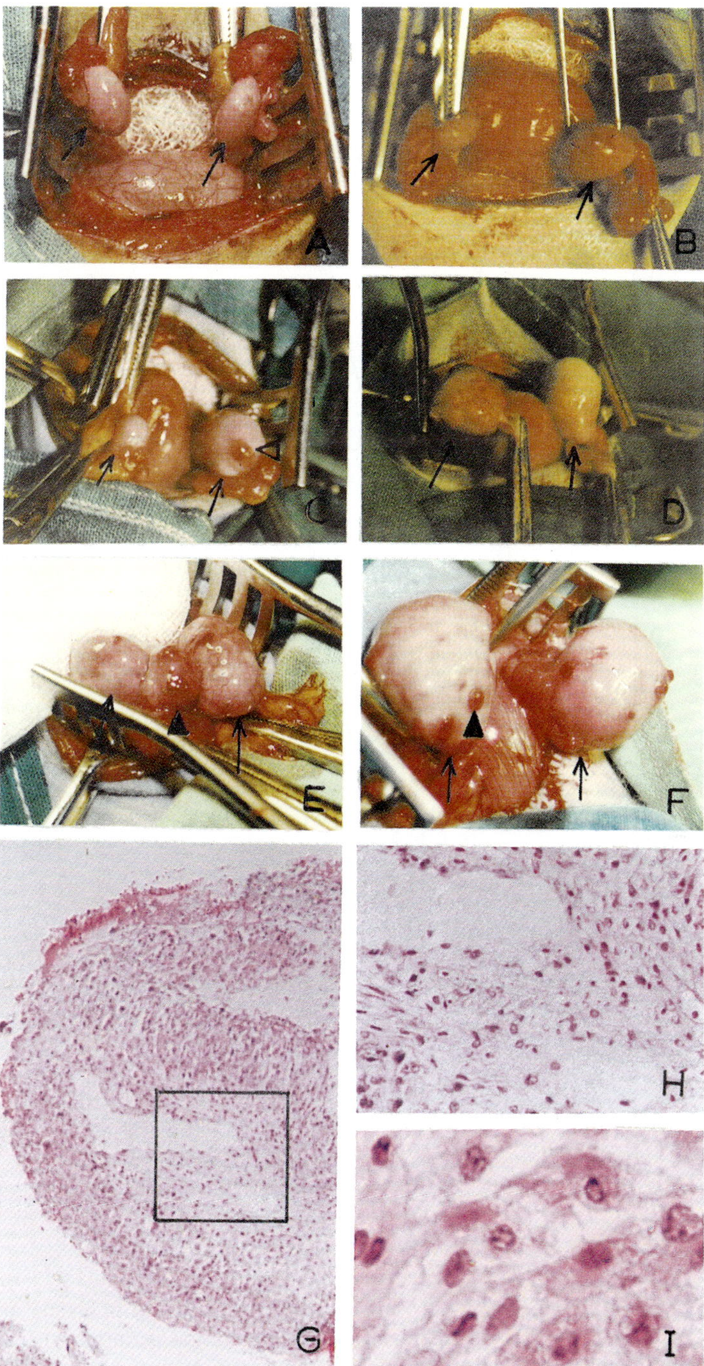

Fig. 2 Appearance of bonnet monkey ovaries following gonadotrophin administration as per protocol-I (A: monkey #46, B: #72) and protocol-II (C: monkey #17, D: #65, E: #66, F: #73). Arrows indicate ovaries; open arrow head: corpus luteum and closed arrow head: hemorrhagic spot. Panels G-I: histological sections of unruptured follicle from monkey #73 showing predominance of luteal cells. Magnification: 10 × (G), 20 × (H) and 40 × (I).

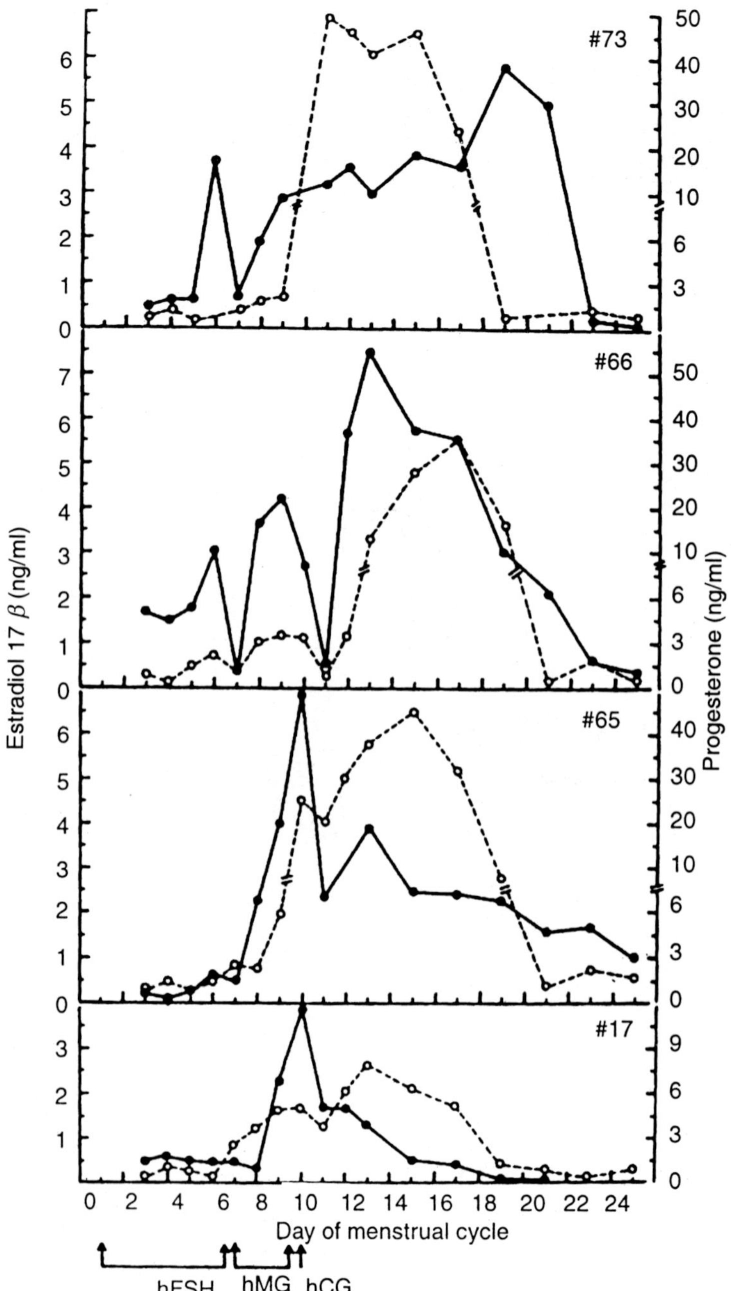

Fig. 3 Serum levels of estradiol (●—●) and progesterone (o—o) in bonnet monkeys subjected to ovarian hyperstimulation using gonadotrophins as per protocol-II. Arrows at the bottom show time and duration of hormone administration. Abbreviations used are as in Fig. 1.

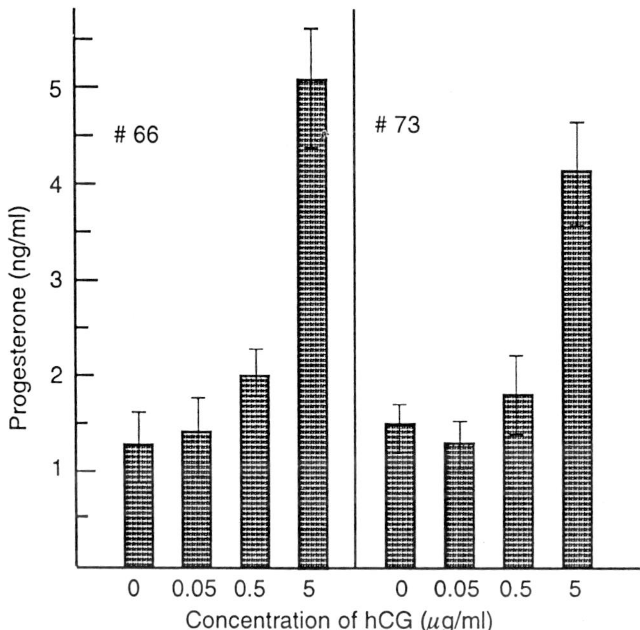

Fig. 4 Progesterone production, in response to human chorionic gonadotrophin (hCG) challenge, by cultured follicular cells from unruptured follicles of monkeys #66 and #73, subjected to ovarian hyperstimulation using gonadotrophins as per protocol-II.

The ovarian and hormone profile features observed with the four bonnet monkeys subjected to protocol-II appear to be similar to those observed in ovarian hyperstimulation syndrome (OHSS) in humans (Rizk and Smitz, 1992; Elchalal and Schenker, 1997). This is one of the most serious iatrogenic conditions with various clinical manifestations caused by intense ovulation induction with clomiphene citrate or highly purified gonadotrophins. It is characterized by cystic enlargement of ovaries, high estradiol levels, hypovolaemia, intense inflammatory reaction, massive accumulation of extracellular exudates and thromboembolism. The pathophysiology of this syndrome is not clearly understood and believed to be linked to inflammatory processes (Rizk and Smitz, 1992; Elchalal and Schenker, 1997; Enskog *et al.*, 1999). In this regard, it would be interesting to exploit the bonnet monkey (protocol II) as the primate model to investigate the cellular and molecular basis of OHSS.

From the foregoing, the first protocol indicates that daily injections of 25 IU per day of hFSH is ineffective and the second protocol shows that twice daily injections of high amounts (50 IU) of hFSH appears to be excessive. It is possible that i.m. injections may not maintain a continuously sustained high levels of gonadotrophins in the body to ensure optimal follicular development, in view of their short half lives. Moreover, it may be unphysiological to expose the ovaries to such frequent, abrupt changes in the levels of gonadotrophins (Wolf *et al.*, 1990). Hence, it would be desirable to have a continuous infusion of hFSH, at a rate of 25 IU per day by subcutaneous implantation of hormone loaded-Alzet mini-osmotic pump on the back of the animal on days 3–9; immediately following pump removal, injections of 25 IU each of hMG (morning) and hCG (evening). Our preliminary data shows that this hormone regimen induces multiple ovulation (up to 4 CLs) with a typical classical response of serum estradiol and progesterone following gonadotrophin administration (data not shown). This however, remains to be verified with more animals. Thus, an ideal protocol for inducing superovulation in bonnet monkeys may involve administering moderate amounts of hFSH by means of osmotic mini pumps.

Conclusion

Daily (i.m.) injections of gonadotrophins, regardless of the dose, are incapable of inducing optimal development of antral follicles and multiple ovulation. Using protocol-II, even though folliculogenesis and estradiol production could be achieved, the incidence of anovulation is high as a consequence of premature luteinization of developing follicles. There appears to be quite different follicular developmental dynamics when high amounts of, or longer exposure to, hFSH and hMG are used. Continuous infusion of constant amount of hFSH 25 IU/day) using osmotic mini-pumps during the follicular phase of menstrual cycle, followed by single injections each of hMG and hCG may be useful in producing multiple ovulation/embryos. This study also indicates that protocol-II, used in bonnet monkeys, could be exploited to produce primate models to investigate cellular and molecular basis of reproductive disorders like OHSS and ULF syndromes.

Acknowledgement

Financial support from the Department of Biotechnology, New Delhi and the Rockefeller Foundation, New York is gratefully acknowledged. The authors thank Dr. S.G. Ramachandra and Mr. V. Ramesh for excellent animal handling and surgical procedures; Mr. H.N. Krishnamurthy for hormone assays; and Ms. M.S. Padmavathy for her help in the preparation of the manuscript.

References

Bavister, B.D., Boatman, D.E., Collins, K., *et al.*, (1984) Birth of rhesus monkey infant following in vitro fertilization and nonsurgical embryo transfer. *Proc. Natl. Acad. Sci.*, **81**, 2218–2222.

Bavister, B.D. and Boatman, D.E. (1993) IVF in nonhuman primates: current status and future directions. *In Wolf, D.P., Stouffer, R.L. and Brenner, R.M (eds.). In vitro fertilization and embryo transfer in primates Springer-Verlag, New York.* pp 30–45.

Chan, A.W., Dominko, T., Luetjens, C.M., *et al.*, (2000) Clonal propogation of primate offspring by embryo splitting. *Science*, **287**, 317–319.

Elchalal, U. and Schenker, J.G. (1997) The pathophysiology of ovarian hyperstimulation syndrome-views and ideas. *Hum. Reprod.*, **12,** 1129–1137.

Enskog, A., Henriksson, M., Unander, M. *et al.*, (1999) Prospective study of the clinical and laboratory parameters of patients in whom ovarian hyperstimulation syndrome developed during controlled ovarian hyperstimulation for in vitro fertilization. *Fertil. Steril.*, **71**, 808–814.

Goodeaux, L.L., Anzalone, C.A., Thibodeaux, J.K., *et al.*, (1990) Successful nonsurgical collection of Macaca mulatta embryos. *Theriogenol.*, **34**, 1159–1167.

Jayaprakash, D., Satish, K.S., Ramachandra, V., *et al.*, (1997) Successful recovery of preimplantation embryos by nonsurgical uterine flushing in the bonnet monkey. *Theriogenol*; **47**, 1019–1026.

McCarthy, J.T., Fotman, J.D., Boice, L.M., *et al.*, (1991) Induction of multiple follicular development and superovulation in the olive baboon, Papio anubis. *J. Med. Primatol.*, **20**, 308–314.

Meng, L., Ely, J.J., Stouffer, R.L. *et al.*, (1997) Rhesus monkeys produced by nuclear transfer. *Biol. Reprod.*, **57**, 454–459.

Moudgal, N.R. Yoshinaga, K, Rao A.J., *et al.* (1991) Perspectives in Primate Reproductive Biology. Wiley Eastern Limited, New Delhi.

Murthy, G.S.R.C., Ramasharma, K., Mukku, V.R., *et al.*, (1979) Reproductive Endocrinology of bonnet monkey. *In Anand Kumar TC, Karger S (eds). Non-human Primate Models for the study of Human Reproduction* pp. 50–55.

Ovadia, J., McArthur, J.W., Smith, O.W. *et al.*, (1971) An individualized technique for inducing ovulation in the bonnet monkey. *J. Reprod. Fert.*, **27**, 13–23.

Rizk, B. and Smitz, J. (1992) Ovarian hyperstimulation syndrome after superovulation using GnRH agonists for IVF and related procedures. *Hum. Reprod.*, **7**, 320–327.

Seshagiri, P.B. and Hearn, J.P. (1993) *In vitro* development of *in vivo* produced rhesus monkey morulae and blastocysts to hatched, attached, and post-attached blastocyst stages: morphology and early secretion of chorionic gonadotrophin. *Hum. Reprod.*, **8**, 279–287.

Seshagiri, P.B., Bridson, W.E., Dierschke, D.J., *et al.*, (1993) Non-surgical uterine flushing for the recovery of preimplantation embryos in rhesus monkeys: lack of seasonal infertility. *Am. J. Primatol.*, **29**, 81–91.

Seshagiri, P.B., Terasawa, E. and Hearn, J.P. (1994) The secretion of gonadotrophin releasing hormone by peri-implantation embryos of the rhesus monkey: comparison with the secretion of chorionic gonadotrophin. *Hum. Reprod.*, **9**, 1300–1307.

Simon, J.A., Danforth, D.R., Hutchison, J.S. *et al.* (1998) Characterization of recombinant DNA derived-human luteinizing hormones in vitro and in vivo. *JAMA.*, **259**, 3290–3295.

Simpson, M.E. and van Wagenen, G. (1958) Experimental induction of ovulation in the macaque monkey. *Fertil. Steril.*, **9**, 386–399.

Simpson, M.E. and van Wagenen, G. (1962) Induction of ovulation with human urinary gonadotrophins in the monkey. *Fertil. Steril.*, **13**, 140–152.

Weston, A.M., Zelinski-Wooten, M.B., Hutchison, J.S., *et al.* (1996) Developmental potential of embryos produced by in-vitro fertilization from gonadotrophin-releasing hormone antagonist-treated macaques stimulated with recombinant human follicle stimulating hormone alone or in combination with luteinizing hormone. *Hum. Reprod.* **11**, 608–613.

Wolf, D.P., Thomson, J.A., Zelinski-Wooten, M.B., (1990) In vitro fertilization-embryo transfer in nonhuman primates: The techniques and its applications. *Mol. Reprod. Dev.*, **27**, 261–280.

Zeliniski-Wooten, M.B., Alexander, M., Cynthia, L., *et al* (1994) Individualized gonadotrophin regimens for follicular stimulation in macaques during in vitro fertilization (IVF) cycles. *J. Med. Primatol.*, **23**, 367–374.

Follicular Growth, Ovulation and Fertilization: Molecular and Clinical Basis
Anand Kumar and Amal K. Mukhopadhyay (Eds.)
Narosa Publishing House, New Delhi, India, 2001

9

Developmental Expression of Estrogen Receptor-α and -β and Androgen Receptor in the Rat Ovary

M. Sar and F. Welsch
Chemical Industry Institute of Toxicology, Six Davis Drive, P.O. Box 12137
Research Triangle Park, NC 27709–2137, USA

Introduction

Estrogens influence the growth, differentiation and maintenance of female reproductive tissues (Clark *et al.*, 1992; Korach *et al.*, 1995). Estrogen is produced by preovulatory follicles by the combined actions of the gonadotropins, FSH and LH (Richards, 1994). Androgen is produced by thecal cells and interstitial cells and its secretion is regulated by LH and chorionic gonadotrophin (hCG) (Erickson *et al.*, 1985). Thecal androgen is converted to estrogen in granulosa cells by the enzyme aromatase which is induced by FSH in granulosa cells (Gore-Langton and Armstrong, 1994). In the ovary estrogens increase growth and differentiation of follicles, and androgens induce mature follicles to undergo atresia (Hsueh *et al.*, 1994). Presumably, estrogen and androgen exert their physiological effects in the ovary through classical nuclear receptors. The effects of estrogen are mediated by at least two different receptors, estrogen receptor-α (ERα) (Koike *et al.*, 1987) and estrogen receptor-β (ERβ) (Kuiper *et al.*, 1996) which are members of the nuclear receptor family (Evans, 1988; Tsai and O'Malley, 1994). In rats, in addition to ERα and ERβ several isofoms of rat ERβ mRNA have been reported (Chu and Fuller, 1997). One of the isoforms known as ERβ_2 is expressed in the ovary, prostate, and other tissues (Peterson *et al.*, 1998). Rat ovaries express both ERα and ERβ mRNA but the ERβ mRNA and protein are localized predominantly in the granulosa cells of growing and mature follicles (Kuiper *et al.*, 1996; Byers *et al.*, 1997). Specific binding of estradiol has been found to granulosa cells (Kudolo *et al.*, 1984), and ligand binding assays have shown that the ERβ protein binds estrogen with an affinity and specificity comparable to that of ERα protein (Kuiper *et al.*, 1997). High affinity binding of androgens to rat ovarian and granulosa cell extracts has been reported (Schreiber *et al.*, 1976; Schreiber and Ross, 1976). Although limited information exists about the specific localization of AR and its mRNA in rat ovary, expression of ERα and ERβ proteins in the ovary has been demonstrated (Power and Sar, 1993; Sar and Welsch, 1999). Furthermore, the temporal and developmental relationships with ERα/ERβ and AR expression have not been known. The present immunocytochemical study descibes the developmental expression of ERα, ERβ and AR in rat ovary using specific ERα, ERβ, and AR antibodies. The results indicate that ERα and ERβ proteins are expressed in different cell types in the ovary while AR is coexpressed with ERα or ERβ in specific cell types.

Materials and Methods

Animals

Timed pregnant Sprague-Dawley rats (Charles River Laboratories, Inc., Raleigh, North Carolina) were fed NIH-07 rodent chow and water ad libitum. Neonatal (1 or 2 days old), postnatal (5–10 days old), immature (21- to 23-day-old), and adult (60 days old) female rats were used for the experiments. The animals were housed in humidity-and temperature-controlled rooms with a 12-h light, 12-h dark photoperiod. The animal experiments were approved by the Institutional Animal Care and Use Committee of the Chemical Industry Instititute of Toxicology, centre for Health Research. Postnatal rats on days 1, 2, 5, and 10 were killed by decapitation, whereas immature and adult female rats were killed by CO_2 asphyxiation. Ovaries were immediately removed, fixed in formalin, embedded in paraffin, and processed for immunohistochemistry.

Immunocytochemistry

Antigen-retrieved paraffin sections were processed for immunostaining by the avidin-biotin peroxidase method as previously described (Sar, 1985; Sar and Welsch, 1999). The sections were incubated overnight at 4°C with ERα monoclonal antibody, ERβ antibody, and preadsorbed ERβ antibody. Monoclonal antibody was used at a concentration of 0.1–0.2 μg/ml, and ERβ antibody was used at a concentration of 5–10 μg/ml in paraffin sections. The optimal working dilution of antibody was determined by incubating sections with various concentrations of antibody ranging from 0.1–10 μg/ml. The secondary antibody (goat-antirabbit IgG or horse-antimouse IgG) and Elite avidin-biotin peroxidase were used at a concentration of either 1:100 or 1:200 for 30–60 min each at room temperature. After a 5 min wash, the sections were treated with liquid diaminobenzedine (DAB) (Biogenex, San Ramon, CA) and then counterstained with hematoxylin. The immunostained slides were evaluated with an Olympus Vanox-S photo microscope.

Antibodies

Clone ID5 of an ERα monoclonal antibody was obtained from Dako Corp., Carpinteria, CA. A rabbit polyclonal antibody (PAI-310) raised against a synthetic peptide corresponding to the C-terminal amino acid residues 467–485 of rat

ERβ was purchased from Affinity Bioreagents Inc., Golden, CO. Preadsorbed ERβ antibody was prepared by incubating 5–10 μg ERβ antibody/ml with 20–40 μg synthetic peptide for 24 h at 4°C. A second ERβ polyclonal antibody raised against a synthetic peptide containing the amino acid sequence of 54–71 of rat and mouse ERβ and amino acids of 46–63 of human ERβ (Upstate Biotechnology, Waltham, MA) was also used at a similar concentration. A rabbit polyclonal anti-AR antibody (PAI-111) raised against a synthetic peptide of 21 amino acid corresponding to the N-terminus of rat and human AR was obtained from Affinity Bioreagents Inc., Golden, CO and used at a concentration of 4 μg/ml. Preadsorbed AR antibody was prepared by incubating 4 μg of AR antibody with 16 μg of synthetic AR peptide. The specificity and immunocharacteristics of the AR antibody have been previously established (You and Sar, 1998).

Results

Immunocytochemistry performed with the monoclonal antibody or polyclonal antibodies revealed nuclear receptor expression in certain cells of the ovary but not in others (Figures 1, 2, 3 and 4). The staining was specific since no positive staining was observed when normal mouse IgG was used for monoclonal antibody immunocytochemistry and preabsorbed ERβ or preabsorbed AR antibody were used for polyclonal antibodies.

In neonatal rats, the germinal epithelium and the differentiating stromal cells showed ERα nuclear immunostaining whereas no specific immunoreactivity (IR) was detected in pregranulosa cells and oocytes (Fig. 1a). In ovaries of 5-and 10-day-old rats, nuclear ERα IR was not previously detected in granulosa cells of the primordial follicles and growing follicles (Sar and Welsch, 1999). In contrast, ERα IR was observed in the nuclei of some stromal cells and germinal epithelium, but weak IR was noticed in a few theca cells. ERβ immunocytochemistry of 1-day-old rat ovaries showed no IR with ERβ antibody in either oocytes or differentiating stromal cells and pregranulosa cells (Sar and Welsch, 1999). In ovaries from 5-and 10-day-old rats, a small number of granulosa cells from growing follicles (primary and secondary stage) showed nuclear IR, although the intensity appeared to be weak to medium (Fig. 1b) No ERβ staining was detected in primordial follicles, but very weak staining was detected in differentiating granulosa cells of the intermediate follicles. Oocytes, stromal cells, and germinal epithelium had no ERβ staining. In neonatal rat ovaries an AR IR pattern was found like that of ERα (Fig. 1c), while ERβ was undetectable. In ovaries of 5- and 10-day-old rats AR IR was expressed in granulosa cells of primary and secondary follicles but no AR IR was detected in the primordial follicles (Fig. 1d). AR was also expressed in connective tissue cells and theca cells similar to ERα localization, but the AR staining was stronger in thecal cells.

In the ovary of immature rats, ERα nuclear IR was observed in thecal cells, interstitial gland cells, and germinal epithelium (Fig. 2a). No ERα staining was detected in granulosa cells of the primary, secondary, and growing follicles

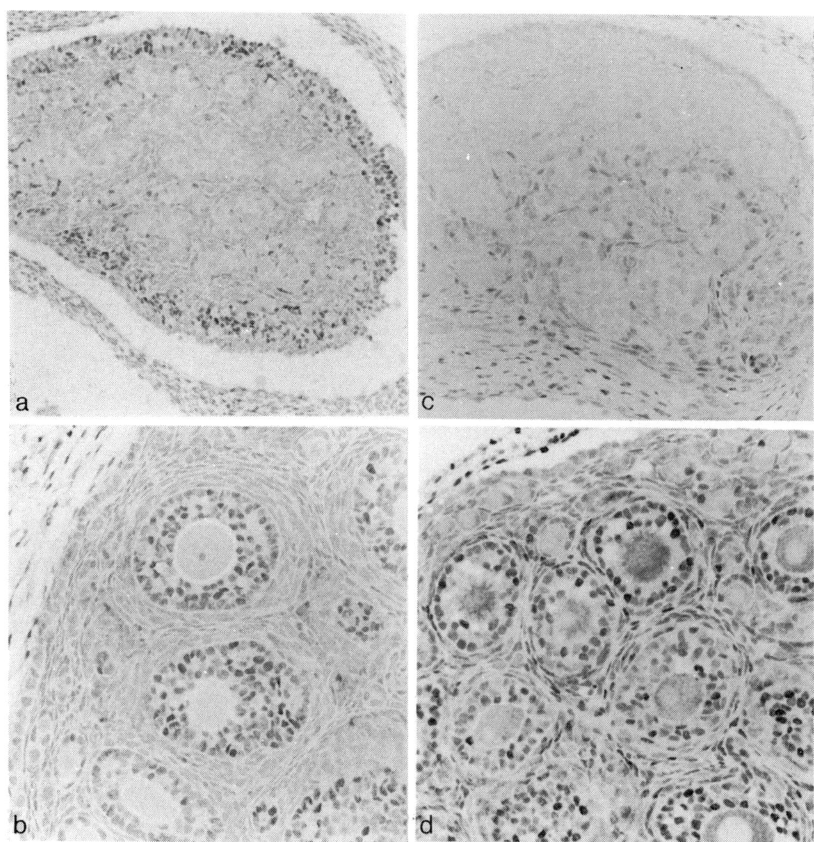

Fig. 1 Localization of ERα(a), ERβ(b), and AR(c-d) in ovary of 1-day-old (a, c) and 10-day-old (b, d) rats. Paraffin sections were immunostained with ERα monoclonal antibody or polyclonal antibodies ERβ and AR. ERα was detected in connective tissue cells, undifferentiated stromal cells and germinal epithelium. ERβ staining was detected in granulosa cells of primary and growing follicles showing variable intensity of IR. Thecal cells and interstitial cells did not show specific staining. AR was detected in granulosa cells, thecal cells and interstitial cells. AR staining was more intense in granulosa cells than ERβ staining. Counterstained with hematoxylin, magnification 290 (a), × 430 (b), × 360 (c), and × 430 (d).

(Fig. 2b). Similarly, in adult rat ovaries granulosa cells did not show ERα staining (Fig. 2c). In contrast, thecal cells, interstitial gland cells and germinal epithelial cells revealed ERα IR but not the corpora lutea cells (Figure 2c–d). In both immature and adult rats ERα was not detected in oocytes.

In both immature and adult rats nuclear ERβ occurred in granulosa cell nuclei of the growing follicles at all stages from primary and secondary to mature follicles, including preantral and antral follicles (Fig. 3a–c). The intensity of ERβ IR was stronger in secondary and mature follicles than in primary follicles. Within the follicles, some granulosa cells did not show nuclear IR with ERβ antibody. The intensity of ERβ IR in granulosa cells of atretic follicles varied considerably. Some cells completely lacked IR, while in others,

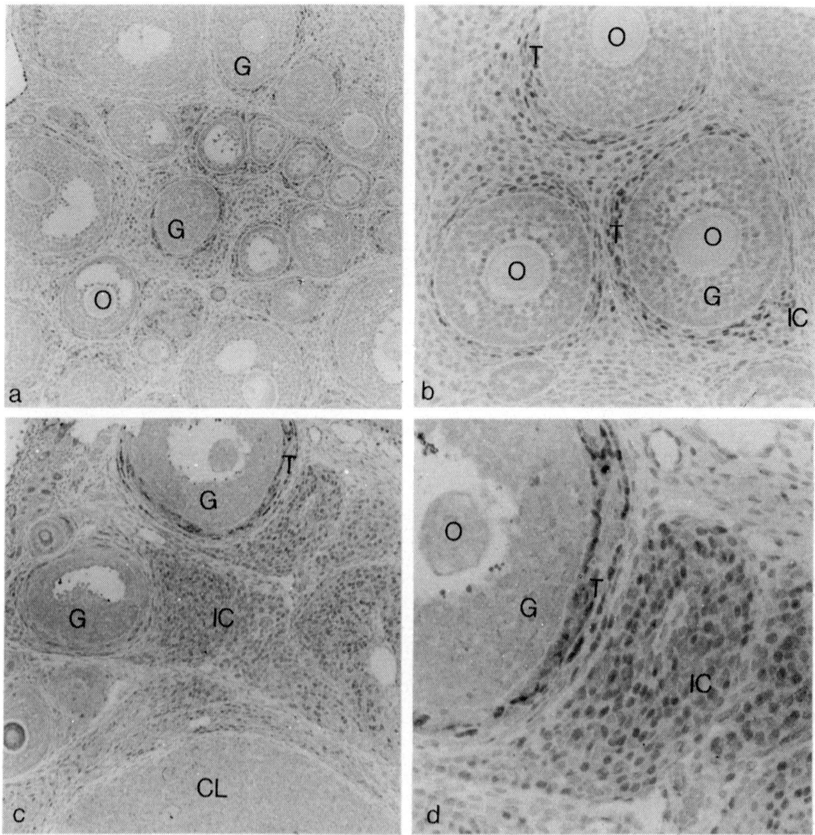

Fig. 2 Localization of ERα in 21-day-old (a–b) and 60-day-old (c–d) rat ovary. Paraffin sections were immunostained with ERα monoclonal antibody. Immunostaining was detected in thecal (T) cells, interstitial gland (IC) cells and germinal epithelium. No staining was detected in granulosa (G) cells of primary, secondary and mature follicles as well as corpus luteum (CL) and oocytes (O). Counterstained with hematoxylin, magnification × 180 (a), × 360 (b), × 210 (c), and × 430 (d).

granulosa cells of the basal cell layers showed IR, and granulosa cells toward the center of the atretic follicle revealed no staining. Thecal cells, interstitial gland cells, oocytes, and germinal epithelium revealed a lack of nuclear ERβ IR (Fig. 3b–c). However, corpora lutea cells showed ERβ IR of weak to medium intensity (Fig. 3d).

In immature and adult rats nuclear AR was detected in granulosa cells of the primary, secondary, and mature follicles (Fig. 4a–c). Thecal cells, interstitial gland cells, and germinal epithelial cells showed nuclear AR staining. In addition, corpora luteal cells revealed positive AR IR while none was detected in oocytes (Fig. 4c).

Discussion

We present here immunohistochemical evidence through the application of

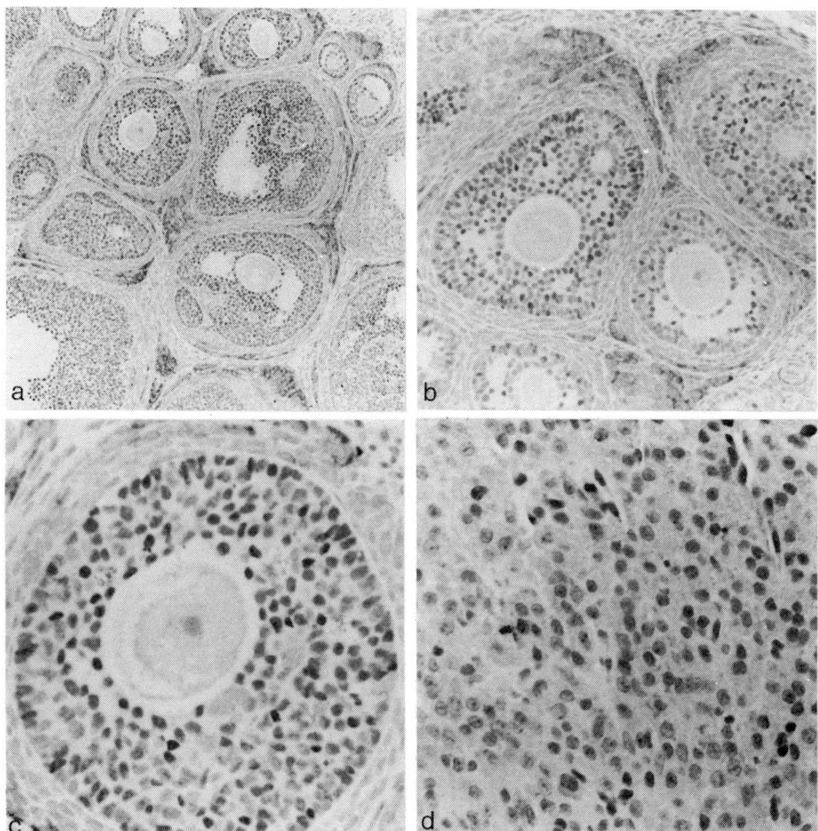

Fig. 3 Localization of ERβ in 21-day-old (a–b) and 60-day-old (c–d) rat ovary. Paraffin sections were stained with polyclonal antibodies either from Affinity Bioreagents (PAI-310) (a–b), or Upstate Biotechnology (c–d) Nuclear staining was in granulosa cells in follicles at different stages of development (a–c). Staining was undectable in thecal cells and inerstitial cells but weak to moderate staining was detected in luteal cells (d). Counterstained with hematoxylin, magnification × 175 (a), × 335 (b), and × 430 (c–d).

specific antibodies that the ERα and ERβ as well as the AR are differentially expressed in rat ovary during postnatal development. ERβ protein was detected exclusively in granulosa cells of primary, secondary, and mature follicles (preantral and antral). The IR of the ERβ protein in granulosa cells showed a parallel distribution with the high ERβ mRNA expression previously demonstrated by in situ hybridization and RT-PCR (Kuiper *et al.*, 1996; Kuiper *et al.*, 1997). In contrast, ERα protein was localized in thecal and interstitial cells and germinal epithelial cells, but not in granulosa cells. Our specific IR observations (Sar and Welsch, 1997; Sar and Welsch, 1999) were recently confirmed by other investigators using Western blot analysis as well as immunocytochemistry (Fitzpatrick *et al.*, 1999; O'Brien *et al.*, 1999). The ERβ and ERα protein IR described here correlates with the reported high expression of ERβ mRNA in granulosa cells and low expression of ERα protein and mRNA in the rat ovary (O'Brien *et al.*, 1999). Western blotting results also showed that granulosa

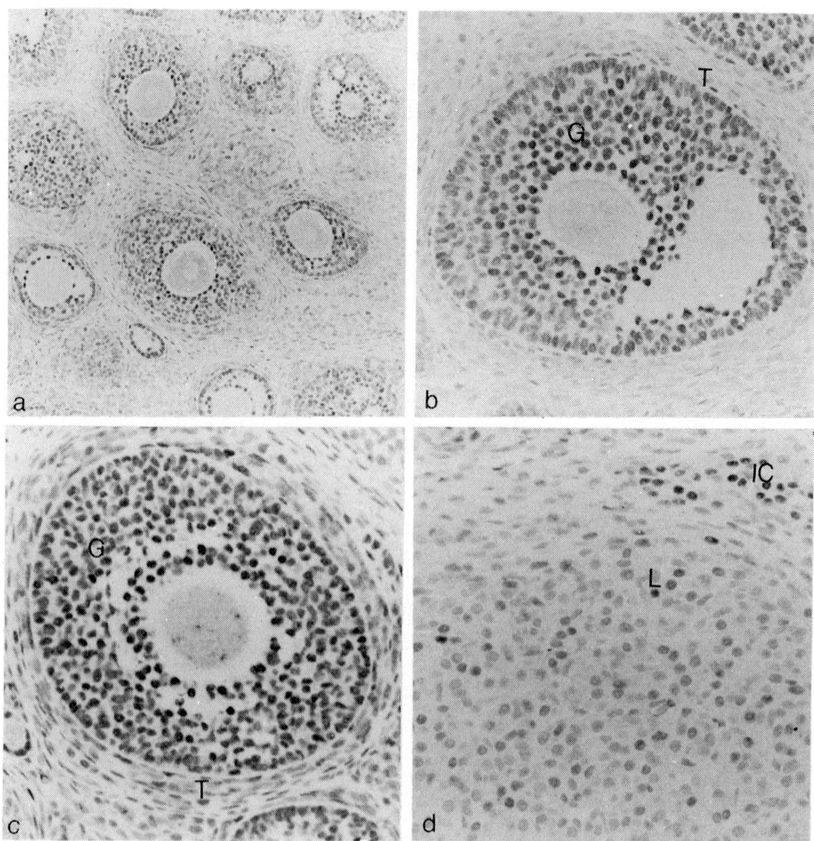

Fig. 4 Localization of AR in 21-day-old (a–b) and 60-day-old (c–d) rat ovary. Paraffin sections were immunostained with a polyclonal antibody (PAI-111). AR IR was detected in granulosa cells (G) in follicles at different stages of developments, theca cells (T) and interstitial gland cells (IC) (a–c), and luteal cells (L) (d). Immunostaining was more intense with AR than ERβ in granulosa cells. Counterstained with hematoxylin, magnification × 175 (a), and × 360 (b–d).

cells contain ERβ-immunoreactive protein(s) approximately 60 kDa to 62 kDa protein by using ERβ transcripts (Fitzpatrick *et al.*, 1999; O' Brien *et al.*, 1999). ERα and ERβ have similar binding affinity for estradiol (Kuiper *et al.*, 1997), and nuclear localization of ^{3}H estradiol has been demonstrated in granulosa cells, theca cells, and interstitial gland cells (Stumpf, 1969; Stumpf and Sar, 1976). The existence of ligand-binding sites and ER in the ovary indicate that estrogen's effect on thecal cells and follicular cells are possibly mediated through ERα and ERβ, respectively. The ERβ protein may be involved in follicular growth and maturation. Recent studies in estrogen receptor α knockout mice showed that lack of the ERα gene does not prevent the growth and maturation of small follicles (Lubahn *et al.*, 1993), which express the ERβ protein but not the ERα protein (Couse *et al.*, 1997; Schomberg *et al.*, 1999). In female ERβ knockout mice the primary reproductive deficit is impaired ovarian function resulting in reduced fertility that has its apparent origin in the ovary (Krege *et al.*, 1998). It appears as if both ERα and ERβ are essential for normal ovarian function. This is clearly evident in female double knockout mice lacking both estrogen ERα and ERβ which exhibit normal reproductive tract development but are infertile (Couse *et al.*, 1999b). Loss of both receptors in these mutants leads to an ovarian phenotype that is distinct from either the ERα or ERβ knockout mice (Schomberg *et al.*, 1999; Couse *et al.*, 1999a; Krege *et al.*, 1998).

In our previous study we demonstrated the developmental expression of ERα and ERβ protein in rat ovary (Sar and Welsch, 1997; Sar and Welsch, 1999). ERβ protein was expressed in granulosa cells of primary, secondary, and growing follicles on postnatal day-5. As follicles grow in size ERβ protein expression appeared stronger which was distinctly observed in 20-day-old rats. In contrast, ERα expression was detected in stromal and interstitial cells of postnatal day 1, 5, and 10, and thecal cells and interstitial cells of prepubertal and adult rat ovary. In the present experiments AR expression was detected in granulosa cells of the primary and secondary follicles similar to ERβ expression and in thecal and interstitial cells similar to ERα expression during postnatal development. However, AR IR was more intense than that for ERα and ERβ. The developmental expression of ERβ and AR IR indicates that both AR and ERβ proteins may be involved in regulation of follicular growth during the early postnatal period. The interaction of ER and AR in follicular growth and differentiation remains to be determined.

Immunohistochemical data showed that AR was expressed in all cell types of the rat ovary including thecal cells, interstitial cells, luteal cells, and granulosa cells of the primary, secondary, and mature follicles. The AR expression level was similar to that of ERα in thecal and interstitial cells and to that of ERβ in granulosa cells, as well as in luteal cells. A similar distribution of AR in human (Horie *et al.*, 1992) and monkey (Hild-Petito *et al.*, 1991) ovaries has been demonstrated. Also high affinity binding of androgens in rat ovarian and granulosa cell extracts has been reported (Schreiber *et al.*, 1976; Schreiber and Ross, 1976). AR transcripts have been detected by Northern blot analysis (Power and Sar, 1993). These data together with AR IR indicate an AR mediated androgen action in the ovary. Androgens induce atresia of granulosa cells but its effect on the corpus luteum is not known. Estrogens, however, mediates luteal cell hypertrophy by stimulating overall protein biosynthesis (Telleria *et al.*, 1998). Estrogen binding sites have also been detected in the rat corpus luteum by binding experiments (McLean *et al.*, 1990) as well as by autoradiography with ^{3}H estradiol (Stumpf and Sar, 1976). Both estrogen receptor subtypes, ERα and ERβ are expressed in the rat corpus luteum of pregnancy (Telleria *et al.*, 1998). The expression of AR and ERβ proteins detected in luteal cells of the rat ovary suggests that both estrogen and androgen are involved in regulation of luteal cell function.

The present results demonstrate that both ERβ and AR were present in granulosa cells of primary and growing follicles where they possibly mediate some of effects of estrogen and androgen action in regulation of growth and development of the follicles. The different expression of ERα and ERβ in specific cell populations of the rat ovary, together with the recent results of ERα and ERβ knockout mice (Schomberg *et al.*, 1999; Couse *et al.*, 1999a; Krege *et al.*, 1998) as well as α, β-ERKO mutant mice (Couse *et al.*, 1999b) suggest that ERα and ERβ are essential for normal biological function of the ovary.

Acknowledgements

We thank Dr. Paul Foster, Dr. Barry McIntyre and Dr. Li You for reviewing the manuscript, Ms. Delorise Williams for tissue preparation, and Ms. Sadie Leak for word processing of the manuscript.

References

Byers, M., Kuiper, G.G.J.M., Gustafsson, J-Å., *et al.*, (1997) Esrogen receptor-β mRNA expression in rat ovary: down regulation by gonadotropins. *Mol. Endocrinol.*, **11**, 172–182.

Chu, S. and Fuller, F.J. (1997) Identification of a splice variant of the real estrogen receptor β gene. *Mol. Cell. Endocrinol.,* **132**, 195–199.

Clark, J.H., Schrader, W.T. and O'Malley, B.W. (1992) Mechanisms of action of steroid hormones. In Wilson, J., Foster, D.W. (eds), *Textbook of Endocrinology*. W.B. Saunders, Philadelphia, pp. 35–90.

Couse, J.F., Bunch, D.O., Lindzey, J., *et al.,* (1999a) Prevention of the polycystic ovarian phenotype and characterization of ovulatory capacity in the estrogen receptor-α knockout mouse. *Endocrinology*, **140**, 5855–5865.

Couse, J.F., Hewitt, S.C., Bunch, D.O., *et al.*, (1999b) Postnatal sex reversal of the ovaries in mice lacking estrogen receptors α and β. Science, **286**, 2328–2331.

Couse, J.F., Lindzey, J., Grandien, K. *et al.*, (1997) Tissue distribution and quantitative analysis of estrogen receptor α (ERα) and estrogen receptor β (ERβ) messenger ribonucleic acid in the wild-type and ERβ knockout mouse. *Endocrinology*, **138**, 4613–4621.

Erickson, G.F., Magoffin, D.A., Dyer, C.A., *et al.*, (1985) The ovarian androgen producing cells: a review of structure/function relationships. *Endocr. Rev.,* **6**, 371–399.

Evans, R.M. (1988) The steroid and thyroid receptor superfamily. *Science* **240**, 889–895.

Fitzpatrick, S.L., Funkhouser, J.M., Sindoni, D.M., *et al.* (1999) Expression of estrogen recetpro-β protein in rodent ovary. *Endocrinology*, **140**, 2581–2591.

Gore-Langton, R.E. and Armstrong, D.T. (1994) Follicular steroidogenesis and its control. In: Knobil, E. and Neill, J.D. (eds) *The Physiology of Reproduction, Second Edition*. Raven Press, Ltd., New York, pp. 571–627.

Hild-Petito, S., West, N.B., Brenner, R.M., *et al.,* (1991) Localization of androgen receptor in the follicle and corpus luteum of the primary ovary durng the menstrual cycle. *Biol. Reprod.* **44**, 561–568.

Horie, K., Takakura, K., Fujiwara, H., *et al.,* (1992) Immunohistochemical localization of androgen receptor in the human ovary throughout the menstrual cycle in relation to oestrogen and progesterone receptor expression. *Human Reprod.,* **7**, 184–190.

Hsueh, A.J.W., Billig. H., and Tsafriri, A (1994) Ovarian follicle atresia: a hormonally controlled apoptotic process. *Endocr. Rev.*, **15**, 707–724.

Koike, S., Sakai, M., Muramatsu, M. (1987) Molecular cloning and characterization of rat estrogen receptor cDNA. *Nuclei Acid Res.*, **15**, 2499–2513.

Korach, K.S., Migliaccio, S., Davis, V.L. (1995) Estrogens. In: Munson,P.L. (ed) *Principles of Pharmacology: Basic Concepts and Clinical Applications.* Chapman and Hall, New York, pp. 827–836.

Krege, J.H., Hodgin, J.B., Couse, J.F., *et al.* (1998) Generation and reproductive phenotypes of mice lacking estrogen receptor β. *Proc. Natl. Acad. Sci. U.S.A*, **95**, 15677–15682.

Kudolo, G.B., Elder, M.G., Myatt, L. (1984) A novel estrogen-binding species in rat granulosa cells. *J. Endocrinol.*, **102**, 83–91.

Kuiper, G.G.J.M., Carlsson, B., Grandien,K., *et al.* (1997) Comparison of the ligand binding specificity and transcript tissue distribution of estrogen receptors α and β. *Endocrinology*, **138**, 863–870.

Kuiper, G.G.J.M., Enmark, E., Pelto-Hnikko, M., *et al.*, (1996) Cloning of a novel estrogen receptor expressed in rat prostate and ovary. *Proc, Natl. Acad. Sci. U.S.A.*, **93**, 5925–5930.

Lubahn, D.B., Moyer, J.S., Golding, T.S. *et al.*, (1993) Alteration of reproductive function but not prenatal sexual development after insertional disruption of the mouse estrogen receptor gene. *Proc. Natl. Acad. Sci. U.S.A.*, **90**, 11162–11166.

McLean, M.P., Khan, I., Puryear, T.K., *et al.*, (1990) induction and repression of specific estradiol sensitive proteins in the rat corpus luteum. *Chin. J. Physiol.*, **33**, 353–366.

O'Brien, M.L., Park,K., In, Y., *et al.*, (1999) Characterization of estrogen receptor-β (ERβ) messenger ribonucleic acid and protein expression in rat granulosa cells. *Endocrinology*, **140**, 4530–4541.

Petersen,D.N., Tkalcevic,G.T., Koza-Taylor, P.H., *et al.*, (1998) Identification of estrogen receptor β2, a functional variant of estrogen receptor β expressed in normal rat tissues. *Endocrinology*, **139**, 1082–1092.

Power, S.G.A. and Sar, M. (1993) Immunohistochemical localization of androgen receptor in ovaries of immature and mature rats. The Endocrine Society: Program and Abstracts, 75th Annual Meeting, p. 104.

Richards, J.S. (1994) Hormonal control of gene expression in the ovary. *Endocr. Rev.*, **15**, 725–751.

Sar, M. (1985) Application of avidin biotin technique for the localization of estrogen receptor in target tissues using monoclonal antibodies. In: Bullock,G.R., Petrusz, P. (eds) *Techniques in Immunochemistry* Vol. 3. Academic Press, New York, pp. 43–54.

Sar, M. and Welsch, F. (1997) Immunocytochemical localization of estrogen receptor β in ovary and prostate of prepubertal and adult Sprague-Dawley rats. NIH/NIEHS Conference on Estrogens on Environment IV (Abstract).

Sar, M. and Welsch, F. (1999) Differential expression of estrogen receptor-β and estrogen receptor-α in the rat ovary. *Endocrinology*, **140**, 963–971.

Schomberg, D.W., Couse, J.F., Mukherjee, A., *et al.*, (1999) Targeted disruption of the estrogen receptor-α gene in female mice. characterization of ovarian responses and phenotype in the adult. *Endocrinology*, **140**, 2733–2744.

Schreiber, J.R., Reid, R., Ross,G.T. (1976) A receptor-like testosterone-binding protein in ovaries from estrogen-stimulated hypophysectomized immature female rats. *Endocrinology*, **98**, 1206–1213.

Schreiber, J.R. and Ross, G.J. (1976) Further characterization of a rat ovarian testosterone receptor with evidence for nuclear translocation. *Endocrinology*, **99**, 590–596.

Stumpf, W.E. (1969) Nuclear concentration of ^{3}H-estradiol in target tissues. Dry-mount autoradiography of vagina, oviduct, ovary, testis, mammary gland and adrenal. *Endocrinology*, **85**, 31–37.

Stumpf, W.E. and Sar, M. (1976) Autoradiographic localization of estrogen, androgen, progestin and glucocorticosteroid in "target tissues" In: Pasqualim, J. (ed) *Receptors and Mechanisms of Action of Steroid Hormones*, Marcel Dekker, New York, pp. 41–81.

Telleria, C.M., Zhong, L., Deb,S., *et al.*, (1998) Differential expression of the estrogen receptors α and β in the rat corpus luteum of pregnancy: regulation by prolactin and placental lactogens, *Endocrinology*, **139**, 2432–2442.

Tsai, M.J. and O'Malley, B.W. (1994) Molecular mechanisms of steroid/thyroid receptor superfamily members. *Ann. Rev. Biochem.*, **63**, 451–486.

You, L. and Sar, M. (1998) Androgen receptor expression in the testis and epididymides of prenatal and postnatal Sprague-Dawley rats. *Endocrine*, **9**, 253–261.

Follicular Growth, Ovulation and Fertilization: Molecular and Clinical Basis
Anand Kumar and Amal K. Mukhopadhyay (Eds.)
Narosa Publishing House, New Delhi, India, 2001

10

Blocking of Ovarian Function Using an Antiprogestin ZK 98.299 in Bonnet Monkeys

C.P. Puri, R.R. Katkam and S.D. Kholkute
Institute for Research in Reproduction (ICMR)
Jehangir Merwanji Street, Parel, Mumbai-400 012, India

Introduction

Progesterone is indispensable for the normal functioning of the female reproductive system. It is essential for folliculogenesis, ovulation induction, preparation of secretory endometrium, implantation and maintenance of pregnancy. In addition, it also maintains the uterus in an quiscent state during pregnancy, increases the viscosity of the cervical mucus and increases basal body temperature.

In a normal ovulatory cycle, granulosa cells in large antral and preovulatory follicles are principle sites for secretion of progesterone and its metabolites. Progesterone biosynthesis in the granulosa cells initially requires follicle stimulating hormone (FSH) and thereafter its action is augmented by luteinizing hormone (LH). In fact, low concentrations of progesterone also have a facilitatory role in the midcycle FSH and LH release followed by ovulation. A transient rise in progesterone levels just prior to ovulation suggests its role in induction of ovulation in primates in which pituitary had been exposed to estrogens for prolonged period.

The action of progesterone in inducing secretory activity in the endometrium is mediated through specific genomic receptors in the uterus. However, the role of progesterone in intra-ovarian regulation of ovarian function is still not clear. Administration of high doses of progesterone during the midfollicular phasc disrupts folliculogenesis and this action is due to impaired release of gonadotrophins. The arrest of folliculogenesis could also be due to direct effects of progesterone on the ovaries as implanting progesterone directly in the ovary also had similar effects. The presence of specific progesterone receptors in the ovary of monkeys and women support the possibility of direct action of progesterone on the ovary [1].

Progesterone is essential for the establishment and maintenance of pregnancy. The action of progesterone on target organs is elicited through binding of progesterone to receptors. Therefore, blocking progesterone receptors might prevent the native hormone from expressing biological effects. With this concept in mind, several antiprogestins like RU 486, ZK 98.299 were synthesized which bind to PR but do not trigger an agonistic hormonal response [2,3].

Although, initially these progesterone antagonists were designed for early abortifacient activity, later studies [4–7] clearly indicated their effects on various other target organs like pituitary and endometrium. It is well established that FSH and LH are required for follicular development [8–10] and at least LH is required for normal progesterone production during the luteal phase of the cycle [11]. In view of this, studies were undertaken in bonnet monkeys to evaluate effects of antiprogestins on follicular development, corpus luteum function and on endometrial maturation. These studies will help in determining the precise role of progesterone in these physiological processes. More knowledge on these aspects may ultimately be useful to develop newer contraceptive strategies.

Effect on Follicular Development and Ovulation

The role of progesterone in folliculogenesis has been suggested by various observations e.g. localization of PR in various ovarian compartments like stromal and thecal cells very early in follicular development [12], high concentrations of intrafollicular progesterone [13], presence of PR in hypothalamus [14] and the synergistic effects of progesterone on hCG-stimulated growth of small antral follicles [15]. The facilitatory effect of progesterone in ovulation is also highlighted by the observations that the levels of progesterone start to increase approximately 12 h before the LH surge [16] and the timing of LH surge is advanced following progesterone treatment during late follicular phase of the menstrual cycle [17]. Progesterone treatment on day six of cycle terminated folliculogenesis in the rhesus monkey [18], presumably due to blockade of the gonadotrophin release indicating both facilitatory and inhibitory effects of progesterone on follicular development. Thus, administration of antiprogestins during various stages of follicular development will not only help in understanding the precise role of progesterone but also would be useful in deciding whether these agents could be used to interfere with follicular development and ovulation.

Our studies in bonnet monkey showed that the follicular development was interrupted following administration of antiprogestins either in early or mid follicular phase. The rise in the circulatory levels of estradiol and the anticipated preovulatory LH surge were either attenuated or completely abolished [19, 20]. Furthermore the effects were dependent on the dose and the frequency of treatment. Administration of ZK 98.299 (20 mg/day s.c.) once daily on days 5 to 8 or days 5 to 15 of the menstrual cycle caused a significant increase in cycle length, which was due to prolongation of follicular phase [9, 20]. The rise in the levels of serum estradiol and bioactive LH was suppressed or completely blocked during the treatment. Following discontinuation of treatment, ovulation was observed as indicated by progesterone levels during luteal phase.

Administration of ZK 98.299 (5, 10 or 20 mg) once a week failed to affect folliculogenesis or ovulation [22]. Although vaginal bleeding was considerably reduced, menstrual cycle length was not affected. Luteal function was normal following 5 or 10 mg treatment (Fig. 1).

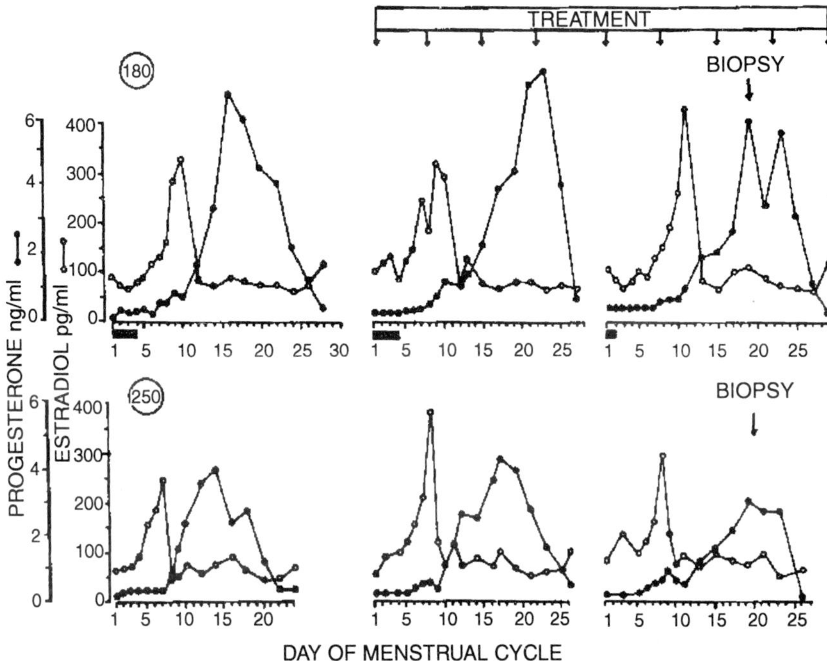

Fig. 1 Serum levels of estradiol and progesterone during the pretreatment and two treatment cycles of two representative bonnet monkeys treated with 5 mg (#250) or 10 mg (#180) ZK 89.299 at weekly intervals. The profile of gonadal hormones in the treatment cycles is similar to that of the pretreatment cycle.

The effects of RU 486, another antiprogestin on folliculogenesis and ovulation, appear to be similar to those of ZK 98.299. Treatment with 25 to 600 mg RU 486, either a single dose or for few days during the mid to late follicular phase arrested folliculogenesis and delayed ovulation in monkeys [23, 41] as well as in women [8, 19, 24]. Oral administration of RU 486, (1, 2 or 5 mg once daily) in women showed dose related effects, a dose of 1 mg failed to consistently block ovulation [25, 26]. Treatment with 5 or 10 mg RU486 once daily throughout the menstrual cycle, suppressed ovulation. Regarding the mode of action on folliculogenesis and ovulation, it appears that gonadotrophin release is impaired following antiprogestin treatment [27]. In bonnet monkeys, ZK 98.734-induced blockade of folliculogenesis was restored (Fig. 2) following hMG (LH and FSH) or by FSH (Metrodin) treatment [28]. In adult male marmosets, estradiol-induced LH release was also successfully suppressed following ZK 98.734 treatment [11]. Treatment with RU 486 (1 mg/day) during mid follicular phase inhibited LH and FSH surge without affecting a subsequent rise in estradiol levels. Furthermore, this effect could be reversed by exogenous administration of progesterone [29].

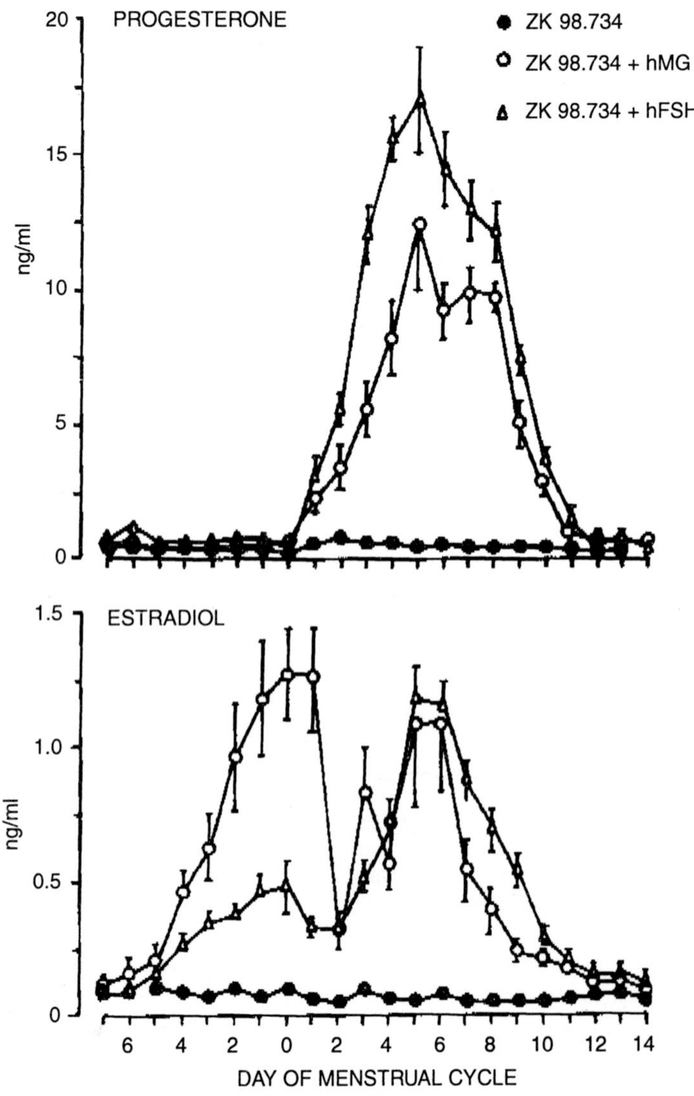

Fig. 2 Reversal of ZK 98.734-induced blockade of folliculogenesis with hCG or hFSH in bonnet monkeys (n = 4 per group). Treatment with hMG or hFSH alone initiated folliculogenesis in ZK 98.734 treated animals.

Apart from impairing gonadotrophin release, antiprogestins may have a direct effect on gonads. Treatment with RU 486 inhibited 3β-hydroxysteroid dehydrogenase, a key enzyme in steroidogenic pathway, in cultured human granulosa cells [30].

Effects on Corpus Luteum Function

Progesterone, secreted by the corpus luteum, prepares the endometrium for implantation of blastocyst and also maintains early pregnancy. Although there are conflicting reports, LH is suggested to be obligatory in maintaining normal luteal function. Reduced availability of LH during luteal phase curtails the luteal function. The presence of LH receptors in the corpus luteum and the luteotropic effects of LH/CG supports the concept that LH is required for normal luteal function. As pointed out earlier, administration of antiprogestins during follicular phase of the cycle impaired LH release, hence it is quite possible that treatment during luteal phase may adversely affect luteal function.

Treatment with ZK 98.299 or ZK 98.734 during luteal phase terminated luteal phase in bonnet monkeys [31, 32] as well as in common marmosets [33], as evidenced by premature decline in circulating levels of progesterone (Fig. 3). In bonnet monkeys, this was also associated with induction of menstruation (Fig. 4). In human, RU 486 showed similar effects [34, 35], however, in nonhuman primates, even higher dosages of RU 486 failed to cause any changes in luteal function [36].

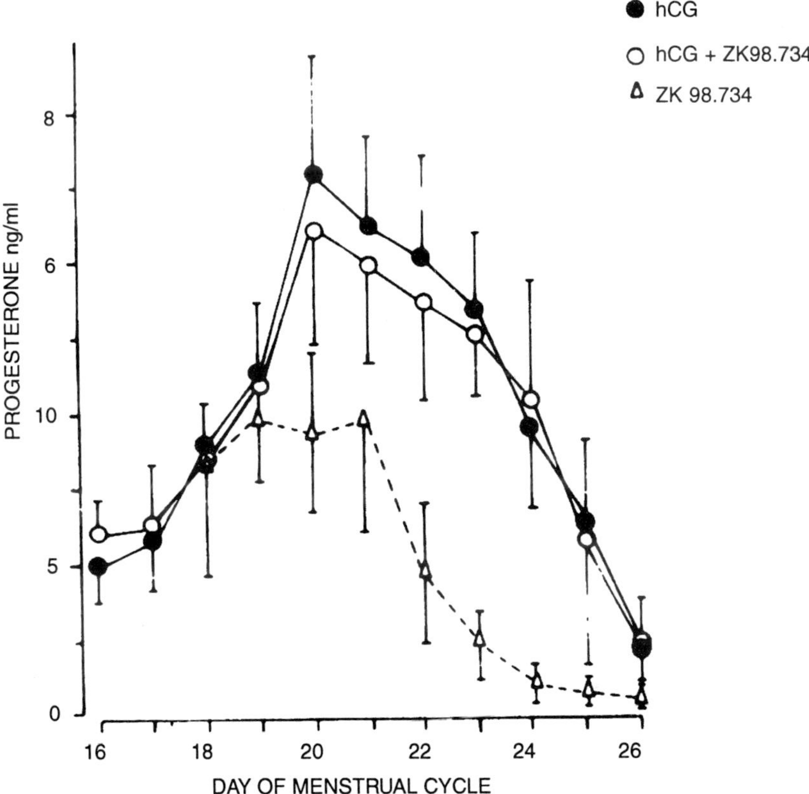

Fig. 3 Effect of ZK 98.734 administered (25 mg/day, s.c.) during the mid-luteal phase (on day 20 to 22 of the menstrual cycle) on serum progesterone levels. Treatment curtailed the luteal function as indicated by the premature decline in serum progesterone levels. ZK 98.734-induced decrease in progesterone levels was reversed by simultaneous treatment with hCG.

Since progesterone production by the corpus luteum in primates is LH-dependent and as antiprogestins have already been shown to impair LH release, it is possible that the luteolytic effects of antiprogestins are primarily mediated via curtailment of gonadotrophin release, which in turn is essential for maintenance of luteal function. Furthermore, antiprogestin-induced decrease in serum progesterone during the luteal phase was successfully prevented by simultaneous administration of hCG supporting the possibility of involvement of gonadotrophin following antiprogestin treatment (Fig. 3). Our studies in bonnet monkeys and marmosets thus suggest effect of antiprogestins in terminating luteal function.

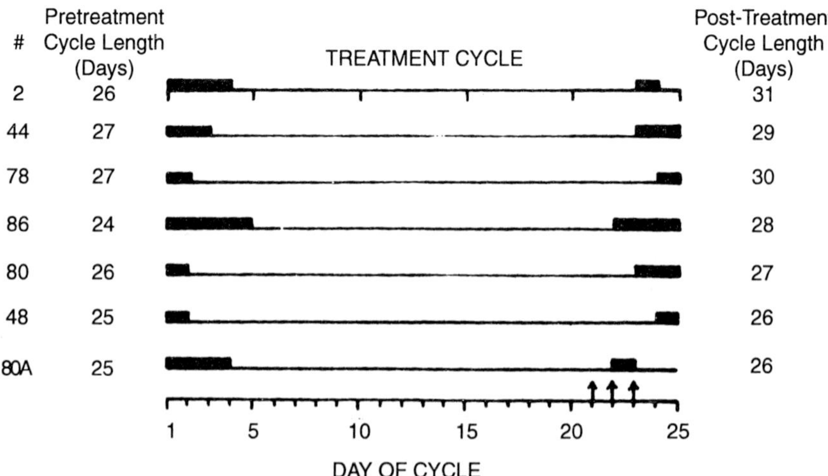

Fig. 4 Effect of ZK 98.299 on menstrual cycle length in cycling bonnet monkeys. ZK 98.299 (30 mg/day, s.c.) was administered once daily on days 21 to 23 of the menstrual cycle. ZK 98.299-induced menstruation in all the animals. Post-treatment cycles were of normal duration.

Effects on Endometrium

Treatment with antiprogestins, depending on the dose, retarded endometrial development, impaired gonadotrophin release and blocked ovulation. However, the hypothalamus, pituitary and endometrium differed in their sensitivity to antiprogestins, endometrium being the most sensitive [5, 37, 38]. Administration of ZK 98.299 (5 or 10 mg once a week, s.c.) failed to show any effect on gonadal hormones, ovulation or menstrual cycle length [22]. However, endometrial development was still impaired suggesting that antiprogestin treatment may selectively interfere with endometrial development sufficient enough to alter its receptivity. In a subsequent study, effects of 2.5 and 5 mg of onapristone (ZK 98.299) administered for 4–7 consecutive cycles were evaluated on various parameters like fertility, cycle length, hormonal profile, endometrial morphology and some of the functional parameters (immuno-histochemical localization of LIF and integrins). Treatment was initiated on day 5 of the first treatment cycle, and thereafter every third day for 4 to 7 consecutive cycles. Control group was also maintained.

All the five animals in the control group, treated with vehicle, became pregnant. However, four animals treated with 2.5 mg onapristone for 17 cycles and another four treated with 5 mg dose for 21 cycles did not conceive. Ovulation occurred in 30 out of 45 treatment cycles. In some of the ovulatory cycles, prolonged treatment suppressed luteal activity, however, in the ovulatory cycles the duration of follicular and luteal phases was not significantly affected as judged by circulating levels of estradiol and progesterone.

Endometrial biopsies were collected on day 8 after estradiol peak, i.e., around the perimplantation period. In treated animals, the endometrial growth and development was retarded and rendered out of phase. In 2.5 mg treated animals, the diameter of endometrial glands was reduced and pseudostratification in the glandular epithelium was observed (Fig. 5b). The stroma also showed condensation. In animals treated with 5 mg onapristone, the glands were partially regressed, secretory activity was significantly reduced and stromal compaction was evident (Fig. 5c) [39].

Ultrastructural studies carried out on endometria from 2.5 mg treated animals showed a general loss of cell organelles compared to vehicle treated control. The presence of intercellular space was increased, with reduced glycogen deposits. The desmosomes were absent with folding of the plasma membrane. Delay in endometrial maturation was also observed, characterized by small round glands. Following treatment with 5 mg ZK 98.299, endometrium showed total loss of cell organelles. Marked vacuolization was also observed. Lumen

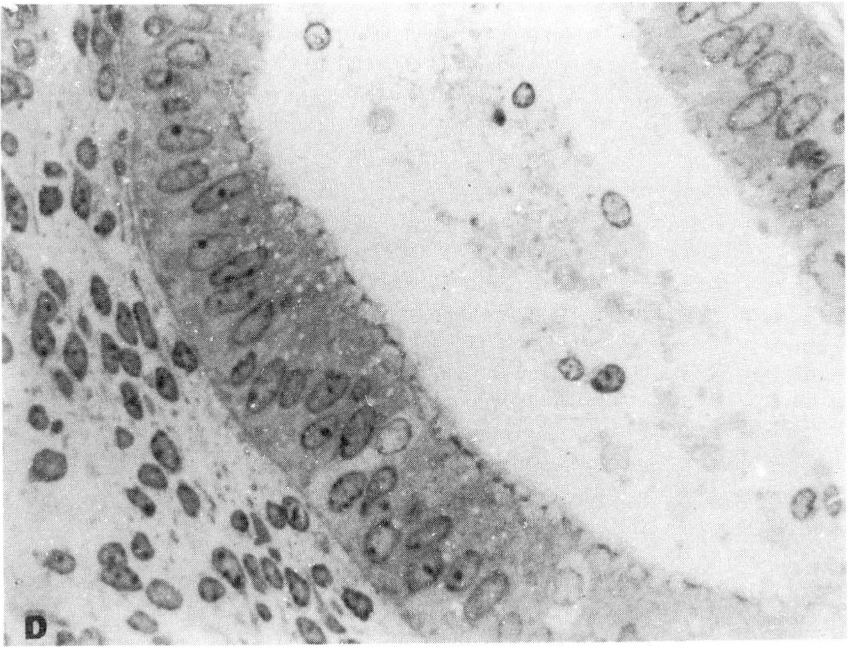

Fig. 5a Control animal showing signs of normal secretory endometrium. Glands are characterized by high columnar epithelium with a few supranuclear vesicles. Stromal cells are dissociated because of edema. (×1250, magnification).

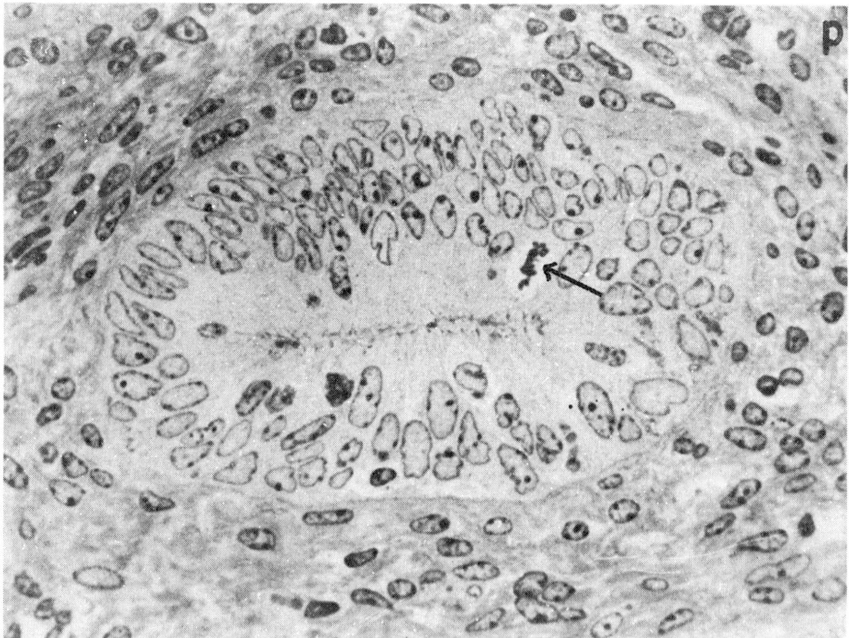

Fig. 5b Animal treated with 2.5 mg of ZK 98.299. Note reduction in diameter of endometrial gland. Mitosis leading to pseudostratification can be observed within gland. Beginning of stromal condensation can be noted (×1250 mignification).

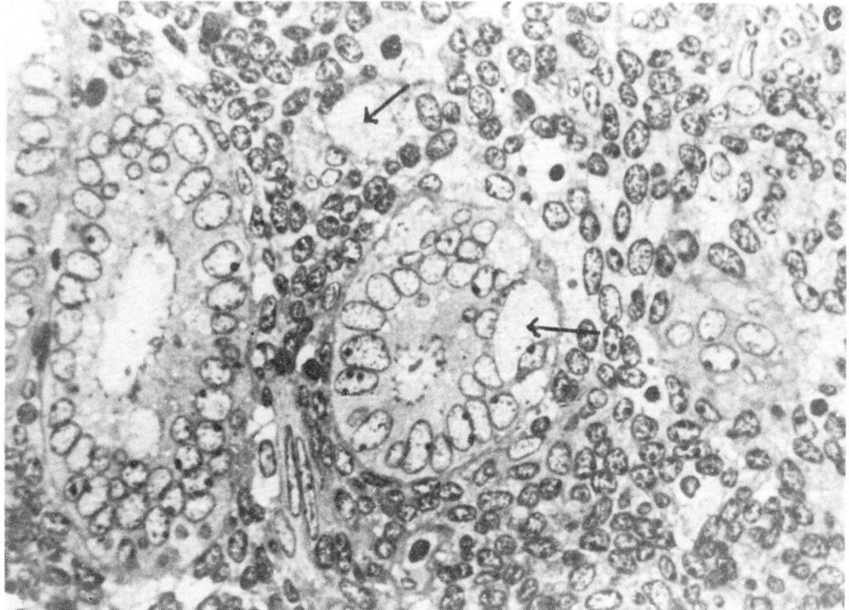

Fig. 5c Animal treated with 5.0 mg of ZK 98.299. Diameter of endometrial glands had decreased further. In shrinking glands, no mitoses was seen. Endometrial stromal tissue between glands was extremely condensed (×1250 mignification).

was drastically reduced with few microvilli. The stromal cells showed accumulation of lipids of the intercellular space. Thus dose-dependent degenerative changes were observed in the endometrium following onapristone treatment for 4 to 7 consecutive cycles [40].

Some of the functional parameters were also studied in the endometria of bonnet monkeys treated with low dose antiprogestin. Leukemia inhibitory factor (LIF) has been suggested to play important role in the preparation of endometrium for implantation in many species including primates [42]. Onapristone treatment caused dose-dependent decrease in the localization of LIF in the endometrium collected on day 8 following estradiol peak. Similarly $\alpha_1 \beta_1$ and $\alpha_v \beta_3$ integrins, which are cycle dependent and regulated by progesterone, were also significantly reduced following onapristone treatment. Thus, it is possible that antifertility effect of low dose onapristone could be due to selective action at the endometrial level leading to aberrant expression of LIF and $\alpha_1 \beta_1$ and $\alpha_v \beta_3$ integrins. This functional alteration may result in non-receptive state thereby preventing implantation.

Conclusions

The results of our studies clearly highlight the effects of antiprogestins on various physiological processes like follicular development, ovulation, corpus luteum function and endometrial maturation. The inhibitory effects of antiprogestins on LH release suggest the possible use of antiprogestins as a contraceptive to block follicular development and ovulation or maintenance of corpus luteum function. The direct and selective effects on endometrial development and maturation open up a new area of "endometrial contraception" wherein the endometrium is rendered out of phase and implantation of blastocyst is prevented. The effects of antiprogestins are dose-dependent. Higher dosages which affect follicular development and inhibit ovulation also retard endometrial development through systemic effect. However, lower doses selectively impair endometrial development without altering the hypothalo-pituitary ovarian axis. The antiprogestins therefore have multifaceted actions on various reproductive processes in female and these in turn may be utilized to inhibit reproduction at various sites.

References

1. Goodman A.L. and Hodgen G.D. (1979). Corpus luteum—conceptus–follicle relationships during the fertile cycle in rhesus monkeys: Pregnancy maintenance despite early luteal removal. *Journal of Clinical Endocrinology and Metabolism* **49**: 469–471.
2. Belanger A., Phibert D., and Teutsch G. (1981). Regio and sterospecific synthesis of 11β-substituted 19-norsteroids. *Steroids* **37**: 361–382.
3. Neef G., Beier S., Elger W., Henderson D. and Weichert R. (1984). New steroids with antiprogestational and antiglucocorticoid activities Steroids, **44**: 349–372.
4. Baulieu E.E. (1989). Contragestion and other clinical applications of RU 486 on antiprogesterone at the receptor. *Science* **245**: 1351–1357.

5. Van Look P.F.A. and Bygdeman M. (1989). Antiprogestational steroids: A new dimension in human fertility regulation. In: Oxford Reviews of Reproductive Biology (ed. S.R. Milligan), pp. 1–61, Oxford University Press, Oxford.
6. Puri, C.P. and Van Look P.F.A. (1991). Newly developed competitive progesterone antagonists for fertility control. *Frontiers of Hormone Research* **19**: 127–167.
7. Thong K.J. and Baird D.T. (1986). Induction of abortion with mifepristone and misoprostol in early pregnancy. *British Journal of Obstetrics and Gynaecology* **99**: 1004–1007.
8. Liu J.H., Garzo G., Morris S., Stuenkel C., Ulmann A. and Yen S.S.C. (1987). Disruption of follicular maturation and delay of ovulation after administration of the antiprogesterone RU 486. *Journal of Clinical Endocrinology and Metabolism* **65**: 1135–1140.
9. Puri C.P., Patil R.K., Elger W.A.G., Vadigoppula A.D. and Pongubala J.M.R., (1989). Gonadal and pituitary response to progesterone antagonist Z.K., 98. 299 during the follicular phase of the menstrual cycle in bonnet monkeys. *Contraception* **39**: 227–243.
10. Collins R.L. and Hodgen G.D. (1986). Blockade of the spontaneous midcycle gonadotropin surge in monkeys by RU 486: progesterone antagonist or agonist? *Journal of Clinical Endocrinology and Metabolism*, **63**: 1270–1276.
11. Kholkute S.D., Patil R.K., Sharma S., Elger W.A.G. and Puri CP. (1980). Suppression of bioactive LH and testosterone by progesterone antagonist ZK 98.734 in adult male common marmosets, *Callithrix Jacchus. Biology of Reproduction* **42**: 808–814.
12. Hild-Petit S., Stouffer R.L. and Brenner R.M. (1988) Immunocytochemical localization of estradiol and progesterone receptors in the monkey ovary throughout the menstrual cycle. *Endocrinology*, **123**: 2896–2905.
13. Westergaard L., Christensen I.J. and Mc Natty K.P. (1986) Steroid levels in ovarian follicular fluid related to follicular size and health status during the normal menstrual cycle in women. *Human Reproduction*, **1**: 227–232.
14. Garris D.R., Villbar R.B., Taboaka Y., White R.S. and Little B. (1981). In situ estradiol and progestin. (R5020). Localization in the vascularly separated and isolated hypothalamus of the rhesus monkey. *Neuroendocrinology* **32**: 202–208.
15. Richards J.S. and Bogovich K. (1982). Effect of human chorionic gonadotropin and progesterone on follicular development in the immature rat. *Endocrinology* **111**: 1429–1438.
16. Haff J.D., Quigly M.E. and Yen S.S.C. (1982). Hormonal dynamics at mid cycle: A evaluation. Journal of Clinical Endocrinology and Metabolism **57**: 792–796.
7. Liu J.H., and Yen S.S.C. (1983). Induction of midcycle gonadotropin surge by ovarian steroids in women: a critical evaluation. *Journal of Clinical Endocrinology and Metabolism*, **57**: 797–809.
18. Dierschke D.J., Yamaji T., Karsch, F.J., Weick R.F., Weick R.G., Weiss G. and Knobil, R.E. (1973). Blockade by progesterone of estrogen-induced L H and FSH release in the rhesus monkey. *Endocrinology* **92**: 1496–1501.
19. Shoupe, D. Mishell D.R. Jr, Page M.A., Madkour H, Spitz I.M. and Lobo R.A. (1987). Effects of the antiprogesterone RU 486 in normal women: II Administration in the late follicular phase. *American Journal of Obstetrics and Gynaecology* **157**: 1421–1426.
20. Puri, C.P., Patil R.K., D'Souza A, Elger W.A.G. and Pongubala J.M.R. (1988). Pharmacokinetics and pharmacodynamics effects of progesterone antagonist ZK 98.299 in female bonnet monkeys. In: Hormone Antagonists for Fertility Regulation (eds. C.P. Puri and P.F.A. Van Look) pp. 123–140, Indian Society for the Study of Reproduction and Fertility, Bombay.

21. Swahn M.L., Johannisson E., Daniore V., de la Torre B and Bygdeman M (1988). The effect of RU 486 administered during the proliferative and secretory phases of the cycle on the bleeding pattern, hormonal parameters and the endometrium. *Human Reproduction* **3**: 915–921.
22. Ishwad P.C., Katkam R.R., Hinduja I.N., Chwalisz K., Elger W. and Puri C.P. (1993). Treatment with a progesterone antagonist ZK 98.299 delays endometrial development without blocking ovulation in bonnet monkeys. *Contraception* **48:** 57–70.
23. Remohi, J., Balmaceda, J.P., Rojas, F.J. and Asch R.H. (1988). The role of preovulatory progesterone in the midcycle gonadotropin surge, ovulation and subsequent luteal phase: studies with RU 486 in rhesus monkeys. *Human Reproduction* **3**: 431–435.
24. Stuenkel C.A., Garzo V.G., Morris S., Liu, J.H. and Yen S.S.C. (1990). Effects of the antiprogesterone RU 486 in the early follicular phase of the menstrual cycle. *Fertility and Sterlity* 53: 642–646.
25. Batista M.C., Cartledge T.P., Zellmer A.W, Merino M.J., Axiotis C, Loriaux D.L. and Neiman L.K. (1992). Delayed endometrial maturation induced by daily administration of the antiprogestin RU 486: a potential new contraceptive strategy. *American Journal of Obsetrics and Gynaecology*, 167: 60–65.
26. Ledger W.L., Sweeting V.M., Hiller H. and Baird D.T. (1992). Inhibition of ovulation by low dose mifepristone (RU 486). *Human Reproduction* 7: 945–950.
27. Croxatto H.B., Salvatierra A.M., Croxatto H.D. Fuentealba B (1993). Effects of continuous treatment with low dose mifepristone throughout one menstrual cycle. *Human Reproduction* **8**: 201–207.
28. Puri C.P., Vadigoppula A. Gopalkrishnan K, Katkam R.R., Elger W.A.G. and Patil R.K. (1991). Contraceptive potential of progesterone antagonist ZK 98. 734. Effect on folliculogenesis, ovulation and corpus luteum in bonnet monkeys In: Perspectives in primate reproductive biology (eds. NR Moudgal, K. Yoshinaga, A.J. Rao and P.R. Adiga), pp. 325–344. Wiley Eastern Limited, New Delhi.
29. Batista M.C., Cartledge T.P., Zellmer A.W., Nieman L.K., Merriam G.R. and Loriaux D.L. (1990). Evidence for a critical role of progesterone in the regulation of the mid cycle gonadotropin surge and ovulation. *Journal of Clinical Endocrinology and Metabolism* **74**: 565–570.
30. Dimattina, M., Albertson, B., Seyler D.E., Loriaux D.L. and Falk R.J. (1986). Effects of the antiprogestin RU 486 on progesterone production by cultured human granulosa cells: Inhibition of the ovarian 3β-hydroxy steroid hydrogenase. *Contraception* **34**: 199–206.
31. Puri C.P., Elger, W.A.G. and Pongubala J.M.R. (1987). Induction of menstruation by antiprogesterone ZK 98.299 in cycling bonnet monkeys. *Contraception* **35**: 409–421.
32. Puri C.P. Katkam R.R., D'Souza A, Elger W.A.G. and Patil R.K. (1990). Effects of a progesterone antagonist lilopristone (ZK 98.734) on induction of menstruation, Inhibition of nidation and termination of pregnancy in bonnet monkeys. *Biology of Reproduction* **43**: 437–443.
33. Puri C.P., Patil R.K., Kholkute S.D., Elger W.A.G. and Swamy X.R. (1989). Progesterone antagonist lipopristone a potent abortificient in the common marmoset. *American Journal of Obstetrics and Gynaecology* **161**: 248–253.
34. Schaison G., George M, Lestrat N, Reinberg, A. and Baulieu E.E. (1985). Effects of the antiprogesterone steroid RU 486 during midluteal phase in normal women. *Journal of Clinical Endocrinology and Metabolism* **61**: 484–489.

35. Neiman L.K., Choate T.M., Chrouses G.P., Healy D.L., Morin M., Ronquist D., Merriam G.R., Sptiz I.M., Bardin C.W., Baulieu E.E. and Loriaux D.L., (1987). The progesterone antagonist RU 486. A potential new contraceptive agent. *New England Journal of Medicine* **316**: 187–191.
36. Asch R.H. and Rojas F.J. (1985). The effects of RU 486 on the luteal phase of the rhesus monkey. *Journal of Steroid Biochemistry* **22**: 227–230.
37. Van Look P.F.A. and Von Hertzen H (1993). Antiprogestogens: perspectives from a global research program. In: Clinical applications of mifepristone (RU 486) and other antiprogestins: assessing the science and recommending a research agenda. Donaldson MS ed. Washington: National Academy Press, 253–277.
38. Spitz I.M. and Bardin C.W. (1993). Clinical pharmacology of RU 486 antiprogestin and antiglucocorticoid. *Contraception* **48**: 403–444.
39. Katkam R.R., Gopalkrishnan, K., Chawalisz K, Schillinger E. and Puri C.P. (1995). Onapristone (ZK 98.299): A potential antiprogestin for endometrial contraception. *American Journal of Obstestrics and Gynecology* **179**: 779–887.
40. Gopalkrishnan K., Katkam RR, Padwal V. Kholkute S.D. and Puri C.P. (2000). Morphometric and ultrastructural studies of endometrium from bonnet monkeys treated with an antiprogestin onapristone. *Contraception:* (Communicated).
41. Collins R.L. and Hodgen G.D. (1986). Blockade of the spontaneous midcycle gonadotropin surge in monkeys by RU 486: A progesterone antagonist or agonist? *Journal of Clinical Endocrinology and Metabolism* **63**: 1270–1276.
42. Vogiagis D. and Salamonsen L.A. (1999). The role of leukaemia inhibitory factor in the establishment of pregnancy. *Journal of Endocrinology* **160**: 181–190.

Follicular Growth, Ovulation and Fertilization: Molecular and Clinical Basis
Anand Kumar and Amal K. Mukhopadhyay (Eds.)
Narosa Publishing House, New Delhi, India, 2001

11

Transforming Growth Factor-β1 and -β2 in Physiological Concentrations Enhances Oestradiol But Not Inhibin Release from Human Granulosa-Luteal Cells in Leukocytes Free Culture

N.E. Schlabritz-Loutsevitch[1], A. Salmassi[2] and L. Mettler[2]
[1]Institute for Hormone and Fertility Research, Hamburg, Germany
[2]Department of Obstetrics and Gynaecology, University of Kiel, Germany

Introduction

The Transforming Growth Factor β (TGF-βs) represent the superfamily of peptide growth factors, which play an important role in the regulation of cell behaviour. The gene superfamily of TGF-βs includes TGF-β family, Mullerian inhibiting substance/antimullerian hormone, Drosophila decapentalgic-products, Xenopus bone morphogenetic factors and the family of activins and inhibins (Burt, 1994).

There are five different isoforms of TGF-β (Roberts and Sporn, 1990). Only TGF-β1, 2 and 3 are present in human tissue. Each TGF-β-isoform has a distinct function in vivo (Miyazono *et al.*, 1988). Biological activities of TGF-β include growth stimulation, growth inhibition, extracellular matrix synthesis, angiogenesis and induction of cell differentiation. TGF-βs play an important role in the physiology of all tissues. Although relative much is known about the presence and biological action of TGF-β isoforms on ovarian cells (Mulheron *et al.*, 1992), knowledge about its function in the human is meager.

Materials and Methods

173 normo-ovulatory women aged 20–42 were recruited to this study: All patients were stimulated according to our, "long protocol" of programmed oocytes retrieval: down-regulation was performed using subcutaneous daily injections of DTRP-6L (Decapeptyl®, Ferring, Germany) of 100–500 mg or the transnasal application of nafarelin acetate (Synarela®, Syntex, Germany) of

800 mg daily, which were administered from day 21 of the natural menstrual cycle until hCG application (Choragon®, Ferring, Germany). Individualized FSH or HMG (Fertinorm HP 75, Serono, Sweden, Menogon®, Ferring, Germany) stimulation started on the second day of the cycle. The SIEMENS ultrasound machine (SONOLINE SL-1) with 5.0 or 7.5 MHz transducers was used. Follicular puncture (FP) was determined 34–36 h after HCG injection. The retrieved oocytes were inseminated 2–5 h later.

The samples of serum and follicular fluid were centrifuged at 1400 g for 10 min to remove cellular fragments and kept frozen at – 70°C until further investigation of inhibin. Inhibin assay was performed by using EASIA (Enzyme Amplified Sensitivity Immuno Assay) methods (Medgenix, Belgium).

Human granulosa cells were obtained from follicular aspirates. Cells were separated from erythrocytes, lymphocytes and macrophages following methods of Best *et al.*, 1994 and Piguette *et al.*, 1994. Cells were plated at 8×10^4/ml and cultured for 24, 48, 72 and 96 hours. The measurement of oestradiol and α-inhibin performed daily.

Student's t-test and Fisher-test were used for statistical analysis.

Results

The Influence of TGF-β1 and-β2 on Estradiol Secretion of Human Granulosa Cells

The evidence in Fig. 1 indicates that TGF-β1 and TGF-β2 enhanced the estradiol production of granulosa cells over 48 h in dose-dependent manner (R=0.96).

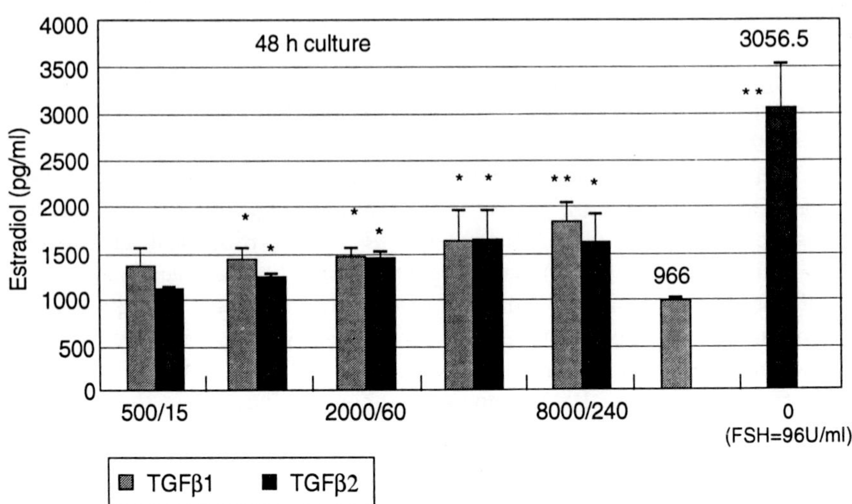

Fig. 1 Estradiol responses on day 2 (48 h culture) to TGF-β1 and TGF-β2
**: Significant difference ($p < 0.01$) between basal levels and TGF-β1- (8000pg/ml), FSH-induced levels.
*: Significant difference ($p < 0.05$) between basal levels and TGF-β1, TGF-β2 induced levels Granulosa cells (4×10^4/ml) obtained from women undergoing IVF-ET-treatment were cultured for 48h with increasing doses of TGF-β1 and TGF-β2.

When the cells were incubated with FSH alone the mean level of estradiol was significantly higher (average increase 2.5 fold) ($p < 0.001$) than with growth factors.

The 72 h granulosa cell culture showed significant increasing of estradiol secretion under the influence of TGF-β1 concentrations (4000 and 8000 pg/ml) and TGF-β2 (240 pg/ml). The FSH-stimulated estradiol production was statistically not significant higher that those under 8000 pg/ml and 240 pg/ml TGF-β1 and TGF-β2 (Fig. 2).

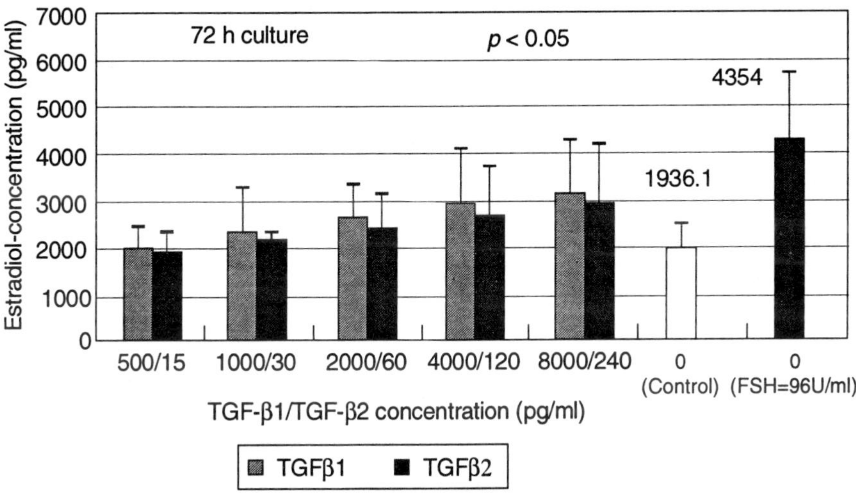

Fig. 2 Estradiol responses on day 3 (72 h culture) to TGF-β1 and TGF-β2
*Significant difference ($p < 0.05$) between basal levels and TGF-β1- (8000pg/ml), TGF-β2- (240 pg/ml) FSH-induced levels.
Granulosa cells (8×10^4/ml) obtained from women undergoing IVF-ET-treatment were cultured for 48h with increasing doses of TGF-β1 and TGF-β2

With the incubation time of 96 h there was also a clear dose-dependent inhancement of estradiol production in the presence of TGF-β1 and TGF-β2. Statistically significant increases in estradiol secretion compared to controls were only seen under the influence of TGF-β1 concentrations 4000 and 8000 pg/ml and TGF-β2 concentrations 120 and 240 pg/ml.

There was also a significantly higher basal oestradiol production at 96 h (2951 $\pm$ 430 pg/ml) compared to 48 h and 76 h in culture (996 $\pm$ 46 and 18523 $\pm$ 607 pg/ml) ($p < 0.01$).

The Influence of TGF- β1 and -β2 on Inhibin Secretion of Human Granulosa Cells

The TGF-β1 and TGF-β2 had no detectable influence on the basal inhibin production in all three cultures (Figs. 3, 4 and 5). Nevertheless, the secretion of inhibin in 96 h culture were significantly higher, as in 48 h and 72 h ($p < 0.05$).

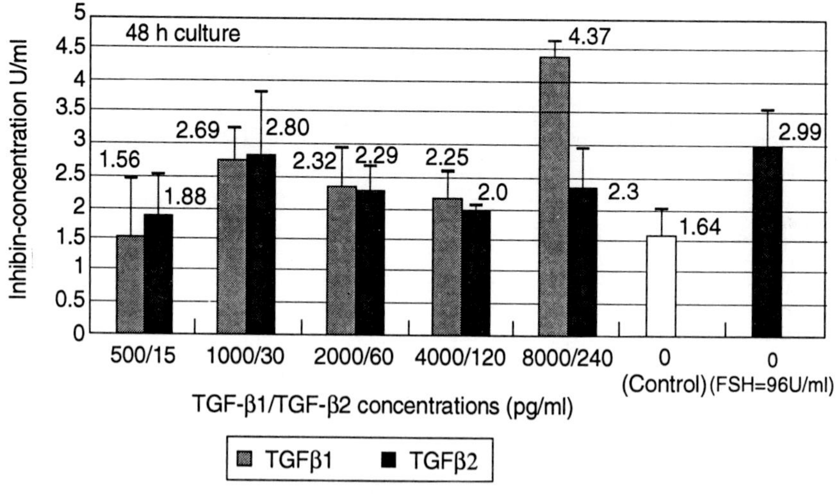

Fig. 3 Inhibin responses on day 2 (48 h culture) to TGF-β1 and TGF-β2.

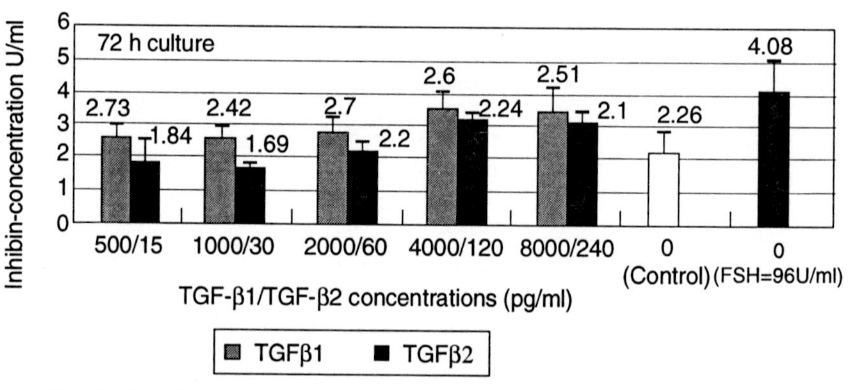

Fig. 4 Inhibin responses on day 3 (72 h culture) to TGF-β1 and TGF-β2.

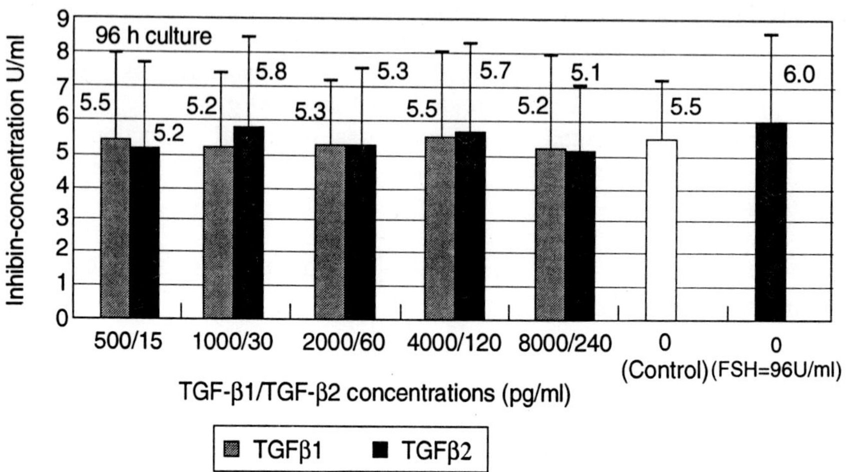

Fig. 5 Inhibin responses on day 4 (96 h culture) to TGF-β1 and TGF-β2

When the granulosa cells were treated with TGF-β1 and TGF-β2 the inhibin response was not significantly different from those in the control. FSH stimulated inhibin production at 72 h ($p < 0.05$), but not on 96 h of incubation. The secretion of inhibin was significantly higher on 96 h compared to 72 h of incubation ($p < 0.05$).

Discussion

As TGF-β1 and -β2 is produced in practically all tissues of the human organism, the influence of oestradiol on the TGF-β secretion– and vice versa- on various cell culture models has already been investigated: breast cancer cells (Zugmayer and Lippman, 1990), osteoclasts (Robinson *et al.*, 1996), trophoblast cells (Matsuraki *et al.*, 1992) and fibroblasts (Colette *et al.*, 1990). The effect of TGF-β on granulosa cell function (hormonal secretion and the ability to proliferate) has been described by Adashi and Resnik (rat cells, 1986), Mondshein and Schomberg (pig cells, 1988) and Roy and Kole (hamster cells, 1995). The results of these studies are contradictory. For example, TGF-β1 stimulated the steroidal synthesis in rat granulosa cells, but inhibited it in pig granulosa cells. Furthermore, the effect of TGF-β has also been dependent on the dosage of gonadotropin used for stimulation (Benahmed *et al.*, 1993). Studies regarding the role of cytokine in the function of human granulosa cells are scarce. One of the reasons for this is the fact that the punctured follicular fluid consists of a mixture of granulosa cells, leucocytes, erythrocytes, lymphocytes and macrophages (Best *et al.,* 1994). The lymphocytes and macrophages the proportion of which—following the usual separation methods—is approximately 16–18% could themselves be the source of cytokine production.

In this study we succeeded in using human granulosa cell culture free of macrophages and lymphocytes (Piquette *et al.*, 1994). If we had used the standard methods of cultivating granulosa cells (Chaffkin *et al.*, 1993), we would not have been successful. The cell density was 8×10^4/well, which is described by Chaffkin *et al.,* (1993) as the density needed for proliferation (up to 48–72 hours of cultivation) or differentiation (more than 96 hours of cultivation) of granulosa cells. This work answers the question raised by Chegini & Flanders (1992), whether TGF-β shows steroidogenic activities. It is the first publication on the positive influence of TGF-β1 and TGF-β2-dependent upon the concentration—on the oestradiol secretion of human granulosa cells. This means that the synthesis of E2 is more complicated than Erickson *et al*, (1990) assumed. Not only FSH, hCG and IGF-I, but TGF-β1 and -β2 are included.

The present work also is the first study on the separate impact of TGF-β1 and -β2 on the production of alpha-inhibin in human granulosa luteal cells. Contrary to Zhang *et al.* (1988), no direct influence of TGF-β1 and TGF-β2 on the secretion of inhibin could be found. These deviating results can be explained by the fact that above authors worked with rat granulosa cells. Besides Erämaa and Ritvos (1996) examined the absence of alpha-inhibin mRNA in human granulosa cells under the influence of TGF-βs, in contrast to β-inhibin, the secretion of which was dependant upon TGF-β. The fact that the basal and FSH, TGF-β1 and -β2 stimulated secretion of inhibin during a cultivation period of 96 hours was significantly higher than during a period of 48–72 hours contradicts the study of Demura *et al.* (1993) on a decreasing inhibin production of differentiated luteal cells. This means that in the human corpus luteum other regulative mechanisms exist than in animals. According to McCartney-Fanscis and Wahl (1994) the paradoxical qualities of TGF-β, leading to an acceleration or inhibition of cell functions, make this cytokine especially interesting for further research projects. One of the most important concepts in therapeutical use of TGF-βs would be: "Substitute, when too little, neutralize, when too much." A prerequisite in following this concept is the knowledge of the actual concentration of TGF-β under different circumstances, as e.g. hormonal treatment, ageing processes, reproductive disorders etc. The present study is a step into this direction.

Acknowledgement

We thank Dr. Olcese for proofreading this article.

References

Adashi, E. Y. and Resnik, C.E. (1986) Antagonistic interactions of transforming growth factors in the regulation of the granulosa cells differentiation. *Endocrinology*, **121 (2)**, 786–792.

Best, C.L., Pudney, J., Anderson, D.J., Hill, J.A. (1994) Modulation of human granulosa cell steroid production in vitro by tumor necrosis factor alpha: implications of white blood cells in culture. *Am. J. Obstet. Gynaecol.*, **84(1)**, 121–127.

Burt D. W. (1994) Evolution of the transforming growth factor β superfamily. *Prog. Growth Factor Res.*, **5(1)**, 99–118

Collette, A. A., Wakfield, L.M., Howell, F.V. *et al.* (1990) Antioestrogens induce the secretion of active transforming growth factor beta from human fetalfibroblast. *Br. J.Cancer*, **62**, 405–409.

Chafikin, L.M., Luciano, A.A., Peluso, J.J. (1993) The role of progesterone in regulating human granulosa cell proliferation and differentiation in vitro. *J. Clin. Endocrinol. Metab.* **76(2)**, 696–700

Chegini, N. and Flanders, K.C. (1992) Presence of Transforming growth factor β and their selective cellular localization in human ovarian tissue of various reproductive stages, *Endocrinology*, **130(3)**, 1707–1715

Demura, R., Suzuki, T., Tajima S. *et al.* (1993) Human plasma free activin and inhibin levels during the menstrual cycle. *J. Clin. Endocrinol. Metab.*, **76**, 1080–1082

Erämaa, M. and Ritvos, O. (1996) Transforming growth factor-β1 and -β2 induce inhibin and activin β_b subunit messenger ribonucleic acid levels in cultured human granulosa-luteal cells. *Fertil. Steril.*, **65(5)**, 954–960

Erickson, G. Nongonadotropic regulation of ovarian function: growth hormone and IGFs. (1994) *In: Ovulation Induction*. Filicori M., Flamingi C. (eds), 73–84 pp

Matsuraki, N., Masuhiro, K., Jo., T., Shimoya., K. *et al.* (1992) Trophoblast-derived transforming growth factor-β1 suppresses cytokine-induced, but not gonadotropin-releasing hormone induced, release of human chorionic gonadotropin by normal human trophoblast. *J.Clin. Endocrinol. Metab.* **74(1)**, 211–216

McCartney-Franscis, N. L. and Wahl S.M. (1994) Transforming growth factor beta: a matter of life and death. *J. Leukoc. Biol.*, **55(3)**, 401–409

Miyazono K., Mellman, C., Werstedt, C., Meldin, C.H. (1988) Latent high molecular complex of TGFβ1, *J. Biol. Chem.,* **263**, 6407–6411.

Mondschein, J. and Schomberg, D.W. (1988) Growth factors modulate gonadotropin receptor induction in granulosa cell culture. *Science*, **211**, 1179–1180.

Mulheron G.W., Bossert N.L., Lapp J.A., Walmer D.K., Schomberg D.W. (1992) Human granulosa-luteal and cumulus cells express transforming growth factor-beta type 1 and type 2 mRNA. *J. Clin. Endocrinol. Metab.* **74**, 458–460.

Piguette, G.N., Simon, C., Danasouri, E.I. *et al.*, (1994) Gene regulation of interleukin-1 receptor type 1 and plasminogen activator inhibitor-1 and 2 in human granulosa-luteal calla. *Fertil. Steril.* **62(4)**, 760–770.

Roberts, A.B. and Sporn M.B., (1990) The transforming growth factors-β. *In Handbook of Experimental Pharmacology, Vol. 95/1. Peptide Growth Factors and their Receptors M.B. Sporn & A.B. Roberts, Eds. Springe-Verlag, Heidelberg*. 419–472.

Robinson, J.A., Riggs, B.L., Spelsberg, T.C. *et al.*, (1996) Osteoclasts and transforming growth factor-β: estrogen-mediated isoform-specific regulation of production. *Endocrinology*, **137(2)**, 615–621.

Roy, S.K. and Kole, A.R. (1995) Transforming growth factor-β receptor type II expression in the hamster ovary: cellular site(s), biochemical properties and hormonal regulation. *Endocrinology*, **136(10)**, 4610–4620.

Zhang, Z.W., Findlay, J.K., Carson, R.D. *et al.,* (1988) Transforming growth factor enhances basal and FSH-stimulated inhibin production by rat granulosa cells in vitro. *Moll. Cell. Endocrinol.*, **58**, 161–169.

Zugmayer, G. and Lippman, M.E. (1990) Effect of TGF-β on normal and malignant mammary epithelium. In: Piez K.A., Sporn M.B. (eds) Transforming growth factors β. *New York Acad. Sci.*, 272–275 pp.

Follicular Growth, Ovulation and Fertilization: Molecular and Clinical Basis
Anand Kumar and Amal K. Mukhopadhyay (Eds.)
Narosa Publishing House, New Delhi, India, 2001

12

Complex Regulation of the Ovarian Prorenin-Renin-Angiotensin-System by Hormones and Growth Factors

Barbel Brunswig-Spickenheier and Amal K. Mukhopadhyay
Institute for Hormone and Fertility Research, University of Hamburg
Grandweg 64, 22529 Hamburg

Introduction

During reproductive life in the female, there is a continuous growth, maturation and demise of ovarian follicles, solely interrupted by the onset of pregnancy. Only a limited number of follicles finally ovulates and differentiates into a corpus luteum, whereas the ultimate fate of the majority of follicles is atresia. There is no doubt that the hormonal microenvironment of each individual follicle determines its final destiny. In this respect, autocrine and paracrine factors that either act alone or modulate gonadotropins action may be of considerable importance.

One such factor is the prorenin-renin-angiotensin-system (PRAS), which is schematically presented in Fig. 1. This mixed enzyme-hormone system has originally been demonstrated to be responsible for the regulation of cardio-vascular homeostasis, but recent studies have emphasized its regulatory role in a number of specialized tissues, including endocrine glands and reproductive

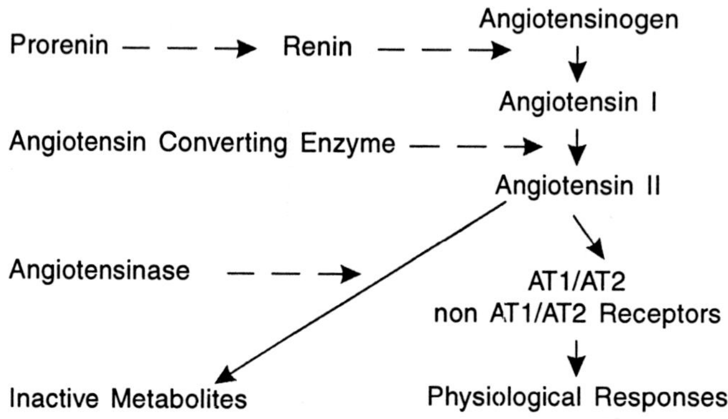

Fig. 1 The prorenin-renin-angiotensin-cascade.

tissues (Ganong, 1995; Pepperell *et al.*, 1995; Mukhopadhyay *et al.*, 1999). Renin, the active enzyme of the PRAS cascade is derived by enzymatic cleavage of the prosegment from its precursor molecule prorenin by a still not fully understood mechanism. Renin is an extremely specific aspartyl protease which cleaves angiotensin I(AI) from its only known substrate, angiotensinogen. By the action of angiotensin-converting enzyme, AI is converted to AII, the biologically active molecule of the PRAS. All is a multifunctional peptide which produces numerous different physiological effects ranging from vasoconstriction and the regulation of adrenal steroidogenesis to activation of neuroendocrine axis, growth promotion and cellular hyperplasia. To elicit its diverse actions, AII binds to a family of seven-transmembrane G-protein coupled receptors, which are cloned and designated AT1 and AT2 (Inagami *et al.*, 1999a). Most known physiological effects of AII are mediated by AT1-receptors which are coupled to adenylate cyclase and phospholipase C. Signaling through AT2 receptors still remains enigmatic; a regulation of cGMP formation and/or phosphotyrosine phosphatase activity however may be involved (Csikos *et al.*, 1998; Marrero *et al.*, 1996). Interestingly, in some cases activation of AT2 receptors may antagonize the action mediated by AT1 receptors (Nakajima *et al.*, 1995; Inagami *et al.*, 1999b). Within the circulation, AII is rapidly split to non-active metabolites by angiotensinase and peptidases.

Whereas the enzymatically inactive prorenin has been reported to be produced and secreted by numerous tissues, only the juxtaglomerular cells of the kidney are capable to secrete renin into the circulation (Yan *et al.*, 1998; Kurtz and Wagner, 1999). In 1986, Glorioso *et al.* described an increase in prorenin activity in human peripheral plasma during the luteal phase of the menstrual cycle and in early pregnancy. These data, together with the observation that prorenin activity was higher in follicular fluid than in peripheral plasma lead to the idea of the existence of an intraovarian PRAS. This was demonstrated shortly thereafter in several species (Schultze *et al.*, 1989; Pepperell *et al.*, 1995; Yoshimura, 1997; Speth *et al.*, 1999). There are numerous hints that in bovine and rat ovaries, the expression of the PRAS may be associated with atresia. Biochemical and morphological data show that prorenin and renin activities in bovine follicular fluid are inversely correlated with the developmental stage of the follicle (Mukhopadhyay *et al.*, 1991) and that the theca layer is the source of prorenin and renin. Experiments performed both in vivo and in vitro lead to the conclusion that the ovarian PRAS is primarily regulated by gonadotropic hormones (Feral *et al.*, 1990; Brunswig-Spickenheier and Mukhopadhyay, 1990; Itskovitz *et al.*, 1992). However since factors produced within the ovary are important modulators of almost all gonadotropic action on ovarian function, we have investigated whether such an interplay may also be involved in the regulation of the PRAS.

Methods

Isolation and Culture of Theca and Granulosa Cells

The isolation, purification and culture of bovine theca and granulosa cells has been described in detail (Brunswig-Spickenheier and Mukhopadhyay, 1990). After aspiration of granulosa cells, theca layers were dissected out of the follicle and dispersed enzymatically. Isolated theca cells were purified over Percoll and cultured on collagen-coated plates for 4 days in serum-free medium (Dulbecco's minimum essential medium/Ham's F12 1:1 with 2 mM L-Glutamine, 0.1% (w/v) BSA, 100 IU/ml penicillin, 100 μg/ml streptomycin sulfate, 5 μg/ml insulin, 5 μg/ml transferrin and 5 ng/ml sodium selenite). Experimental treatments were performed for 24 h as indicated and prorenin activity was measured in the supernatant. To obtain granulosa cell conditioned medium, these cells were cultured for 48 h in the above mentioned medium.

Assay for Prorenin Activity

To convert inactive prorenin into renin, culture was subjected to a limited proteolysis with TPCK-trypsin and renin activity was measured by the capacity to release angiotensin I (AI) from exogenous angiotensinogen. AI was determined by radioimmunoassay and expressed as ng AI/250,000 cells/h. Since exclusively prorenin is secreted by theca cells, the renin activity measured in the medium can be taken as prorenin.

AII Binding Studies

Theca cell membranes or cultured theca cells were incubated with different amounts of ^{125}I-AII in the absence or presence of an excess of either unlabelled AII or receptor-subtype specific antagonists. After separation from free ^{125}I-AII, cell-bound radioactivity was measured in a γ-counter. Details are given elsewhere (Brunswig-Spickenheier and Mukhopadhyay, 1992).

Results

Effects of Gonadotropins on Prorenin Secretion by Bovine Theca cells in Culture

In preliminary experiments, we had demonstrated that the highest concentrations of prorenin were present in fluid from atretic follicles (Mukhopadhyay *et al.*, 1991). Therefore it would have been reasonable to use theca cells from atretic follicles to study the prorenin production. However, since it was not possible to isolate sufficient amounts of viable theca cells from these follicles, we have pooled theca cells from follicles of different developmental stages for the present investigations. We have previously demonstrated that prorenin production by theca cells in vitro could be stimulated by LH, 8Br-cAMP and Forkolin (Brunswig-Spickenheier and Mukhopadhyay, 1990). These data demonstrate that the LH-induced increase in prorenin production is a cAMP-mediated process. In the following experiments, we wanted to determine whether locally produced regulatory factors could modify this effect of gonadotropin.

Modulation of LH-Stimulated Prorenin Secretion by Intraovarian Factors

At the outset we have examined whether granulosa cell-derived factors could influence theca cell prorenin production by adding increasing amounts of conditioned medium from granulosa cell culture to theca cells in the absence or presence of LH. The addition of granulosa-cell conditioned medium reduced the LH-stimulated prorenin secretion (125 ± 5.6 ng AI/250,000 cells/h in LH-stimulated cells without addition of granulosa-cell conditioned medium vs. 71 ± 3.8 ng AI/250,000 cells/h in LH-stimulated cells with addition of 500 µl/ml granulosa-cell conditioned medium). The basal unstimulated prorenin secretion was not affected (3.5 ± 0.3 ng AI/250,000 cells/h in the absence of granulosa-cell conditioned medium vs. 2.9 ± 0.7 ng AI/250,000 cells/h in the presence of 500 µl/ml granulosa-cell conditioned medium). Thus granulosa cells appear to secrete a factor(s) capable of suppressing prorenin production by theca cells in response to LH. Consequently, in a growing follicle with fully active granulosa cells, these factor(s) may keep a check on prorenin production by theca cells, inspite of a sustained level of LH. However, the nature of the inhibitory factor(s) still remains to be identified.

Steroids, growth factors and cytokines are prominent among several biologically active factors produced by granulosa cells. We therefore examined the effect of steroids (estradiol, progesterone, androstendione) on prorenin production. Incubation of theca cells with steroids (1–1000 ng/ml) did not alter either basal or gonadotropin-stimulated prorenin production (data not shown).

Next, we examined the effects of growth factors like transforming growth factor α, TGFα, fibroblast growth factor-2, FGF-2, and cytokines like tumor necrosis factor α, TNFα on LH-stimulated prorenin production. All of these factors suppressed the LH-stimulated prorenin production in a dose-dependent manner, however none of thee agonists had an effect on unstimulated cells (Table 1).

To define whether these factors exerted their inhibitory effect on LH-stimulated prorenin production by altering LH binding or more distal signal transduction pathways, we examined the effects of these factors on 8Br-cAMP-stimulated prorenin production. The result obtained was identical to that observed after stimulated of the cells with LH, as TNFα, FGF-2 and TGFα inhibited 8Br-cAMP-induced prorenin production in a dose-dependent way (Table 1).

Within the ovary-granulosa cells have been identified as a major source of FGF-2 and TNFα (Neufeld et al., 1987; Roby and Terranova, 1989); whereas TGFα is also produced by theca cells (Lobb *et al.,* 1989; Reeka *et al.*, 1998; Singh and Armstrong, 1995). Therefore it appears plausible that the inhibition produced by granulosa cell conditioned medium on LH-stimulated production is related to the presence of these factors in the medium. However this contention can only be substantiated by proper analysis of the medium and identification of the factors therein.

In contrast to the inhibitory effect observed after incubation with TNFα, FGF-2 and TGFα, treatment of the cells with transforming growth factor-β, TGF-β resulted in a further significant augmentation of both gonadotropin and 8Br-cAMP-stimulated prorenin production. The basal unstimulated prorenin production remained unchanged (Table 2). A number of reports have confirmed

Table 1 Effect of different growth factors on LH-stimulated prorenin production by theca cells

	Prorenin (*ng AI*/250,000 cells/h)		
	No addition	*LH* (10 *ng/ml*)	8 *Br-cAMP* (1 *mM*)
No growth factors	2.1 ± 0.4	95.0 ± 2.4[a]	104.0 ± 9.2[a]
+ FGF-2 (0.1 ng/ml)	2.5 ± 0.6[ns]	93.1 ± 1.5[ns]	87.6 ± 1.6*
+ FGF-2 (1 ng/ml)	3.5 ± 1.2[ns]	85.5 ± 4.9*	68.8 ± 2.5**
+ FGF-2 (10 ng/ml)	3.1 ± 0.7[ns]	62.7 ± 6.1**	20.8 ± 2.9***
+ FGF-2 (100 ng/ml)	2.8 ± 0.3[ns]	58.9 ± 1.9***	16.7 ± 4.2***
+ TGFα (0.1 ng/ml)	4.1 ± 0.8*	70.3 ± 2.8***	91.5 ± 7.1[ns]
+ TGFα (1 ng/ml)	2.4 ± 1.0[ns]	41.8 ± 0.5***	70.0 ± 4.9**
+ TGFα (10 ng/ml)	1.9 ± 0.6[ns]	30.4 ± 3.1***	39.5 ± 2.6***
+ TGFα (100 ng/ml)	2.7 ± 0.7[ns]	26.6 ± 6.8***	29.1 ± 3.3***
+ TNFα (0.1 ng/ml)	3.6 ± 0.8*	105.5 ± 5.4*	120.0 ± 8.4[ns]
+ TNFα (1 ng/ml)	1.9 ± 0.4[ns]	66.8 ± 7.2**	80.1 ± 5.3*
+ TNFα (10 ng/ml)	3.2 ± 0.7[ns]	49.2 ± 1.6***	48.4 ± 4.1***
+ TNFα (100 ng/ml)	2.6 ± 0.6[ns]	38.7 ± 1.2***	16.0 ± 0.9***

Theca cells were treated for 24 h with the indicated agonists and the amount of prorenin secreted into the medium was measured. Data represent mean±S.D. of triplicate determinations.
[ns]not significant
[a]$p < 0.01$ compared to basal values
*$p < 0.05$; **$p < 0.01$; ***$p < 0.001$ to "no growth factors".

Table 2 Dose-dependent stimulation of LH/8Br-cAMP-induced prorenin production by TGFβ

	Prorenin (*ng AI*/250,000 *cells/h*)		
	No addition	*LH* (10 *ng/ml*)	8 *Br-cAMP* (1 *mM*)
No TGFβ	4.3 ± 1.2	102 ± 4.3[a]	112 ± 19.2[a]
+ TGFβ (1 pg/ml)	4.5 ± 0.8[ns]	145 ± 7.9**	96 ± 5.8[ns]
+ TGFβ (10 pg/ml)	5.2 ± 0.9[ns]	189 ± 14.2***	161 ± 14.0*
+ TGFβ (100 pg/ml)	3.4 ± 1.2[ns]	280 ± 9.6***	211 ± 19.1**
+ TGFβ (1 ng/ml)	4.1 ± 0.8[ns]	312 ± 21.2***	315 ± 7.9***
+ TGFβ (10 ng/ml)	5.1 ± 1.0[ns]	275 ± 10.9***	326 ± 11.4***

Theca cells were treated with LH or 8Br-cAMP in the absence or presence of incresing amounts of TGFβ and prorenin was measured in the supernatant. All data represent mean ± S.D. from triplicate determinations.
[ns]Not significant.
[a]$p < 0.01$ compared to basal values.
*$p < 0.05$, **$p < 0.01$, ***$p < 0.001$ compared to in the absence of TGFβ.

that in the ovary TGFβ is mainly produced by theca cells (Teerds and Dorington, 1992; Gangrade and May, 1990) and the present observation suggests that TGFβ is acting as an autocrine factor which upregulates the gonadotropin-stimulated prorenin production. The precise mechanism involved in this upregulation still remains elusive.

Regulation of Ovarian AII-Receptors

Generally, biological actions of the PRAS are considered to be mediated by AII via specific membrane-bound receptors. To test the ovarian distribution of AII receptors and therefore identify the putative site for AII action, we have incubated granulosa, theca and luteal cells with ^{125}I-AII in the absence or presence of unlabelled peptide and have detected that only theca cells could bind labelled AII in significant amounts. A Scatchard-Plot of the cells revealed a single class of high-affinity, low capacity binding sites (K_d: 0.3 nM, B_{Max}: 67 fmol/mg protein) which were highly specific for AII (unrelated peptides like oxytocin, GnRH, VIP, ANP did not compete for the binding sites). AII exerts its action through different receptor subtypes (Inagami, 1995). Molecular characterization of at least subtype AT1a, AT1b and AT2 has been accomplished (Sandberg, 1994). Pharmacological characterization using different subtype specific antagonists of AII binding sites to theca membrane clearly shows that bovine ovarine are occupied mainly by AT2 receptors. The distribution of AT2 receptors in bovine ovarian cells differs from the rat and human, where granulosa cells are occupied with AT2 receptors (Pucell *et al.*, 1991; Tanaka *et al.*, 1995; Johnson *et al.*, 1997). Since in the rat, these receptors could be downregulated by in vitro treatment with FSH, we have incubated theca cells in the absence or presence of LH and determined the AII binding capacity. Treatment of the cells resulted in a doubling of receptor number (20,000 receptors per cell after LH-stimulation vs. 10,000 receptors in unstimulated cells) without affecting binding capacity of the individual receptors (K_d: 0.08 nM in unstimulated vs. K_d: 0.1 nM in LH-stimulated cells).

Taken together, our results show that LH enhances prorenin production as well as AT2 receptor expression in bovine theca cells. The effect of LH is reduced by simultaneous incubation with granulosa- and theca cells derived growth factors and cytokines (TGFα, FGF-2, TNFα) and is enhanced by theca-cell derived growth factor like TGFβ.

Discussion

Gonadotropic hormones are considered to be key regulators of follicular growth and differentiation. The observation however, that during each reproductive cycle a number of follicles start growing but only few finally ovulate, emphasizes the importance for paracrine and autocrine factors to finely modulate gonadotropin's action. Creation of a unique microenvironment composed of steroids, proteins and peptides is a prerequisite for the determination of an individual follicle's fate.

An important feature of the developmental stage of an individual follicle is the expression of the PRAS, which at least in the cow and in the rat is tightly associated with atresia. Significantly higher amounts of prorenin can be detected in the fluid of atretic follicles in cow (Mukhopadhyay *et al.*, 1991).

The expression of AII receptors in the rat granulosa cell is inversely correlated with the expression of FSH receptors (Pucell *et al.*, 1988) and these granulosa cells already exhibit morphological features of atresia (Obermuller *et al.*, 1998). High expression of the PRAS is not only a consequence of follicular atresia but also appears to be a cause thereof. In vitro treatment of hamster ovaries with AII resulted in a shift of the androgen/estradiol level to a more androgenic milieu (Kitzman and Hutz, 1992). Since a high level of estradiol in connection with FSH is obligatory for the induction of granulosa cell LH receptors and consequently for ovulation, this AII-induced change in steroid production may mediate the progression of follicular atresia. Recently it has been demonstrated, that treatment of rat granulosa cells with AII inhibited FSH-action like e.g. prevention of DNA fragmentation, estrogen production and LH receptor content (Kotani *et al.*, 1999). Taken together, these data strongly argue for a role of an intraovarian PRAS in the development of follicular atresia (Speth *et al.*, 1999). In the light of these observations it appears evident that in a follicle destined for ovulation, the PRAS has to be suppressed. Since all follicles may gain access to LH from the circulation, locally produced factors in individual follicles must be involved in the fine regulation of local PRAS expression which in turn may influence growth or demise of an individual follicle or a group of follicles. Of outstanding importance as modulators of ovarian function are growth factors and cytokines, which are produced and secreted by follicular cells and which act either as modulators of the differentiation state of an individual follicle or as factors that stimulate or inhibit cellular proliferation. This article demonstrated that FGF-2, TGF-α, TGF-β and TNF-α modulate expression of the PRAS and may thereby represent a link determining the developmental stage of an individual follicle.

Healthy, preovulatory follicles are equipped with a large number of biochemical active granulosa cells. These granulosa cells synthesize and secrete factors like FGF-2 and TNF-α, which may act on the theca cell to keep the PRAS suppressed. This effect can be furthermore enhanced by the action of theca cell derived TGF-α, which not only inhibits LH-stimulated prorenin production, but also acts as a mitogen for granulosa cells (Fig. 2).

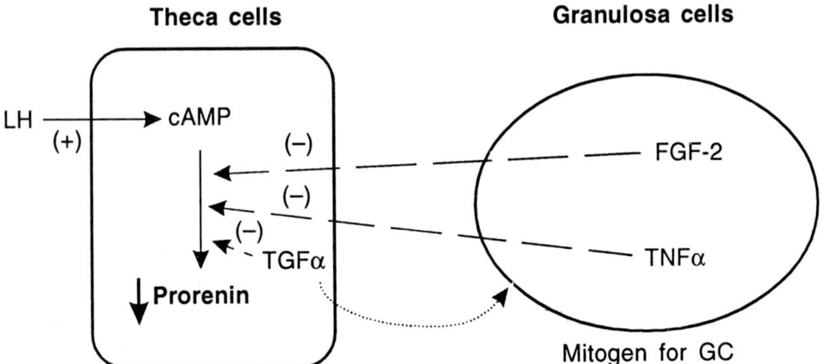

Fig. 2 Model for the granulosa and theca cell induced suppression of LH-stimulated prorenin production in healthy preovulatory follicles.

The onset of atresia is characterized by a rapid degeneration of the granulosa cells accompanied by a hypertrophy of theca cells (Greenwald and Roy, 1994). In this case, the inhibitory influence of granulosa cell derived factors on LH-stimulated prorenin production is attenuated. This effect is furthermore supported by theca cell derived TGF-β which on the one hand is antagonistic to granulosa cell growth and on the other hand may act in an autocrine manner to synergistically enhance LH-induced prorenin production. The resulting increase in prorenin may lead to follicle demise (Fig. 3).

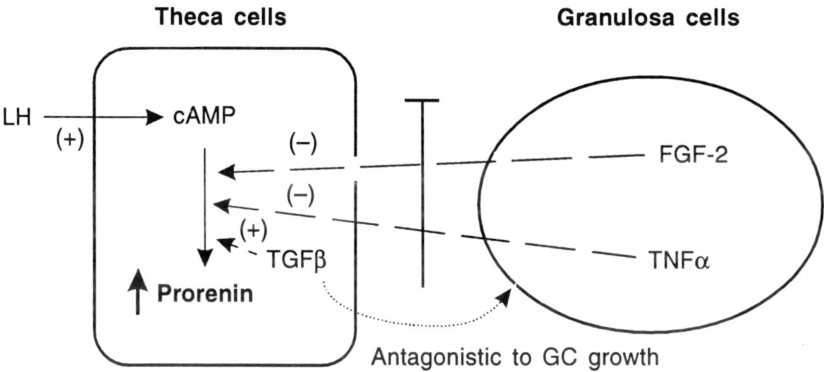

Fig. 3 Proposed mechanism of TGFβ-enhanced stimulation of theca cell prorenin production in follicles at the verge of atresia. In this case, the inhibitory effects of FGF-2 and TNFα observed in healthy follicles are abolished since granulosa cells are already degenerated.

Recent observations by Hayashi *et al.*, (2000) and Acosta *et al.*, (1999) have added further evidence for the complexity of the PRAS in ovarian function. These authors demonstrated that PRAS is not only a regulator of follicular cell function but that endothelial cells of the ovarian microvasculature may also be subject to regulation by ovarian PRAS. Furthermore, the PRAS appears not only to consist of the classical renin-angiotensin pathway, but may also be tightly connected to the regulation of endothelin and atrial natriuretic peptide.

Defining the role of intraovarian factors like the PRAS, cytokines and growth factors in follicular growth and differentiation is one of the main challenges to today's reproductive endocrinology.

References

Acosta, T.J., Berisha, B., Ozawa, T., Sato, K., Schams, D. and Miyamoto A. (1999) Evidence for a local endothelin-angiotensin-atrial natriuretic peptide system in bovine mature follicles in vitro: effects on steroid hormones and prostaglandin secretion. *Biol. Reprod.*, **61**, 1419–1425.

Brunswig-Spickenheier, B. and Mukhopadhyay, A.K. (1990) Inhibitory effects of a tumor-promoting phorbol ester on luteinizing hormone-stimulated renin and prorenin production by cultured bovine theca cells. *Endocrinology*, **127**, 2157–2165.

Brunswig-Spickenheier, B. and Mukhopadhyay, A.K. (1992) Characterization of angiotensin-II receptor subtype on bovine thecal cells and its regulation by luteinizing hormone. *Endocrinology,* 1**31**, 1445–1452.

Csikos, T., Chung, O. and Unger, T. (1998) Receptors and their classification: focus on angiotensin II and the AT2 receptors. *J. Hum. Hypertens.*, **12**, 311–318.

Feral, C., Reznik, Y., LeGall, S., Mahoudeau, J., Corvol, P. and Leymaire, P. (1990) Stimulation by hCG of ovarian inactive renin synthesis in rabbit preovulatory theca cells. *J. Reprod. Fertil.*, **89**, 407–414.

Gangrade, B.K. and May, J.V. (1990) The production of transforming growth factor-beta in the porcine ovary and its secretion in vitro. *Endocrinology,* **127**, 2172–2180.

Ganong, W.F. (1995) Coda. Tissue renin-angiotensin-systems, 1994. *Adv. Exp. Med. Biol.*, **377**, 435–440.

Glorioso, N., Atlas, S.A., Laragh, J.H., Jewelewicz, R. and Sealey, J.E. (1986) Prorenin in high concentrations in human ovarian follicular fluid. *Science,* **233**, 1422–1424.

Greenwald, G.S. and Roy, S.K. (1994) Follicular development and its control. Knobil, E. and Neill, J (eds) *The physiology of reproduction*. Raven Press, New York, 629pp.

Hayashi, K., Miyamoto, A., Berisha, B., Kosmann, M.R., Okuda, K. and Schams, D. (2000) Regulation of angiotensin II production and angiotensin receptors in microvascular endothelial cells from bovine corpus luteum. *Biol. Reprod.*, **62**, 162–167.

Kitzman, P.H., and Hutz, R.J. (1992) In vitro effects of angiotensin II on steroid production by hamster ovarian follicles and on ultrastructure of the theca internal *Cell and Tissue Research,* **268**, 191–196.

Inagami, T. (1995) Recent progress in molecular and cell biological studies of angiotensin receptors. *Curr. Opin. Nephrol. Hypertens.*, **4**, 47–54.

Inagami, T., Kambayashi, Y., Ichiki, T., Tsuzuki, S., Eguchi, S. and Yamakawa, T. (1999a) Angiotensin receptors: molecular biology and signalling. *Clin. Exp. Pharmacol. Physiol.* **26**, 544–549.

Inagami, T., Eguchi, S., Numaguchi, K., Motley, E.D., Tang, H., Matsumoto, T, and Yamakawa, T. (1999b) Cross-talk between angiotensin II receptors and the tyrosine kinases and phosphatases. *J. Am. Soc. Nephrol.*, **11**, S57–S61.

Itskovitz, J., Bruneval, P., Soubrier, F., Thaler, I., Corvol, P. and Sealey, J.E. (1992) Localization of renin gene expression to monkey ovarian theca cells by in situ hybridization. *J. Clin. Endocrinol. Metab.*, **75**, 1374–1380.

Johnson, M.C., Vega, M., Vantman, D., Troncoso, J.L. and Devoto, L. (1997) Regulatory role of angiotensin II on progesterone production by cultured human granulosa cells. Expression of angiotensin II type-2 receptor. *Mol. Hum. Reprod.,* **8**, 663–668.

Kotani E., Sugimoto, M., Kamata, H., Fujii, N., Saitoh, M., Usuki, S., Kubo, T., Song, K., Miyazaki, M., Murakami, K. and Miyazaki, H. (1999) Biological roles of angiotensin II via its type 2 receptor during rat follicle atresia. *Am. J. Physiol.,* **276**, E25–E33.

Lobb, D.K., Kobrin, M.S., Kudlow, J.E. and Dorrington, J.H. (1989) Transforming growth factor-alpha in the adult bovine ovary: identification in growing follicles. *Biol. Reprod.*, **40**, 1087–1093.

Kurtz, A. and Wagner, C. (1999) Cellular control of renin secretion. *J. Exp. Biol.*, **202**, 219–225.

Marrero, M.B., Paxton, W.G., Schieffer, B., Ling, B.N. and Bernstein, K.E. (1996) Angiotensin II signalling events mediated by tyrosine phosphorylation. *Cell. Signal,* **8**, 21–26.

Mukhopadhyay, A.K., Holstein, K., Szkudlinski, M., Brunswig-Spickenheier, B. and Leidenberger, F.A. (1991) The relationship between prorenin levels in follicular fluid and follicular atresia in bovine ovaries. *Endocrinology*, **129**, 2367–2375.

Mukhopadhyay, A.K., Cobilanschi, J. and Brunswig-Spickenheier, B. (1999) The prorenin-renin-angiotensin-system in human semen: possible relevance for the regulation of male fertility. Rajalakshmi, M. and Griffin, P.D. (eds:) *Male contraception: present and future*. New Age International Ltd., New Delhi, 125pp.

Nakajima, M., Hutchinson, H.G., Fujinaga, M., Hayashida, W., Morishita, R., Zhang, L., Horiuchi, M., Pratt, R.E. and Dzau, V.J. (1995) The angiotensin-II type-2 (AT_2) receptor antagonizes the growth effects of the AT_1 receptor: gain-of function study using gene-transfer. *Proc. Natl. Acad. Sci., USA*, **92,** 10663–10067.

Neufeld, G., Ferrara, N., Schweigerer, L., Mitchell, R. and Gospadarowicz, D. (1987) Bovine granulosa cells produce basic fibroblast growth factor. *Endocrinology,* **121**, 597–603.

Obermuller, N., Schlamp, D., Hoffmann, S., Gentili, M., Inagami, T., Gretz, N. and Weigel, M. (1998). Localization of the mRNA for the angiotensin II receptor subtype 2 (AT2) in follicular granulosa cells of the rat ovary by nonradioactive in situ hybridization. *J. Histochem. Cytochem.*, **46**, 865–870.

Pepperell, J.R., Yamada, Y., Nemeth, G., Palumbo, A. and Naftolin, F. (1995) The ovarian renin-angiotensin system. A paracrine-intracrine regulator of ovarian function. *Adv. Exp. Med. Biol.,* **377**, 379–389.

Pucell, A.G., Bumpus, F.M. and Husain, A. (1988) Regulation of angiotensin receptors in cultured rat ovarian granulosa cells by follicle stimulating hormone and angiotensin II. *J. Biol. Chem.*, **263**, 11954–11961.

Pucell, A.G., Hodges, J.C., Sen, I., Bumpus, F.M. and Hussain, A. (1991). Biochemical properties of the ovarian granulosa cell type-2 angiotensin II receptor. *Endocrinology*, **128**, 1947–1959.

Reeka N., Berg, F.D. and Brucker, C. (1998) Presence of transforming growth factor alpha and epidermal growth factor in human ovarian tissue and follicular fluid. *Hum. Reprod.* **13**, 2199–2205.

Roby, K.F. and Terranova, P.F. (1989) Localization of tumor necrosis factor (TNF) in the rat and bovine ovary using immunocytochemistry and cell blot: evidence for granulosa cell production. Hirshfield, A.N. (ed) *Growth Factors and the Ovary*. Plenum Press, New York, 273 pp.

Sandberg, K. (1994) Structural analysis and regulation of angiotensin II receptors. *Trends Endocrinol. Metab.* **5**, 1–8.

Schultze, D., Brunswig, B. and Mukhopadhyay, A.K. (1989) Renin and prorenin-like activities in bovine ovarian follicles. *Endocrinology*, **124**, 1389–1398.

Singh, B. and Armstrong. D.T. (1995) Transforming growth factor alpha gene expression and peptide localization in porcine ovarian follicles. *Biol. Reprod.,* **53**, 1429–1435.

Speth, R.C., Daubert, D.L. and Grove, K.L. (1999) Angiotensin II: a reproductive hormone too? *Regul. Peptides*, **79**, 25–40.

Tanaka, M., Ohnishi, J., Ozawa, Y., Sugimoto, M., Usuki, S., Naruse, M., Murakami, K. and Miyazaki, H. (1995) Characterization of angiotensin II receptor type 2 during differentiation and apoptosis of rat ovarian cultured granulosa cells. Biochem. *Biophys. Res. Commun.*, **207**, 593–598.

Teerds, K.J. and Dorrington, J.J. (1992) Immunohistochemical localization of transforming growth factor-β1 and -β2 during follicular development in the adult rat ovary. *Mol. Cell. Endocrinol.* **84**, R7–R13.

Yan, Y., Chen, R., Pitarresi, T., Sigmund, C.D., Gross, K.W., Sealey, J.E., Laragh, J.H. and Catanzaro, D.F. (1998) Kidney is the only source of human plasma renin in 45-kb human renin transgenic mice. *Circ. Res.,* **83**, 1279–1288.

Yoshimura, Y. (1997) The ovarian renin-angiotensin-system in reproductive physiology. *Front. Neuroendocrinol*, **18**, 247–291.

Follicular Growth, Ovulation and Fertilization: Molecular and Clinical Basis
Anand Kumar and Amal K. Mukhopadhyay (Eds.)
Narosa Publishing House, New Delhi, India, 2001

13

Expression of P450 SCC Gene in Ovaries by Reverse Transcription Polymerase Chain Reaction: A Non-isotopic Riboprobe Based Assessment

Vrushali S. Tanavde, Heena U. Shirwalkar and Anurupa Maitra
Institute for Research in Repoduction, Jehangir Merwanji Street
Parel, Mumbai 400012, India

Introduction

Cytochrome P450 cholesterol side chain cleavage (P450 scc) is now known to be the enzyme that catalyzes the first irreversible step in steroidogenesis viz., cleavage of the cholesterol side-chain complex that occurs in the inner mitochondrial membrane of a steroidogenic cell (Miller, 1988). Gene encoding this P450 scc enzyme is thus believed to be a key factor in regulating steroidogenesis by way of controlling the utilization of cholesterol into the steroidogenic pathway. A cell specific regulation of the gene has also been reported (Voutilainen and Miller, 1987). Thus, although the gene is now known to be a key locus for steroidogenic control, its developmental expression particularly in the ovarian cells, during follicular growth and luteinization has not been studied in detail. The studies had been difficult until the advent of the PCR-based techniques. Advent of reverse transcription polymerase chain reaction (RT/PCR) has now provided a standard method for detection of these low abundance mRNAs in tissues as well as cells in culture (Delidow *et al.*, 1989). Purpose of our present study was to develop such a RT/PCR based system for assessing the expression of P450 scc gene at the mRNA level, with particular emphasis on cells of the ovary. The method uses an endogenous internal standard or "housekeeping" gene for a relative quantitation of the mRNAs. Relative quantitation can be achieved by densitometric analysis of hybridized duplexes.

The method was validated and applied for a preliminary assessment of the basal expression of P450 scc gene in corpus luteum tissue, as against granulosa cells, using a porcine ovary model.

Materials and Methods

RNA Isolation

Porcine ovaries, freshly obtained from a local abattoir and transported to laboratory on ice were used for the study. Corpus luteum tissue was freshly excised from a few corpora lutea, homogenised in guanidium solution and total RNAs isolated by a modified single step method of Chomozynski et al., (1987). Granulosa cells were isolated from follicular fluid obtained through fine needle aspiration of the follicles. The cells were pelleted by centrifugation, washed in MEM medium and subjected total RNA extraction by the single step procedure. RNA pellets were ethanol precipitated, resuspended in Tris EDTA buffer (pH 8.0) and quantitated spectrophotometrically at 260/280 nm. Integrity of the preparations were also checked by running part of the RNAs on formaldehyde denaturing agarose gel. 1 μgm qunatities of the RNA samples were then subjected to reverse transcription polymerase chain reaction.

Reverse Transcription Polymerase Chain Reaction

Based on published sequences for the P450 scc gene, primers were designed that would be complementary to 3'-termini of the porcine sequence and spanning 376 bp including the primers (porcine base positions 196 to 219 and 549 to 571). A cDNA template was first synthesized by reverse transcriptase reaction from 1 μg of the total RNAs isolated from porcine corpus luteum tissue. Sample was first preheated at 25°C for 10 min and then reverse transcription carried out with AMV Reverse Transcriptase at 42°C for 60 min. Denaturation of the reverse transcriptase at the end of reaction was then carried out at 99°C for 5 min. Amplification of the cDNA template was carried out by polymerase chain reaction (PCR) using a DNA amplification kit (Roche Diagnostics, Germany). The amplification was carried out for 35 cycles using following temperature profile: Denaturation at 96°C for 1 minute, annealing at 55°C for 1 minue and polymerisation at 72°C for 2 min followed by a final extension at 70°C for 5 min. Primer concenration used was 1 μm in a reaction volume of 100 μl. As an internal control, another set of primers were designed for the constitutive gene cyclophilin. Primer sequences complementary to 3'-termini of the porcine cyclophilin sequence and spanning 113 bp (base positions 105 to 123 and 197 to 217) were used. Conditions of the RT/PCR were identical to that described for the target P450 scc gene.

Visualisation of the two amplified products were carried out in a ethidium bromide stained 1.5% agarose gel.

Characterisation and Validation of the RT/PCR System

Restriction Analysis of Amplified cDNA Fragments

Based on expected sequences of the amplified fragments of procine P450 scc and cyclophilin genes, restriction sites were identified, using a computer databse (Gene Pro). Three such restriction enzymes were identified for the 376 bp P450 scc fragment: Hpa II yielding expected size products of 286 bp and 90 bp, Alu I yielding expected products of 262 bp and 114 bp size and Mbo II yielding expected products of 150 bp, 127 bp and 99 bp size. The RT/PCR amplified P450 scc fragment generated from the procine corpus luteum tissue was digested with the three enzymes and product sizes compared.

For the 113 bp cyclophilin gene fragment, a unique enzyme Bgl I was identified. However, the products yielded are of 108 and 5 bp sizes. For a better confirmation therefore, a second RT/PCR amplification was carried out of a larger segment (358 bp) of the gene that flanks and includes the 113 bp fragment. The primers used for this were complementary to 3'-termini of the porcine base positions 13 to 30 and 352 to 370. Expected sizes of the products yielded from this larger fragment after Bgl I digestion correspond to 200 bp and 158 bp.

Southern Hybridization

The RT/PCR amplified fragments, immobilized on a nylon membrane (Hybond+, Amersham, UK) were subjected to hybridization with digoxigenin labelled probes. A complementary RNA (cRNA) probe generated for P450 scc and a cDNA probe generated for cyclophilin were used. A specific 1.2 kb P450 scc cDNA cloned into transcription vector pGEM3 and a 700 bp rat cyclophilin cDNA cloned into pUC-19 vector were available for the purpose. Both the plasmids were obtained as a gift from the laboratory of Dr. J.D. Veldhuis (University of Virginia, USA), through kind courtesy of Dr. Thomas Wise (Nebraska, USA) and Dr. Anthony Zeleznik (Pittsburg, USA) respectively.

Generation of Digoxigenin Labelled cRNA and cDNA Probes

Digoxigenin labelled antisense cRNAs for P450 scc, approximately 1.2 kb were generated from the pGEM plasmid using the DIG RNA Labelling kit (Boehringer Mannheim, Germany). Plasmid was linearised with Hind III enzyme and "in vitro" transcription carried out for 2 hours at 37°C in presence of T7 polymerase and DIGdUTP. A 700 bp DIG labelled cDNA probe for cyclophilin was generated from the pUC-19 plasmid by random primed labelling using the DIG DNA labelling kit (Boehringer Mannheim, Germany). The cDNA insert was isolated from the palsmid by digestion with enzymes Bam H1 and Hind III and gel purified. The purified cDNA was denaured by boiling and then subjected to random primed labelling with Klenow enzyme, in presence of a mixture of deoxyribonucleotides containing DIG-11-dUTP. This reaction was carried out overnight at 37°C.

Detection of the labelled probes was carried out on a X-ray film by a chemiluminescent method using CSPD substrate (DIG luminescent detection kit, Boehringer Mannheim, Germany).

The probes were quantified by dot blot analysis against standard labelled probes. Hybridization of the southern membrane was carried out at 50°C overnight, using 100 ng/ml of the cRNA probe and 25 ng/ml of the cDNA probe. Exposure of the X-Ray film for about 30 minutes to 1 hour at room temperature was required for optimum detection of the hybridized duplexes.

Results

Fig. 1 shows restriction analysis of the 376 bp P450 scc gene and the 113 bp and 358 bp fragments of the cyclophilin gene amplified from porcine corpus luteum RNA. Hpa II, Alu I and Mbo II digestion of P450 scc yielded the expected major bands of 286 bp, 262 bp and 150 bp respectively. Smaller fragments of < 100 bp size were however not distinctly visible on gel. As for cyclophilin, Bgl I digestion of the 358 bp fragment led to the two distinct products corresponding to 200 bp and 158 bp sizes. Digestion of the 113 bp cyclophilin fragment also led to the major product of 108 bp, which was however not distinguishable. Following this confirmation, RT/PCR was further carried out from corpus luteum RNA, to amplify individually the two cyclophilin fragments (113 bp and 358 bp) as well as the 376 bp P450 scc fragment. Fig. 2 shows the specific amplification products obtained with all these reactions. The products were then immobilised on a nylon membrane and

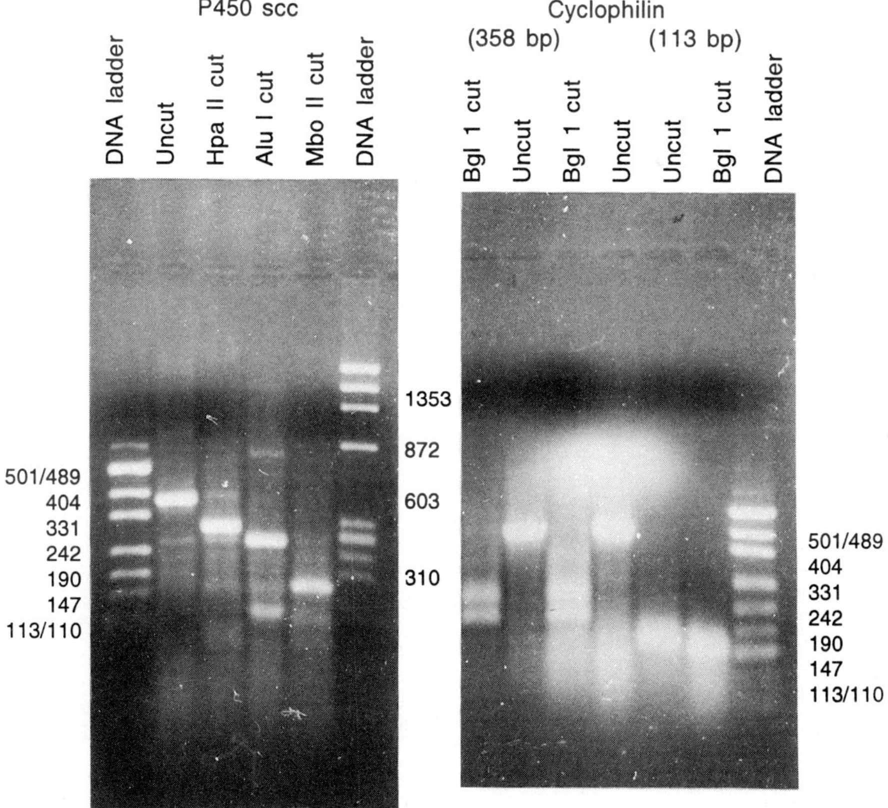

Fig. 1 Restriction analysis of RT/PCR amplified fragments of P450 scc and cyclophilin genes.

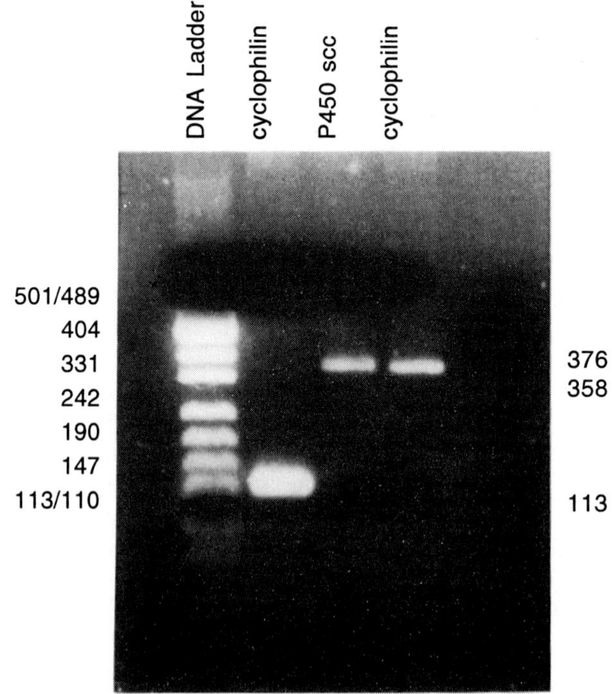

Fig. 2 RT/PCR amplification of P450 scc and cyclophilin gene fragments from porcine corpus luteum tissue.

subjected to hybridization with the DIG-labelled P450 scc cRNA and cyclophilin cDNA as described under Materials and Methods. Specific signals corresponding to the three amplified products, were obtained confirming their specificity (Fig. 3).

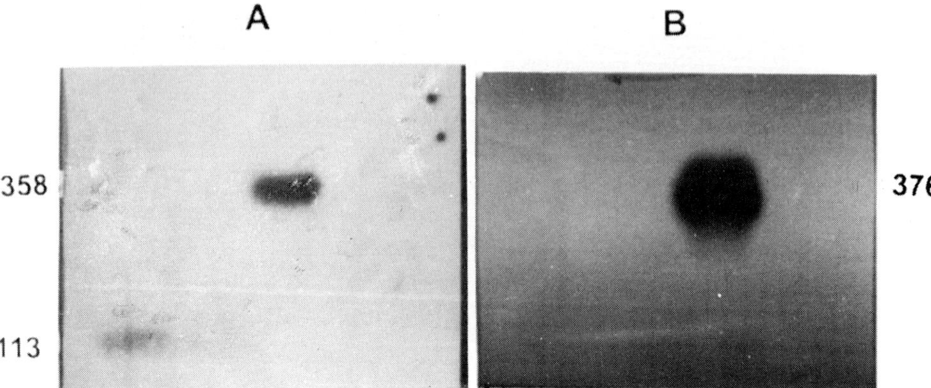

Fig. 3 Southern hybridization of the RT/PCR amplified fragments with specific DIG-labelled probes. A) cyclophilin was hybridized with DIG-labelled cDNA whereas B). P450 scc was hybridized with DIG labelled cRNA. Detection of hybridized duplexes carried out by a chemiluminescent method.

Using this validated system, a preliminary analysis was then carried out to assess the basal expression of P450 scc gene in the corpus luteum tisuse as against granulosa cells and also to assess the pattern of P450 scc expression in the CL at varying stages of development. Total RNAs were isolated from different corpora lutea obtained from abattoir procine ovaries and designated as young, mature and regressing, on the basis of their gross appearance (size, colour, presence or absence of blood clots) described by Ireland et al., (1980). Simultaneously, RNAs were also obtained from granulosa cells of the preovulatory follicles. Co-amplification of P450 scc (376 bp) and cyclophilin (113 bp) was then carried out in these samples. Fig. 4 shows the amplified bands corresponding to the respective transcripts, as visualized on a 1.5% agarose gel. Results clearly indicated a higher expression of the gene in the corpus luteum tissues as compared to granulosa cells. Expression of cyclophilin was consistent in all the tissues. A southern hybridization was then carried out with the two probes and subjected to chemiluminescent detection as described before. Results

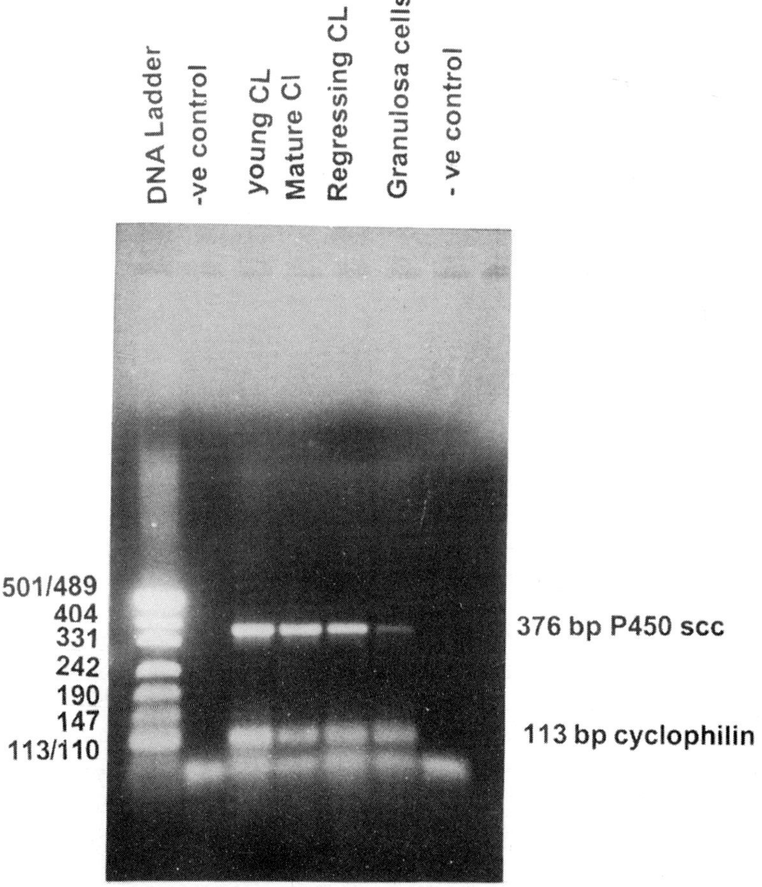

Fig. 4 Relative expression of P450 scc gene in corpus luteum tissue and granulosa cells, as against the internal control gene cyclophilin.

presented in Fig. 5 confirm a higher basal expression of the P450 scc gene in all the three corpora lutea selected, as compared to granulosa cells. No difference was however seen in the expression within the CL specimens selected. Cyclophilin gene yielded consistent signals in all the samples. The system is thus suitable for a densitometric assessment of the relative gene expression by way of normalization of the target gene signals against the standard gene.

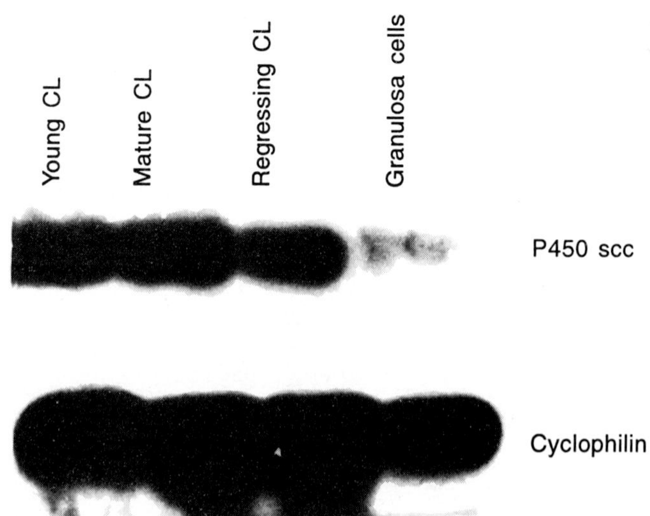

Fig. 5 Chemiluminescent detection of the relative expression of P450 scc gene as against cyclophilin in corpus luteum tissue and granulosa cells.

Discussion

Detection of low abundance mRNAs like those for the steroid enzyme genes, have been carried out conventionally by methods such as Northern blot analysis or RNAse protection assays (Maitra *et al.*, 1995). Advent of reverse transcription polymerase chain reaction (RT/PCR) has now provided a standard technique to assess these messages specifically even from small biopsies and cell numbers (Tarnuzzer et al., 1996). Exquisite sensitivity and relative simplicity of the technique has also stimulated search for application of the technique to quantify the mRNAs.

Use of a house-keeping gene in this regard has been accepted as a standard technique for relative quantitation. Some assay variations however, may arise due to actual variations in the levels of internal standard within the compared samples. However, this does not seem to be the case with the method described here. Also, as the method is hybridization based, a high degree of sensitivity and/or specificity is provided by the technique. As for quantitation of the PCR and RT/PCR products a need has been felt over the years for development of sensitive non-isotopic methods as an alternative to the radioactive systems conventionally used. Techniques based on fluorescence, colorimetry, chemiluminescence, electrochemiluminescence or bioluminescence for detection of hybridized products have been developed by various workers (Xiao L *et al.*, 1996). The chemiluminescence solid phase hybridization system described by us here for detection and quantitation of the RT/PCR products has the additional advantage of a high degree of sensitivity provided by RNA probes. The method thus combines the efficiency of RT/PCR with the sensitivity of digoxigenin labelled chemiluminescent system for quantification of the mRNA levels.

The incresed expression of P450 scc gene in corpus luteum tissue, as compared to granulosa cells seen in our study corroborates with the finding of LaVoie et al (1997). In a study based on RNAse protection assay they conclude that luteal progesterone synthesis in the pig model is associated with an increase in P450 scc enzyme mRNA above preovulatory levels. A 2 to 3 fold increase in P450 scc enzyme mRNA over preovulatory follicles was observed by them in CL tissue whereas corpora albicantia showed a dramatic decline. Pescador et al in 1996 carried out a study along similar line in the bovine model, in which the P450 scc message abundance was compared in developing, early and late mid cycle CL as well as in regressed CL. Using Northern blot analysis they observed no significant differences between the developing and early and late mid cycle CL. In regressed CL, the message reached undetectable levels. A similar trend was observed in our study with the pig model. The three CL specimens, approximately graded as young, mature and regressing showed no apparent differences in the abundance of P450 scc message. However, the study is only preliminary and more confirmatory evidence needs to be obtained for any definite conclusions. The results, nonetheless, add to the validation of the RT/PCR based method developed as a non-isotopic alternative for sensitive quantification of the P450 scc gene expression.

Acknowledgements

We gratefully acknowledge the encouragement and support received from Dr. H.S. Juneja, Director, IRR, for carrying out the work. Typographical assistance from Ms. Doris D'Souza in preparing the manuscript is also being acknowledged.

References

Chomozynski, P. and Sacchi, N., (1987). *Anal Biochem.,* **162**: 156–159.

Delidow B.C., Peluso J.J. and White B.A., (1989). Quantitative measurement of mRNA by polymerase chain reation. *Gene Anal. Tech.*, **6**: 120–124.

Ireland J.J., Murphree R.L., Coulson B.P. (1980). Accuracy of predicting stages of bovine estrus cycle by gross appearance of the corpus luteum Jour. *Dairy Sci.*, **63**: 155–160.

LaVoie H.A, Benoit A.M, Garmey JC., *et al.*, (1997). Co-ordiante developmental expression of genes regualting sterol economy and cholesterol side-chain cleavage in the porcine ovary. *Biol. Reprod.* **57**: 402–407.

Maitra A., LaVoie H.A, Day R.N. et al., (1995). Regulation of procine granulosa cell 3, hydroxy-3-methylglutaryl co-enzyme, A reductase by insulin and insulin like growth factor I: Synergism with follicle stimulating hormone or protein kinase A agonist. *Endocrinology* **136**: 5111–5117.

Miller W.L. (1988). Molecular biology of steroid hormone synthesis (Review) *Endocr. Rev.* **9**: 295–318.

Pescador N., Soumano K., Stocco D.M. et al. (1996). Steroidogenic acute regulatory protein in bovine corpora lutea. *Biol. Reprod.* **55**: 485–491.

Tarnuzzer R.W., Macauley W.G., Farmerie S. et al., (1996). Competitive RNA templates for detection and quantitation of growth factors, cytokines, exracellular matrix components and matrix metallo proteinases by RT/PCR. *Biotechniques* **20**: 670–674.

Voutilainen R. and Miller W.L. (1987). Coordinate tropic hormone regualtion of mRNA for insulin like GF II and cholesterol side chain cleavage enzyme P450 scc in human steroidogenesis. *Proc. Natl., Acad. Sci. USA.* **84**: 1590–1594.

Xiao L., Yang C., Nelson CO et al. (1996). Quantitation of RT-PCR amplified cytokine mRNA by aequorin-based bioluminescence immunoassay. *Jour. Immunol. Methods,* **199**: 139–147.

Follicular Growth, Ovulation and Fertilization: Molecular and Clinical Basis
Anand Kumar and Amal K. Mukhopadhyay (Eds.)
Narosa Publishing House, New Delhi, India, 2001

14

Regulation of Folliculogenesis by Intraovarian Peptide

Tarala D. Nandedkar, Perinaaz Wadia, Smita Mahale, S.B. Moodbidri and K.S. Iyer
Institute for Research in Reproduction, Parel, Mumbai 400 012, India

Introduction

Gonadotrophins have been known to control ovarian follicular maturation. However it is now well documented that autocrine/paracrine factors present within the ovarian follicle can modulate gonadotrophin action or directly regulate folliculogenesis. Inhibin-related peptides are examples of such paracrine factors that alter the levels of pituitary FSH or act directly on granulosa cells (Table 1). Besides these, cytokines, growth factors and other non steroidal factors are also involved in the paracrine regulation of folliculogenesis.

Follicular fluid from various animal species has been shown to contain substances that inhibit the binding of FSH to its receptor. One such FSH binding inhibitor (FSHBI) from bovine, porcine and human follicular fluid was reported by Reichert and his colleagues (Darga and Reichert., 1978). Our group has been engaged in the purification of ovarian follicular fluid peptide (OFFP) from human follicular fluid. Studies carried out with the partially purified peptide of sheep origin have revealed the ability of the peptide to inhibit FSH-binding to granulosa cells in vitro (Shahid *et al.*, 1991). In mice the peptide suppressed ovulation, induced follicular atresia and decreased plasma progesterone levels (Nandedkar *et al.,* 1988). The partially purified peptide of human origin impaired fertility in post partum marmosets (Nandedkar *et al.*, 1989) and decreased circulating steroid hormone levels leading to the suppression of ovulation in bonnet monkeys (Nandedkar *et al.*, 1992a). Further, administration of the peptide in mice induced apoptosis in granulosa cells thus triggering follicular atresia (Nandedkar *et al.*, 1996; 1998).

In the present study FSHBI from human ovarian follicular fluid has been purified and characterised. This peptide when injected in Swiss female mice has resulted in the suppression of ovulation on estrus day. In view of the above observation as well as previous results showing a decline in progesterone synthesis (Nandedkar *et al.*, 1988), we have studied effect of the peptide on 3β hydroxysteroid dehydrogenase (3β HSD), a key enzyme in progesterone biosynthesis, in the murine ovary.

Table 1 Intraovarian nonsteroidal factors that act as paracrine regulators of folliculogenesis

	Regulator	*Molecular weight*	*Localisation in cell type*	*Autocrine/ Paracrine action in vitro*	*References*
1.	Follicular Regulatory Protein (FRP)	15	GC	E_2	di Zerega et al, 1983
2.	Oocyte Maturation Inhibitor (OMI)	2	GC	Oocyte meiosis	Tsafriri et al, 1976
3.	Inhibin A/B	32	GC	E_2	Ying et al, 1986
4.	Activin A/AB/B	28	GC	E_2, P_4	Hutchinson et al, 1987
5.	Follistatin	32–43	GC	E_2	Ueno et al, 1987; Ying, 1988
6.	FSH Binding Inhibitor	57	GC	E_2	Darga and Reichest, 1978
	"	2.5	GC	E_2	Sluss et al, 1987; Lee et al, 1993
	"	< 4	GC	P_4	Nandedkar et al, 1988
7.	Gn RH-like peptide/ binding inhibitor	2	GC	P_4	Aten et al., 1987; Aten and Berhman, 1989
8.	Mullerian Inhibiting	70	GC	–	Takahashi et al, 1986
9.	Growth Differentiation Factor-9 Substance	–	oocyte/GC	P_4	Elvin et al., 1999

The peptide has been shown to have a significant role in the induction of apoptosis in granulosa cells (Nandedkar *et al.*, 1996;1998). In addition to this, our aim has been to identify the molecular mechanisms activated by the peptide to generate apoptosis. Our focus has been on the Bcl_2 family of proteins known to be apoptotic as well as anti-apoptotic (Chao and Korsmeyer, 1998), especially the expression of Bcl_{xs} and Bcl_2 proteins in the mouse ovary.

Materials and Methods

Purification of hFSHBI

Follicular fluid was aspirated from the ovaries of women recruited for an *in vitro* fertilization programme. The fluid was pooled and centrifuged at 2000 rpm for 10 min to clear out cell debris. The supernatant was subjected to ultrafiltration on Amicon PM-10 filter (cut off < 10 kDa). The filtrate was further processed on a Sephadex G-25 column (2.7 × 35 cm) with 0.2M acetic acid as eluting buffer and absorbance read at 280 nm. The active fraction (Peak 3) was loaded on a preparative reversed phase HPLC, RP-18 column (21.5 × 250 mm) developed with a gradient of 50% acetonitrile in 0.1%

trifluoroacetic acid and absorbance read at 280 nm. In order to assess its purity, the active peak was then loaded on analytical reversed phase HPLC, RP-18 column (4 × 250 mm) developed with a linear gradient of 100% acetonitrile and 0.1% trifluoroacetic acid, absorbance read at 220 nm.

Amino Acid Analysis

Hydrolysis of 25 μg of the active Prep. RP HPLC fraction was carried out using 6N HCI containing 0.5% phenol at 110°C for 24 h. Composition of amino acids was analysed using a Shimadzu Amino Acid Analyser.

Radioreceptor Assay

Immature female Swiss mice (21–23 days old) bred in our animal colony were administered 10iu pregnant mare's serum gonadotropin (PMSG) and sacrificed 48 h later (Shahid *et al.*, 1991). Granulosa cells were harvested from the ovaries of these mice and the cell number adjusted to 20×10^6 cells/ml. Human FSH obtained from Dr. Leo Reichert Jr., Albany USA was radioiodinated by the method of Reichert and Abou-Issa (1977) and the assay was carried out as described by Darga and Reichert (1978). Briefly, granulosa cells were incubated with ^{125}I labelled hFSH (20,000 cpm, specific activity 20 μCi/μg) for 2h at 37°C. The reaction was terminated by the addition of cold buffer followed by centrifugation of the tubes at 3000rpm for 15 min. The radioactivity of the pellet was counted on a Kontron gamma counter. The unlabelled hFSH (1–100 ng/tube) was used to obtain a standard curve. Samples of various fractions were incubated with the cells prior to the addition of radiolabelled FSH to determine percent inhibition of FSH binding to granulosa cells.

Biological Action of FSHBI on Female Mice

The estrous cycle of female Swiss mice bred in our animal colony were monitored daily by vaginal smears. Mice in the metestrus stage were injected with FSHBI (Treated) or saline (Control) followed by a second injection in the diestrus stage. On the estrus day of the cycle the injected mice were sacrificed. The oviducts were observed under the microscope to detect the presence of oocytes whereas the ovaries were sectioned on the cryostat and stained with haematoxylin and eosin stains to observe the various stages of follicular development.

Histochemical Localisation of 3βHSD in Murine Follicles

Immature female mice (21–23 days old) were administered 10 iu PMSG(D_0). The mice were sacrificed at different time intervals in order to obtain ovarian follicles at various stages of growth and atresia. The control group (N) was sacrificed 48 h (D_2) following PMSG injection and the atretic group (A) sacrificed 72 h (D_3) after this injection. In the treated group (T) 5μg of FSHBI (RP-HPLC fraction) was administered 24 h (D_1) following the PMSG injection and sacrificed 24 h (D_2) later. The ovaries of all the three groups were snap frozen and sectioned on a Leitz cryostat at –20°C and sections (5μ) were stored at –70°C to be later used for localization of 3β hydroxysteroid dehydrogenase (3βHSD) activity (Dickman and Dey, 1974).

Immunohistochemical Localisation of Bcl_2 Family

The ovarian sections of groups N, A and T were processed for immunohistochemical localisation studies. The procedure used was as reported earlier (Nandedkar et al., 1992b). In brief, 5 μ frozen sections were fixed in acetone for 10 mins and incubated with 1% H_2O_2 in methanol for 20 min. to remove the endogenous peroxidase. Nonspecific blocking was carried out by incubation with 1% normal rabbit serum in 0.01 M PBS (pH 7.2). Primary monoclonal antisera to Bcl_2 and Bcl_{xs} at a concentration of 1:50 were incubated overnight at 4°C. Rabbit second antibody to mouse IgG conjugated to HRP at a dilution of 1:500 was employed. Using 0.01% 3,3′-diaminobenzidine-HCl and 0.002% H_2O_2 in PBS, a positive reaction was visualized as brown colour. Culture supernatant (SP_2) at a dilution of 1:50 was used as a negative control.

Results

The active fraction on preparative RP-HPLC showed a homogenous peak (Fig. 1) when subjected to analytical RP-HPLC which was further characterized and used for biological studies.

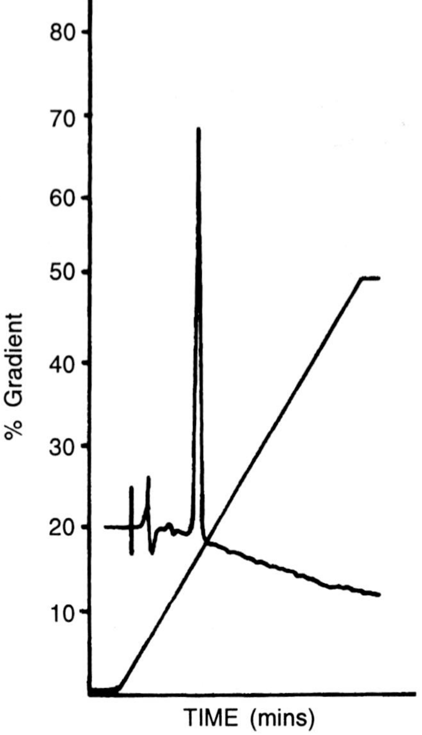

Fig. 1 Analytical RP-HPLC profile of active material purified on prep RP-HPLC. 25μg detected at 220 nm, 0.32 AUFS.

Amino acid analysis of this fraction revealed predominantly aspartic acid, glutamic acid and glycine suggesting hydrophylic nature (Table 2).

Table 2 Amino acid composition of FSHBI obtained after acid hydrolysis

Asp	3	Val	1
Thr	1	Ile	4
Ser	4	Leu	1
Glu	4	His	2
Gly	7	Arg	1
Ala	1	Lys	1

Radioreceptor Assay

About 5-7% binding of ^{125}I-hFSH was inhibited by cold hFSH at concentrations 1ng-100ng. The RP-HPLC fraction inhibited FSH binding by 50% at concentration of 1.88μg. A dose related response was observed.

Biological Action on Female Mice

Table 3 shows the oviducts of saline injected controls containing 7–8 oocytes on estrus day. Mice injected with the peptide at various levels of purity showed a significant decrease in oocyte number in the oviducts. Maximum suppression of ovulation (88%) was observed in mice injected with the RP–HPLC active fraction.

Table 3 Ovulation inhibition in mice induced by FSHBI at varying levels of purity

Sample	*No. of mice*	*No. of oocytes/ mouse ± SEM*	*Inhibition of ovulation*	*Sample showing 50% inhibition*
Saline	14	7.28 ± 0.63	–	–
Human Follicular Fluid (250 μg)	10	3.8 ± 1.26	47.84%	261.28 μg
PM-10 filtrate (40 μg)	8	3.62 ± 1.41	50.25%	39.8 μg
SG-25 Fraction III (40 μg)	10	3.4 ± 0.91	53.33%	37.5 μg
Prep RP HPLC Fraction VII (10 μg)	6	0.83 ± 0.76	88.57%	5.64 μg

Effect of FSHBI on Localisation of 3β HSD in Murine Follicles

In the ovaries of PMSG injected control mice (N) an intense localisation of 3βHSD activity was observed in the interstitial gland tissue (IGT) whereas the large follicles were weakly stained (Fig. 2a). In the A group as well as in the FSHBI treated (T) group the granulosa cell arrangement was disturbed, the

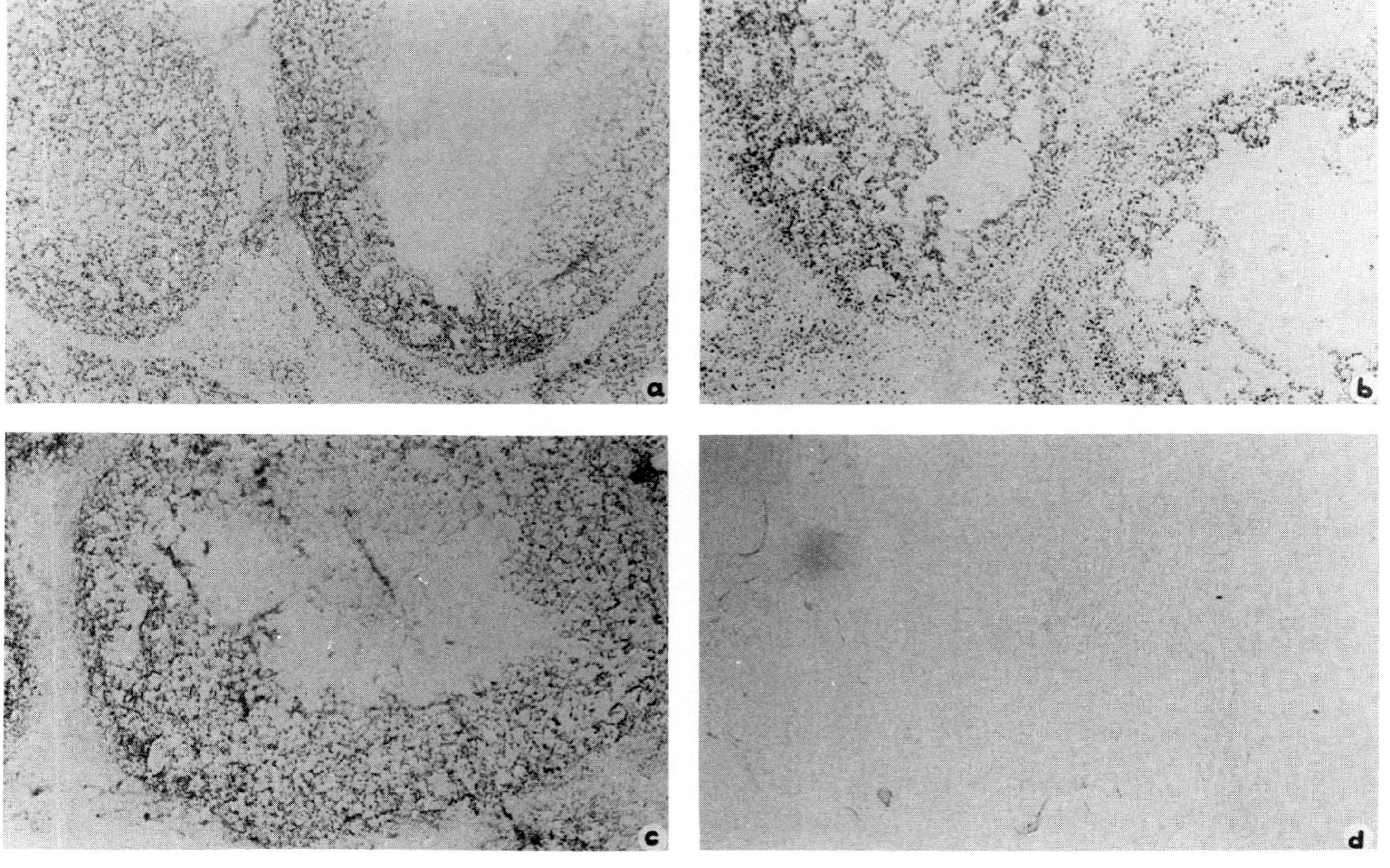

Fig. 2 Histochemical localization of 3βHSD on mouse ovarian sections (a) normal group, (b) atretic group, (c) treated group, (d) negative control-in absence of substrate.

cells were moderately stained showing mild 3β HSD activity while IGT cells showed weak or moderate localisation of 3β HSD activity (Fig. 2b, c). No staining was seen in the negative control (Fig. 2d).

Effect of FSHBI on Localisation of Bcl_2 and Bcl_{xs}

In the control group (N) weak localisation of Bcl_2 was observed in granulosa cells (Fig. 3a) while no staining was observed in atretic follicles of groups A and T (Fig. 3b, c). However, IGT showed weak to moderate staining for Bcl_2 in all the three groups.

Weak localisation of Bcl_{xs} was observed in normal preovulatory follicles of control (N) mouse ovaries (Fig. 3d). In the A and T groups the early atretic follicles showed weak staining while the late atretic follicles were intensely stained (Fig. 3e, f). This agrees with the apoptotic role of bcl_{xs}.

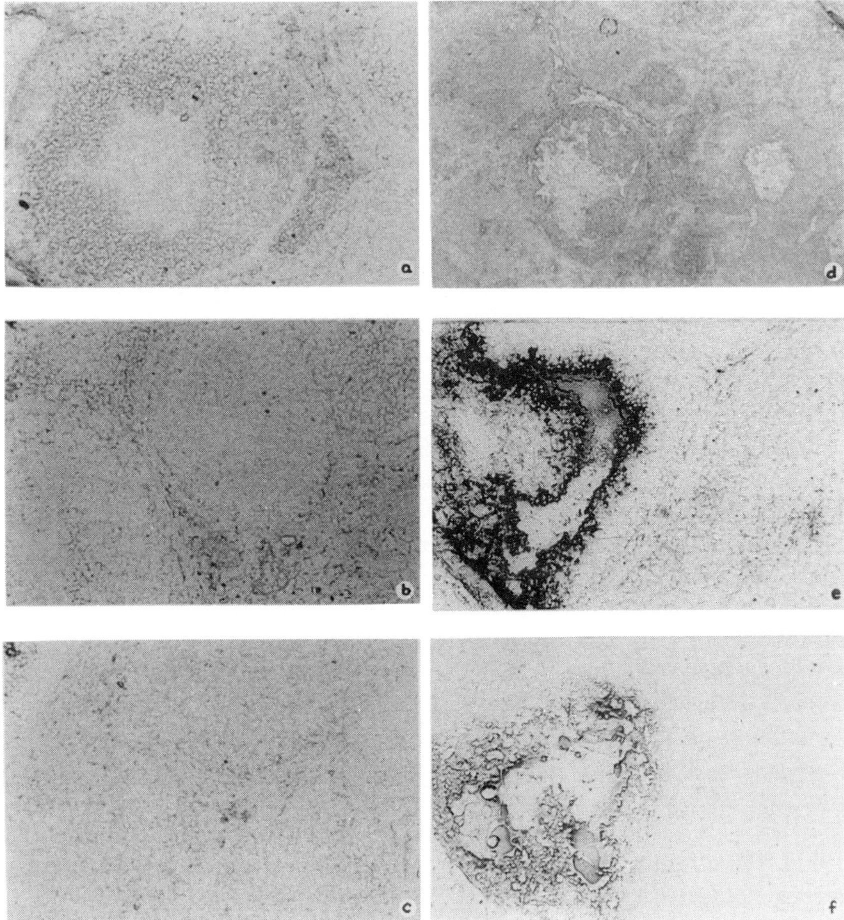

Fig. 3 Immunohistochemical localisation of Bcl_2 and Bcl_{xs} in mouse ovarian sections (a) Bcl_2-normal group (b) Bcl_2-atretic group, (c) Bcl_2-treated group, (d) Bcl_{xs}-normal group (e) Bcl_{xs}-atretic group, (f) Bcl_{xs}-treated group.

Discussion

FSHBI purified from human ovarian follicular fluid is a low molecular weight peptide (< 5kDa), of a hydrophilic nature and predominantly contains aspartic acid, glutamic acid and glycine. This peptide is distinct from FSH as well as FSHBI identified by Reichert and his colleagues. They have identified a low and high molecular weight human FSHBI which showed FSH antagonist and FSH agonist action respectively (Lee *et al.*, 1990; 1993). A decapeptide (BI–10) selected randomly from the amino acid sequence of the high molecular weight FSHBI also revealed FSH antagonist action. Both BI--10 and low molecular weight –FSHBI were shown to inhibit FSH– stimulated estradiol synthesis in rat Sertoli cell cultures (Lee *et al.*, 1993; 1995). On the other hand, human FSHBI isolated by our group has earlier been shown to decrease plasma progesterone levels in marmosets and bonnet monkeys (Nandedkar *et al.*, 1989;1992a).

The decrease of plasma progesterone levels, in mice and monkeys, caused by the administration of partially purified FSHBI (Nandedkar *et al.*, 1988; 1989; 1992a) may be due to its effect on 3β HSD activity as this is the key enzyme in progesterone biosynthesis. But, the histochemical localisation of 3β HSD in ovarian follicles of control and treated mice was found to be similar and did not differ. Thus, the detection of 3β HSD activity has not yielded significant change possibly due to the semi-quantitative nature of this histochemical technique.

Administration of FSHBI to female mice blocked the action of FSH on growing follicles. The non availability of FSH to these follicles, prevented further growth and subsequently ovulation. These follicles therefore became atretic.

Follicular atresia induced in mice by FSHBI is infact due to the onset of apoptosis or programmed cell death. This has been indicated by the localisation of fragmented chromosomal DNA in the nucleus of granulosa cells (Nandedkar *et al.*, 1998) and also by using the TUNEL technique (Nandedkar and Dharma In communication). Apoptosis is activated by a number of molecular mechanisms. The mechanism of our interest is the Bcl_2 family proteins which are both apoptotic and antiapoptotic. In the present study, immunohistochemical localisation of Bcl_2 is weak in normal follicles while it is absent in atretic follicles. This corroborates with the antiapoptotic role of bcl_2 (Chao and Korsmeyer, 1998). However, in the rat ovary Goodman *et al.*, 1998 has reported localisation of bcl_2 in atretic follicles.

On the other hand, localisation of Bcl_{xs} is weak in the early atretic follicles and intense in the late atretic follicles while absent in the normal follicles. This agrees with the apoptotic role of bcl_{xs} (Chao and Korsmeyer, 1998). These results demonstrated for the first time that bcl_{xs} may have a role in granulosa cell apoptosis in atretic follicles induced by FSHBI.

Recently, Matzuk and his colleagues have developed transgenic mice that ectopically overexpress human FSH from multiple tissues. These female transgenic mice develop hemorrhagic and cystic ovaries (Rajendra Kumar *et al.*, 1999).

In a previous study we have reported a high concentration of FSHBI in follicular fluid from cystic follicles of sheep (Nandedkar, 1995) and women (unpublished data) as compared to fluid from atretic and normal follicles. It is therefore postulated that high levels of FSHBI in cystic fluid may be the factor inhibiting FSH action in the transgenic animals.

In conclusion, FSHBI isolated from human ovarian follicular fluid has been shown to block the availability of FSH to growing follicles. It may thus, hamper a cascade of events triggered by FSH such as steroidogenesis, LH receptor formation and mitosis (Fig. 4) ultimately triggering atresia/apoptosis in these follicles.

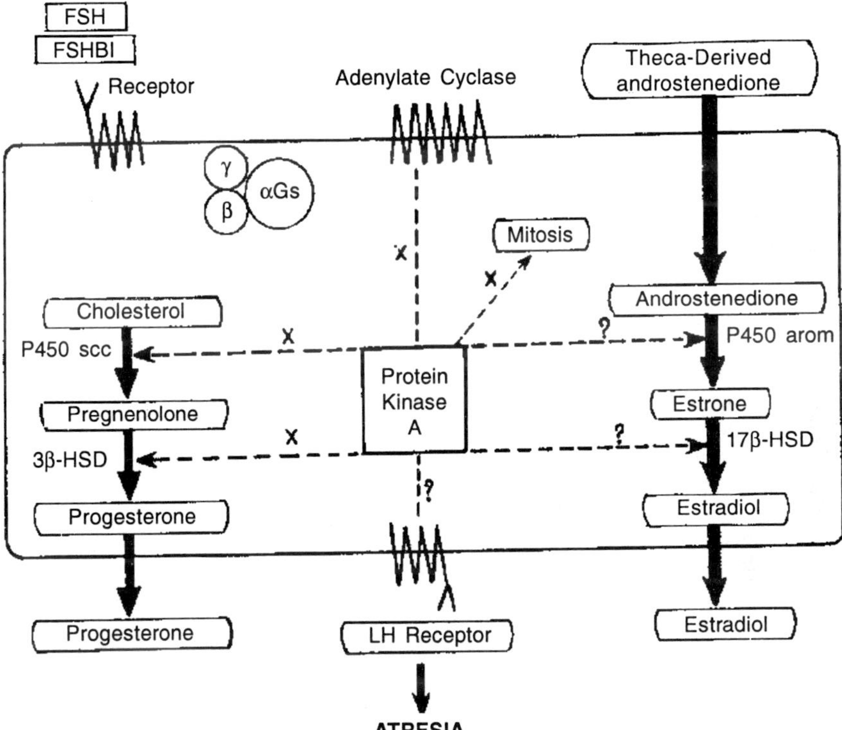

Fig. 4 A hypothetical representation wherein FSHBI may block the cascade of events triggered by FSH in granulosa cells finally leading to follicular atresia.

Acknowledgement

The authors acknowledge Dr. L. E. Reichert Jr., Albany, USA for the supply of human FSH and the Hormone Bank of IRR for the PMSG used in the above studies. We are grateful to Dr. Mehroo Hansotia and Dr. Sadhna Desai, Fertility Clinic, Mumbai for providing human ovarian follicular fluid. We also thank Dr. H. S. Juneja Director, I.R.R. for his constant support and encouragement throughout the work. Technical assistance of Mr. S. T. Ghanekar and Mr. S. M. Rewadekar and typographical assistance of Mrs. Doris D'Souza is appreciated.

References

1. Aten, R.F., Polan, M.L., Bayless, R., Behrman, H.R. (1987) A GnRH-like protein in human ovaries: similarity to the GnRH-like ovarian protein of the rat. *J. Clin. Endocrinol.*, **64**, 1288–1293.
2. Aten, R.F., Behrman, H.R. (1989) A GnRH binding inhibitor protein from bovine ovaries: Purification and identification as histone H2A. *J. Biol. Chem.*, **264**, 11065–11071.
3. Chao, D.T., Korsmeyer, S.J. (1998) Bcl-2 family: regulators of cell death. *Ann. Rev. Immunol.*, **16**, 395–419.
4. Darga, N.C., Reichert, L.E.Jr. (1978) Some properties of the interaction of follicle stimulating hormone with bovine granulosa cells and its inhibition by follicular fluid. *Biol. Reprod.*, **19**, 235–240.
5. Dickman, Z., Dey, S.K. (1974) Steroidogenesis in rat embryo and its possible influence on morula-blastocyst transformation and implantation. *J. Reprod. Fertil.*, **37**, 91–93.
6. di Zerega, G.S., Campeau, J.D., Ujita, E.L., King, O.R., Marrs, R.P., Lobo, R., Nakamura, R.M. (1983) The possible role of a follicular protein in the intraovarian regulation of folliculogenesis. *Sem. Reprod. Endocrinol.*, 1, 309–322.
7. Elvin, J.A., Clark, A.T., Wang, P., Wolfman, N.M., Matzuk, M. (1999) Paracrine actions of growth differentiation factor-9 in the mammalian ovary. *Mol. Endocrinol.*, **13**, 1035–1048.
8. Goodman, S.B., Kugu, K., Chen, S.H., Preutthipan, S., Tilly, J.L., Dharmarajan, A.M. (1998) Estradiol mediated suppression of apoptosis in the rabbit corpus luteum is associated with a shift in expression of bcl-2 family members favouring cellular survival. *Biol. Reprod.*, **59**, 820–827.
9. Hutchinson, L.A., Findlay, J.K., de Vos, F.L., Robertson, D.M. (1987) Effects of bovine inhibin, transforming growth factor-β and bovine activin-A on granulosa cell differentiation. *Biochem. Biophys. Res. Commun.*, **146,** 1405–1412.
10. Lee, D.W., Sheldon, R.M., Reichert, L.E.Jr. (1990) Identification of low and high molecular weight follicle stimulating hormone (FSH) receptor binding inhibitors in human follicular fluid. *Fertil. Steril.*, **53**, 830–835.
11. Lee, D.W., Grasso, P., Dattatreyamurty, B., Dezeil, M.R., Reichert, L.E.Jr. (1993) Purification of a high molecular weight follicle-stimulating hormone receptor-binding inhibitor from human follicular fluid. *J. Clin. Endocrinol. Metab.* **77**, 163–168.
12. Lee, D.W., Grasso, P., Leng, N., Izquierdo, R., Grabb, J.W., Deziel, M.R., Reichert, L.E.Jr. (1995) A decapeptide corresponding to the partial amino acid sequence of a high molecular weight FSH receptor binding inhibitor is a specific inhibitor of FSH binding. *Pep. Res.*, **8**, 171–177.
13. Nandedkar, T.D., Kadam, A.L., Moodbidri, S.B. (1988) Control of follicular maturation in the mouse by a non steroidal regulator from sheep follicular fluid. *Int. J. Fertil.*, **33**, 52–59.
14. Nandedkar, T.D., Kholkute, S.D., Shahid, J.K., Moodbidri, S.B. (1989) Luteal deficiency caused by ovarian follicular fluid peptide in post partum marmosets, *Callithrix jacchus. Contraception*, **40**, 101–109.
15. Nandedkar, T.D., Shahid, J.K., Kholkute, S.D., Darpe, M.B., Moodbidri, S.B. (1992a) Interference with ovulation and luteal function by human ovarian follicular fluid peptide in bonnet monekys, Macaca radiata. *Contraception*, **45**, 379–385.
16. Nandedkar, T.D., Shahid, J.K., Mehta, R., Moodbidri, S.B., Hegde, U.C., Hinduja, I. (1992b) Localisation and detection of ovarian follicular fluid protein in follicles of human ovaries. *Ind. J. Exp. Biol.*, **30**, 271–275.

17. Nandedkar, T.D. (1995) Detection of ovarian follicular fluid peptide and albumin in sheep ovary by Enzyme-Linked Immunosorbant Assay (ELISA). *J. Reprod. Biol. Comp. Endocrinol.*, **7**, 83–91.
18. Nandedkar, T.D., Parker, S.G., Iyer, K.S., Mahale, S.D., Moodbidri, S.B., Mukhopadhyaya, R.R., Joshi, D.S. (1996) Regulation of follicular maturation by human ovarian follicular fluid peptide. *J. Reprod. Fertil., Supplement,* 50, 95–104.
19. Nandedkar, T.D., Rajadhyaksha, M.S., Mukhopadhyaya, R.R., Rao, S.G.A., Joshi, D.S. (1998) Apoptosis in granulosa cells induced by intrafollicular peptide. *J. Biosciences,* **23**, 271–277.
20. Rajendra Kumar, T., Ganesh, P., Wang, P., Woodruff, T.K., Boime, I., Byrne, M.C., Matzuk, M.M. (1999) Transgenic models to study gonadotropin function: the role of follicle stimulating hormone in gonadal growth and tumorigenesis. *Mol. Endocrinol.*, **13**, 851–865.
21. Reichert, L.E.Jr., Abou-Issa, H. (1977) Studies on a low molecular weight testicular factor which inhibits binding of FSH to receptor. *Biol. Reprod.,* **17**, 614–621.
22. Shahid, J.K., Nandedkar, T.D., Iyer, K.S.N., Mahale, S.D. (1991) Effect of ovarian follicular fluid peptide (OFFP) on FSH and EGF binding to granulosa cells. *Horm. Metab. Res.*, **23**, 456–458.
23. Sluss, P.M., Shneyer, A.L., Franke, M.A., Reichert, L.E.Jr. (1987) Porcine follicular fluid contains both FSH agonist and antagonist activities. *Endocrinol.*, **120**, 1477–1481.
24. Takahashi, M., Koide, S.S., Donahue, P.K. (1986) Mullerian inhibiting substance as oocyte meiosis inhibitor. *Mol. Cell. Endocrinol.*, **47**, 225–234.
25. Tsafriri, A., Pomerantz, S.H., Channing, C.P. (1976) Inhibition of oocyte maturation by porcine follicular fluid: partial characterization of the inhibitor. *Biol. Reprod.,* **14**, 511–516.
26. Ueno, N., Ling, N., Ying, S.Y., Esch, F., Shimasaki, S., Guillemin, R. (1987) Isolation and partial characterization of follistatin - a novel Mr 35000 monomeric protein which inhibits the release of follicle stimulating hormone. *Proc. Natl. Acad. Sci. USA.,* **84**, 8282–8286.
27. Ying, S.H., Becker, A., Ling, N., Ueno, N., Guillemin, R. (1986) Inhibin and beta type transforming growth factor (TGF beta) have opposite modulating effects on the follicle stimulating hormone (FSH) induced aromatase activity of cultured rat granulosa cells. *Biochem. Biophys. Res. Commun.,* **136**, 969–975.
28. Ying, S.Y. (1988) Inhibins, activins and follistatins : gonadal proteins modulating the secretion of follicle-stimulating hormone. *Endocr. Rev.,* **9**, 267–293.

Follicular Growth, Ovulation and Fertilization: Molecular and Clinical Basis
Anand Kumar and Amal K. Mukhopadhyay (Eds.)
Narosa Publishing House, New Delhi, India, 2001

15

Bacterial Modulations of Granulosa Cell Steroidogenesis

Anand Kumar and Nishat Ahmad
Department of Reproductive Biology, All India Institute of Medical Sciences, New Delhi–110029, India

Introduction

The bacteria and endocrine systems have a dynamic and mutually regulatory relationship. A wealth of data is available on the effect of sex steroids on bacterial growth and pathogenicity. Certain infections are exacerbated during a particular phase of the reproductive cycle under the influence of ovarian steroids. In *in vivo* conditions gonadal steroids show effects on bacterial growth in a species specific manner. However there is a paucity of literature on the effect of microbes on endocrine and reproductive functions. This review examines the potential role of ovarian steroids on bacterial pathogenesis. We have also presented some of our preliminary findings on the effect of *Escherichia coli* and *Mycobaterium tuberculosis* on ovarian cell steroidogenesis in rat, buffalo (*Bubalus bubalis*) and human.

Effect of Hormones on Bacterial Growth and Metabolism

Hormonal variations during the reproductive cycle affect the host-resistance against several bacterial infections. High steroidal states such as pregnancy in women, increase their susceptibility to fungal infections (Buck & Hasenclever, 1963 and Drutz *et al*, 1981), bacterial infections (Luft and Ramington, 1982 and Holmes et al, 1971) and viral infections (Amstey, 1975 and Aycock, 1941). The high susceptibility to infections during pregnancy could be partly explained on the basis of immunological depression due to raised concentration of steroids (Birkeland & Kristofferson, 1980 and Chaout & Voisin, 1979, Kasakura, 1971).

Bergstrom (1987) reported a high incidence of urinary tract infections (UTI) in elderly postmenopausal women. The prevalence of UTI in Swedish women over 75 years of age is 5% as compared to 1% in women of all age groups. Estrogen appears to have a protective role against *E. coli* infection in post menopausal women (Parsons and Schmidt, 1982). An association between estrogen and *Escherichia coli* infection in premenopausal women has been reported by Hooton et al (1996). Females are more susceptible to this infection between 8–15 days after the onset of the menstrual cycle. Oopherectomized rabbits and bitches have a high incidence of urinary tract infection by *E. coli* (Mulholland et al, 1982). This increase in infection is attributed to the decrement in the secretion of protective lining of mucopolysaccharides in the bladder. The production of mucin is controlled by estrogen and its secretion by progesterone (Zachariae, 1958 and Greenwald, 1958). The mucus production which is a natural defence mechanism of host against the reproductive tract infections, is clearly under hormonal control in human females (Elstein, 1978). Lactoferrin present in mucosal secretions has been shown to inhibit the growth of bacteria requiring iron for normal metabolism, in addition to having a direct bactericidal effect. Vaginal lactoferrin concentration varies dramatically during the menstrual cycle with the lowest levels being detected just before menstruation (Arnold *et al*, 1977). Progesterone has been shown to increase the survival time of *E. coli* in the uterus of rabbit and estradiol potentiates the effect (Mastuda *et al*, 1985). Estradiol prevents *E. coli* induced purulent endometritis in the uterus of rat whereas progesterone has been shown to inhibit the phagocytic activity of the uterine luminal leucocytes. Binding of estrogen has been demonstrated in *E. coli* (Nishikawa and Baba, 1990). Estradiol shows binding to other common bacteria such as *Pseudomonas aeruginosa, Staphylococcus aureus, Streptococcus pyogenes and Streptococcus haemolyticus* also (Sugarman and Mummaw, 1990).

The load of *Bacteroides melaninogenicus* increases in the subgingival microflora duirng pregnancy when the levels of the gonadal steroids are high (Kornman and Loesche, 1982). *B. melaninogenicus* binds estradiol and progesterone both (Kornman and Loesche, 1982). Estrogen and progesterone have been shown to promote the growth of this bacilli in the culture media like Vitamin K (Kornman and Loesche, 1982). Estradiol induces cessation of the estrus cycle of mouse at the estrus stage and predisposes the animal to genital infection with *Ureaplasma urealyticum* and *Mycoplasma hominis* (Furr and Taylor-Robinson, 1989a&b). *Mycoplasma pulmonis,* primarily a respiratory pathogen, is capable of colonising the mouse vagina after treatment with progesterone. However, the adminstration of estradiol during the course of *M. pulmonis* infection leads to rapid elimination of organisms from the genital tract (Furr and Taylor-Robinson, 1993a). It is interesting to note that some mycoplasmas depend upon estrogen for their growth and the others on progesterone. The mycoplasmas which depend either on progesterone or estradiol show different metabolic characteristics. Progesterone dependent mycoplasmas primarily metabolize glucose and the estrogen dependent strains metabolize arginine or arginine and glucose both. Both the strains also show differences in their mechanism of binding to the host cells (Furr and Taylor-Robinson, 1993b). These observations suggest the presence of receptros for mycoplasma on vaginal epithelial cells and these receptors appear to be under hormonal control.

Gonococcal pelvic infection and disseminated gonococcal infection (DGI) both exacerbate at the time of menstruation (Sweet *et al*, 1986, Britigan *et al,* 1985). Consumers of oral contraceptives containing androgenic progestins are prone for gonococcal infection (Eschenbach *et al,* 1977 and Louv *et al*, 1989). In mice injected with estrogen, gonococcal bacteria develops within 12 hours after inoculation and the mice die within 18 hours. Progesterone supplementation leads to inhibition of bacteremia (Kita *et al,* 1985). Intraperitoneal administration of gonococci enchances the infiltration of polymorphonuclear leukocytes (PMN) cells in the peritoneal cavity. Treatment of mice with estradiol significantly reduces the PMN cell infiltration in response to gonococcal inoculation (Kita *et al*, 1985). Estradiol has been shown to stimulate the grwoth of *N. gonorrhoeae* whereas progesterone inhibits the growth of gonococci in vitro (Saht, 1982). Gonococci invade the non-ciliated cells in the mucosa of cultured fallopian tubes. HCG in high concentration blocks the mucosal binding of gonococci (Gorby *et al,* 1991). The strength of adherence of *N. gonnorrhoeae* with vaginal epithelial cells varies during the menstrual cycle attaining the maximum adherence just before menstruation (Forslin *et al,* 1979).

Similar to gonococci, chlamydial infections also occur at the time of menstruation (Sweet *et al*, 1986). Estradiol has been reported to enhance the infection of HeLa cells by *Chlamydia trachomatis* (Bose and Goswami, 1986). Using a human endometrial epithelial cell system, Maslow *et al*, 1988 has reported that the degree of chlamydial attachment is significantly increased when the cells are obtained from women in the early phase of the menstrual cycle and cultured in the presence of physiological doses of estrogen. When epithelial cells are obtained from women in the later phases of the menstrual cycle and cultured in the presence of estrogen and progesterone both, a significant reduction in chalamydial attachment to cells is observed. The users of steroidal oral contraceptives have higher prevalence of chlamydial infection (Elgaali *et al*, 1994). The antichlamydial activity of the cervical secretions is reduced in the users of oral contraceptives (Elgaali *et al*, 1994). However, Patton *et al*, 1994 could not demonstrate any effect of oral contraceptives on the development of chlamydial acute salpingitis in monkeys.

Endocrinal Modulations

As we have seen that sex steroids could modify the growth and infectivity of the bacteria, so could the bacteria affect the metabolism and functions of reproductive hormones. Bacteria produce and metabolize hormones and hormone like materials, including androgens and progestins. A fecal species of clostridium has been shown to convert cortisol to androgens. This strain has a constitutive desmolase that cleaves the side chain of cortisol to form 11β-hydroxy-4-androstene-3, 17-dione, 3α, 1β-dihydroxy-5β-androstane-17-one; and 3α, 11β, 17β-trihydoxy-β-androstane (Bokkenheuser *et al*, 1984). The growth of mycobacterium strain DP (a marine mycobacterium) is enhanced by cholesterol. It also converts cholesterol to androgens. The strain DP degrades cholesterol to cholestenone, androstendione (AD), androstadiendione (ADD), testosterone and dehydrotestosterone (DHT) (Smith *et al*, 1993). In comparison, rapid cholesterol oxidation has been observed in mycobacterium strain NRRL B-3683. The oxidized products are androstendione, androstadiendione (Marsheck *et al*, 1972). Chipley *et al* 1975, has reported that culture supernatant fractions of Mycobacterium ATCC 19652 degrades significant quantitites of cholesterol within 24 h to androstendione and testosterone. This degradation has been attributed to the increased cholesterol oxidase activity caused by the culture supernatants of the mycobacteria. *Mycobacterium aureum* a commensal of ear wax produces progesterone from cholesterol in fermentation cultures (Horhold and Bohme, 1990).

Effect of Bacteria on Granulosa Cell Steroidogenesis

In order to evaluate the direct action of bacteria on ovary, we studied the effects of the sonicated lysates of *M. tuberculosis* (pathogenic, H37Rv and non-pathogenic, H37Ra) and *E. coli* (ATCC 25922) on the cultures of granulosa cells from various species like human, buffalo and rat.

Effect of *E. coli*

E. coli is one of the common causes of non-sexually transmitted pelvic infections. The variations in the pattern of its pathogenesis during pre- and post-menopausal stages in women has been described earlier. The sonicated lysate of *E. coli* directly affects the granulosa cell steroidogenesis. The lysate of *E. coli* has been prepared by ultrasonicating 3×10^8 bacteria (equivalent to 1 Mcferland) in phosphate buffered saline (PBS) for 1 hour. The live bacteria have been removed by centrifugation and sterile filtration by 0.2 µ pore size filters. The filtrate has been cultured on peptone agar for 48 h to ensure the completion of lysis and sterility of the lysate.

Studies on Human Granulosa cells

Granulosa cells have been isolated from women undergoing follicular puncture for isolation of oocytes under IVF programme at the University Hospital, Bonn (Germany). 10^6 human granulosa cells have been cultured in gamete medium (modified HTF medium from Scandinavian IVF Science) in a total volume of 1 ml which included 0.1 ml of lysate or PBS for 18 h at 37°C under 5% CO_2 in air. The viability of the cells has been assessed by trypan blue exclusion method. The granulosa cells are treated with and without 100 µl, 50 µl, 25 µl of lysate of a clinical isolate of *E. coli* and 5 UhCG (supramaximal dose). All the treatments have been done in quadruplicates. The dose of 100 µl inhibits basal progesterone production. The hCG-stimulated progesterone production is significantly blocked by doses of 100 µl, 50 µl and 25 µl (Fig. 1). The inhibitory activity is retained by the lystate boiled for one hour. The boiled as well as unboiled lysate have totally abolished the stimulatory action of db-cAMP (Fig. 2). The chloroform extract of the boiled lysate has shown a slight but significant stimulation of basal secretion of progesterone by the human granulosa cells (unpublished observation). No effect on the viability of the cells has been detected by any of the treatments.

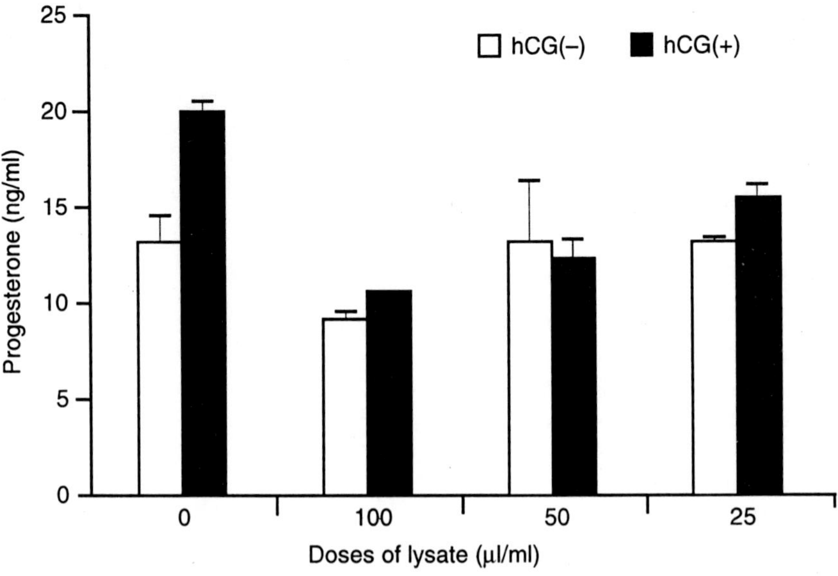

Fig. 1 Effect of unboiled lysate of *E.coli* on basal and hCG (1IU)-stimulated progesterone production by human granulosa cells. The values are mean ± sd of four replicates. The hCG-stimulated progesterone production is significantly inhibited by all the doses of the lysate ($p < 0.05$). The basal steroid production is inhibited significantly only by a dose of 100µl/ml ($p < 0.05$). The experimental details are given in the text.

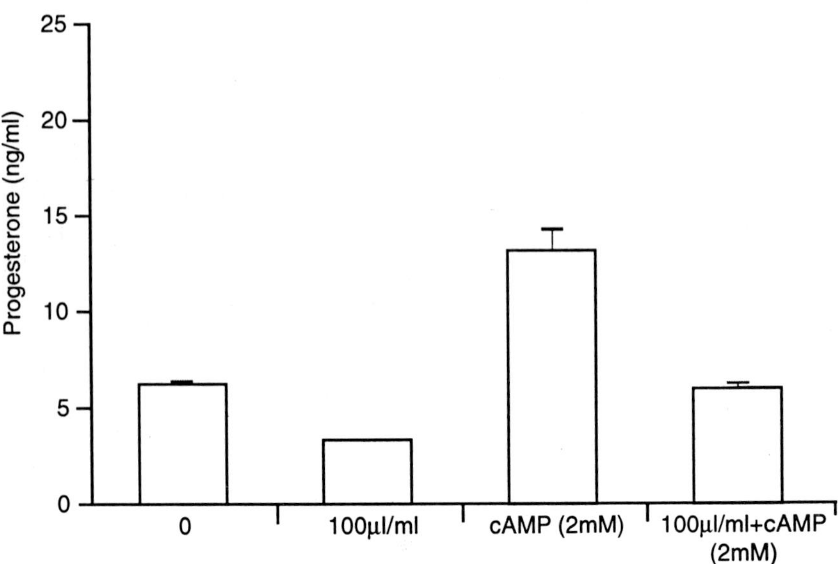

Fig. 2 Effect of boiled lysate of *E.coli* on db-cAMP (2 mM)-stimulated progesterone production by human granulosa cells. The values are mean ± sd of four replicates. The db-cAMP (2mM) has significantly stimulated the progesterone production from basal value ($p < 0.05$). The lysate in a dose of 100µl/ml significantly inhibits the basal as well as cAMP-stimulated progesterone production ($p < 0.05$). The experimental details are given in the text.

Studies on Buffalo Granulosa Cells

Granulosa cells aspirated from medium-sized follicles (5–9 mm diameter) produce maximum amount of progesterone. One million viable cells have been cultured in McCoy's 5A tissue culture medium in presence or absence of 1 IU hCG (supramaximal dose) with or without the bacterial lysate containing 320 μg, 32 μg, 3.2 μg and 0.32 μg protein in 0.1 ml PBS or equivalent volume of PBS for 8 h in a total volume of 1 ml at 37°C under 5% CO_2 in air. The lysate, unboiled and boiled both, have shown no effect either on basal or on hCG-stimulated progesterone production by the granulosa cells. The data is not presented here.

Granulosa cells obtained from small follicles of 1–4 mm diameter produce maximum amount of estradiol. One million cells have been cultured in McCoy's 5A tissue culture medium in presence or absence of 1 IU FSH (maximal dose) with or without lysate containing proteins in above mentioned doses or PBS for 24 h in a total volume of 1 ml. Testosterone dissolved in the tissue culture medium has been added to each well to achieve the concentration of 2 ng/ml. The lysate, unboiled and boiled both, has not shown any effect eiher on basal or on FSH-stimulated estradiol production by the granulosa cells. The data is not given here.

Effect of *Mycobacterium tuberculosis*

Nogales-Ortiz et al 1979, have shown that pure involvement of ovarian parenchyma in genital tuberculosis is seen only in 11% of the cases. Active pulmonary tuberculosis without demonstrable lesions or structural involvement of either the ovaries or the genital tract has also been shown to be associated with amenorrhoea and infertility in human female (Shaefer, 1972). About 30 percent of women who suffer from tuberculosis also suffer from amenorrhoea and other menstrual irregularities (unpublished observation). The mechanism of amenorrhoea and infertility in the patients of extragenital tuberculosis (with no involvement of ovary or genital tract) has not been understood.

Mycobacteria lie in close proximity of ovaries in genital tuberculosis and may even enter the ovarian macrophages, which may then alter ovarian function in a paracrine manner. The biological role of several cytokines on ovarian functions has been described (Brannstrom and Norman, 1993). It is however not known if *M. tuberculosis* has any direct action on the ovarian cells. The lystae of *M. tuberculosis* (H37Rv) and (H37Ra) have been prepared by ultrasonicating 3×10^8 bacteria in 1ml PBS (equivalent to 1 Mcferland) for 30 min in five strokes. The live bacteria has been removed by centrifugation and sterile filtration by 0.2 μ pore size filters. The smear of sterile lysate has been examined microscopically by Zeihl-Nelson staining. The lysate has also been cultured Lowenstein-Jensen medium for six weeks to confirm the completion of lysis.

Studies on Rat Granulosa Cells

5×10^5 cells have been cultured in McCoy's 5 A tissue culture medium in presence and absence of 250 mIU hCG (maximal dose), with and without 100 μl of the lysate or PBS in a final volume of 1 ml for 24 h. Progesterone is measured by radioimmunoassay in the spent medium. The sonicated mycobacterial lysate has inhibited the basal as well as hCG-stimulated progesterone production by the rat granulosa cells without affecting the viability of the cells. The observation rules out any possibility of cytotoxic effect of the lysate (Kumar and Rattan, 1997, Fig. 3). The observation suggests the direct involvement of mycobacterial components/toxins on the granulosa cell function.

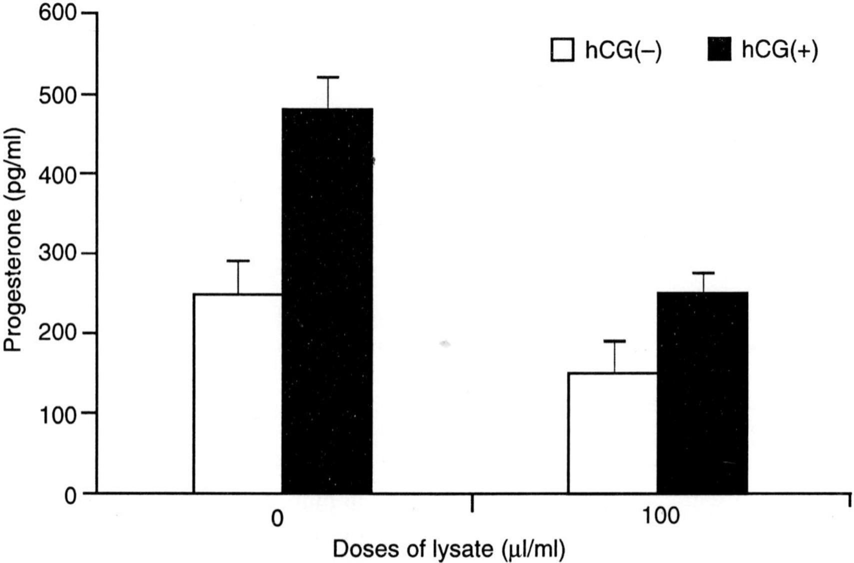

Fig. 3 Effect of unboiled lysate of *M. tuberculosis* on basal and hCG (250 mIU)-stimulated progesterone production by rat granulosa cells. The values are mean ± sd of four replicates. The basal as well as hCG-stimulated progesterone production is significantly inhibited by a dose of 100 μ/ml of the lysate ($p < 0.05$). The experimental details are given in the text.

Studies on Buffalo Granulosa Cells

Granulosa cells aspirated from medium-sized follicles, produce much larger amount of progesterone in comparison to small-sized follicles. One million granulosa cells from medium-sized follicles have been cultured in McCoy's 5A tissue culture medium for 18 hours with or without the lysate in PBS or PBS alone in presence or absence of 1 IU hCG (supramaximal dose) in a total volume of 1 ml. The lysate containing 47 μg, 4.7 μg, 0.47 μg and 0.047 μg of protein in 0.1 ml PBS not only blocks the hCG-stimulated progesterone production, but also significantly suppresses the basal production of the steroid in a dose dependent manner (Fig. 4). The lysate boiled for one hour, has no significant effect on the progesterone production by the cells in all the

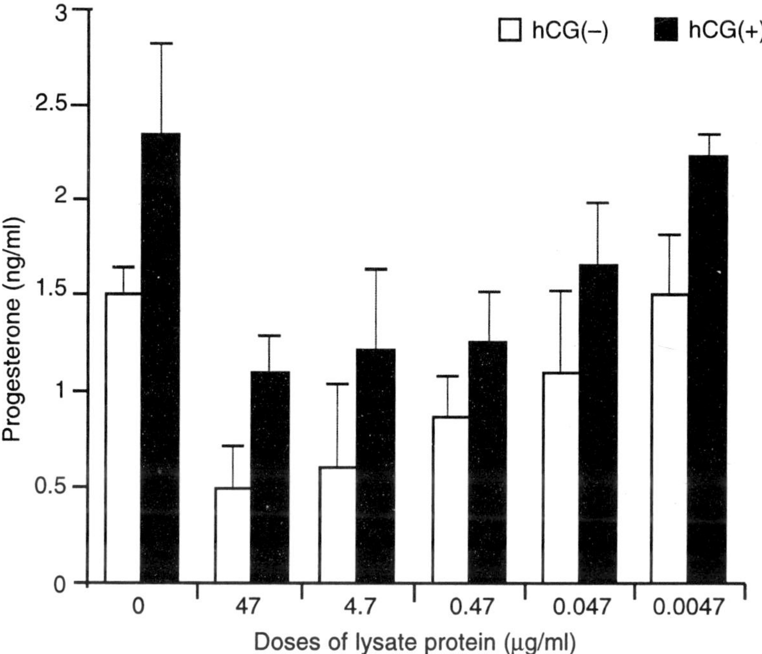

Fig. 4 Effect of unboiled lysate of *M.tuberculosis* on basal and hCG (1IU)-stimulated progesterone production by buffalo granulosa cells. The values are mean ± sd of four replicates. The doses of lysate in protein concentration of 47mg/ ml, 4.7μg/ml, 0.47μg/ml and 0.047μg/ml has significantly inhibited basal as well as hCG-stimulated progesterone production by the granulosa cells ($p < 0.05$). The doses in the lower protein concentrations have shown no significant effect on either basal or hCG-stimulated progesterone production. The experimental details are given in the text.

above doses. The particulate and non-particute fraction of the lysate have been separated by ultracentrifugation at 105,000g for 1 h. The supernatant containing the non-particulate fraction and the pellet containing the particulate fraction has been isolated. The cell particulate fraction of the lysate containing 47 and 4.7 μg of protein has also inhibited both the basal and the hCG-stimulated progesterone production in contrast to the non-particulate fraction which has no effect on the steroid production (unpublished observation). The chloroform extract of the lysate also has no significant effect on the production of this steroid (unpublished observation). This observation rules out any role of lipid fraction of the lysate on the modulation of the granulosa cell steroidogenesis. The lysate of non-pathogenic strain of *M. tuberculosis* (H37Ra) in the above doses exhibits no effect on progesterone production by the granulosa cells (unpublished). This inhibitory action of *M. tuberculosis* is linked to the virulence of the bacilli.

One million granulosa cells obtained from small follicles have been cultured in McCoy's 5A tissue culture medium in presence or absence of 1 IU FSH with or without lysate containing proteins in 0.1 ml PBS in the various mentioned doses or equivalent volume of PBS for 24 h in a total volume of 1 ml. Testosterone in the dose of 2 ng has been added to each well. The lysate proteins, unboiled and boiled both, have shown no effect on basal as well as FSH-stimulated estradiol production by the granulosa cells (Fig. 5). Also the lysate proteins of non-pathogenic strain of *M. tuberculosis* (H37Ra) in similar doses have no effect on estradiol production by the granulosa cells (unpublished observation). The viability of the cells has been assessed using MTT (3-{4,5-Dimethyl thiazol-2-y1}-2,5-diphenyl tetrazolium bromide). Following the completion of culture the spent medium has been removed and replaced by MTT containing medium. The cells have been further incubated for 4 h and crystal formed has been disolved in DMSO and dull blue color developed has been read at 550 nm in the spectrophotometer.

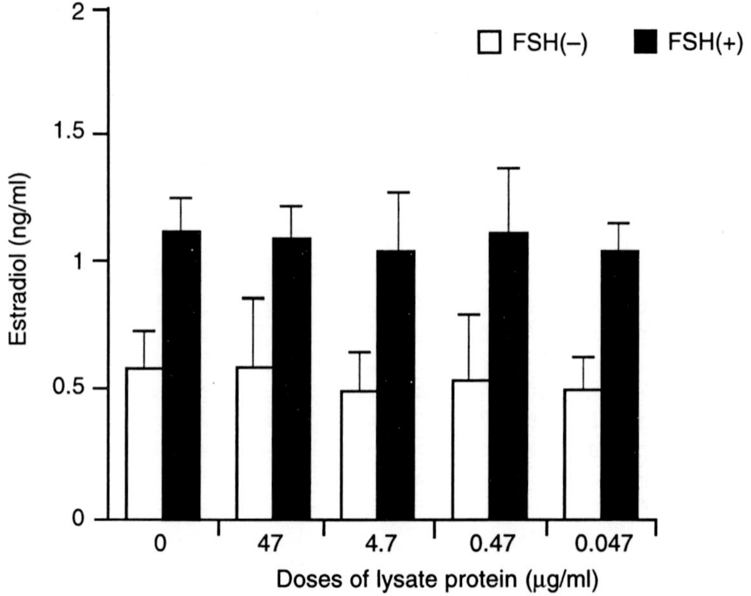

Fig. 5 Effect of unboiled lysate of *M. tuberculosis* on basal and FSH (1IU)-stimulated estradiol production by buffalo granulosa cells. The values are mean ± sd of four replicates. None of the doses of lysate has any effect on either basal or FSH-stimulated progesterone production by the granulosa cells. The experimental details are given in the text.

The cytotoxity of these inhibitor(s) have been ruled out as none of the inhibitory doses of the lysate has any effect on the viability of granulosa cells (Fig. 6). The mycobacterial protein(s) may possibly act through the inhibition of steroidogenesis at a post-cAMP site of progesterone biosynthesis and thus inhibit the differentiation and luteinization of granulosa cells. The work is being carried out to elucidate the possible mechanism of action of these mycobacterial proteins on bovine granulosa cell steroidogenesis.

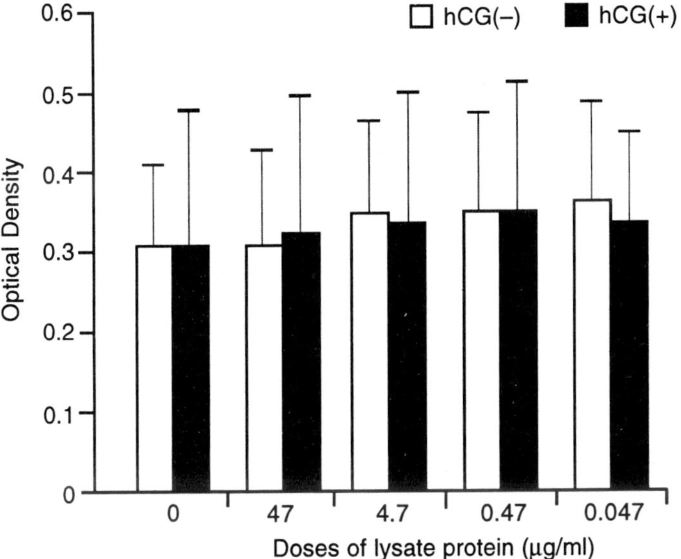

Fig. 6 Effect of lysate of *M. tuberculosis* on the viability of buffalo granulosa cells. The values are mean ± sd of four replicates. The doses of lysate shows no significant effect on the viability of granulosa cells. The experimental details are given in the text.

References

1. Amstey, M.S., (1975). Genital herpes virus infection. *Clin. Obst. Gynaecol.* **18**: 80–100.
2. Arnold, R.R, Cole, M.F., McGhee J.R. (1977). A bactericidal effect for human lactoferrin. *Science.* **197**: 263–265.
3. Aycock W.L., (1941). The frequency of poliomyelitis in pregnancy. *N. Eng. J. Med.* **225**: 405–408.
4. Bergstrom H., (1987). Estriol in urinary tract infection in elderly women. *Acta. Obst. Gynecol. Scand. Suppl.* **140**: 79–83.
5. Birkeland SA, Kristofferson, K., (1980). The fetus as an allograft: a longitudinal study of normal human pregnancies studied with mixed leucocytes cultures between mother-father and mother-child. *J. Immunol.* **11**: 311–320.
6. Bokkenheuser, V.D., Morris, G.N., Ritchie, A.E., Holderman, L.V., Winter, J., (1984). Biosynthesis of androgen from cortisol by a species of Clostridium recovered from human fecal flora. *J. Inf. dis.* **149**: 489–494.
7. Bose, S.K., Goswami, P.C., (1986). Enhancement of adherence and growth of Chlamydia trachomatis by estrogen tretment of HeLa cells. *Infect. Immun.* **53**: 646–650.
8. Brannstrom M., Norman R.J., (1993). Involvement of leucocytes and cytokines in the ovulatory process and corpus luteum function. *Human Reproduction.* **8**: 1762–1775.
9. Britigan B.E., Cohen M.S., Sparling P.F. (1985). Gonococcal infection: A model of molecular pathogenesis. *N. Eng. J. Med.* **313**: 1683–1684.
10. Buck A.A., Hasenclever H.F., (1963). Epidemiogical studies of skin reactions and serum agglutinins to *Candida albicnas* in pregnant women. *Am. J. Hyg.* **78**: 232–240.

11. Chaout G., Voisin G.A., (1979). Regulatory T. cell subpopulation in pregnancy: In evidence of suppressive activity in the early phase of MLR. *J. Immunol.* **122**: 1383–1388.
12. Chipley J.R., Dreyfuss M., Smucker R., (1975). Cholesterol metabolism by mycobacterium. *Microbios*, **12**: 199–207.
13. Drutz D.J. Huppert M., Sun S.H., McGuire, W.A. (1981). Human sex hormones stimulate growth and maturation of *Coccidioides immitis*. Infect. Immun. **32:** 897–907.
14. Elgaali A., Mahmoud E.A., Hamad E.E., Olsson S.E., Mardh P-A (1994). Anti-chlamydial activity of cervical secretion in different phases of the menstrual cycle and the influence of hormonal contraceptives. *Contraception;* **49**: 265–274.
15. Elstein M., (1978). Functions and physical properties of mucus in the female genital tract. *Br. Med. Bull.* **34:** 83–87.
16. Eschenbach D.A., Harnisch J.P., Holmes K.K., (1977). Pathogenesis of acute pelvic inflammatory disease: role of contraception and other risk factors. *Am. J. Obst. Gynecol.* **128**: 838–850.
17. Forslin L., Danielsson D., Falk V., (1979). Variations in attachment of Neisseria gonorrhoeae to vaginal epithelial cells during the menstrual cycle and early pregnancy. *Med. Microbiol. Immunol.* **167**: 231–238.
18. Furr P.M., Taylor-Robinson D. (1989a). The establishment and persistence of *Ureaplasma urealyticum* in estradiol-treated female mice. *J. Med. Microbiol.* **30**: 233–236.
19. Furr P.M., Taylor- Robinson D. (1989b). Estradiol-induced infection of genital tract in female mice by *Mycoplasma hominis*. **135**: 2743–2749.
20. Furr P.M., Taylor- Robinson D. (1993a). The contrasting effects of progesterone and estrogen on susceptibility of mice to genital infection with *Mycoplasma pulmonis*. **38**: 160–165.
21. Furr P.M., Taylor- Robinson D. (1993b). Factors influencing the ability of different mycoplasmas to colonize the genital tract of hormone- treated female mice. *Int. J. Exp. Path.* **74**: 97–101.
22. Gorby G.L., Clemens C.M., Barley L.E., McGee Z.A. (1991). Effect of human chorionic gonadotrophin (hCG) on Neisseria gonorrhoeae invasion and of IgA secretion by human fallopian tube mucosa. *Microbial Pathogenesis.* **10**: 373–384.
23. Greenwald G.S., (1958). Endocrine regulation of mucin in the tubal epithelium of the rabbit. *Anatomical Record.* **130**: 477–479.
24. Holmes K.K., Counts G.W., Beaty H.N., (1971). Disseminated gonococcal infection. *Ann. Intern. Med.* **74**: 979–993.
25. Hooton T.M., Winter C., Tiu F., Stamm W.E. (1996). Association of acute cystitis with the stage of the menstrual cycle in young women. *Clin. Infect. Dis.* **23**: 635–636.
26. Horhold C., Bohme K.H. (1990). Formation of progesterone and 1-dehydroprogesterone from cholesterol in fermentation cultures of *Mycobacterium aureum. J. Steroid Biochem.* **36**: 181–183.
27. Kasakura, S. (1971). A factor in maternal plasma during pregnancy that suppresses the reactivity of mixed leukocytes cultures. *J. Immunol.* **107**: 1296–1301.
28. Kita E, Takahashi S, Yasui K, Kashiba S., (1985). Effect of estrogen (17betaestradiol) on the susceptibility of mice to disseminated gonococcal infection Infect. *Immun.* **49**: 238–243.
29. Kornman K.S., Loesche W. (1982). Effects of estradiol and progesterone on *bacteroides melaninogenicus* and *Bacteroides gingivalis. Inf. Immun.* **35**: 256–263.
30. Kumar A., Rattan, A. (1997). Antigonadotrophic effect of *Mycobacterium tuberculosis. Hormon. Metab. Res.* **29**: 504–506.

31. Louv W.C., Austin H., Perlman J, Alexander W.J. (1989). Oral contraceptive use and the high risk of chlamydial and gonococcal infections. *Am. J. Obst. Gynecol.* **160**: 396–402.
32. Luft B.J., Ramington J.S. (1982). Effect of progesterone resistance to *Listeria monocytogens* and *toxoplasa* infections in mice. *Infect. Immun.* **38**: 1164–1171.
33. Marsheck W.J., Kraychy S, Muir R.D. (1972). Microbial degradation of sterols. *Appl. Microbiol.* **23**: 72–77.
34. Maslow A.S., Davis C.H., Choong J., Wyrick P.B. (1988). Estrogen enhances attachment of Chlamydia trachomatis to human endometrial epithelial cells in vitro. *Am. J. Obst. Gynecol.* **159**: 1006–1114.
35. Mastuda H., Okuda K., Fukui K, Kamata Y (1985). Inhibitory effect of estradiol-17 beta and progesterone on bactericidal activity in uteri of rabbits infected with Escherichia coli. *Infect. Immun.* **48**: 652–657.
36. Mulholland S.G., Qureshi S.M., Fritz R.W., Silverman H. (1982). Effect of hormonal deprivation on the bladder defense mechanism. *J. Urology*, **127**: 1010–1013.
37. Nishikawa Y., Baba T., (1985). Effects of ovarian hormones on manifestation of purulent endometritis in rat uteruses infected with *Escherichia coli. Infect. Immun.* **47**: 311–317.
38. Nogales-Ortiz F, Tarancon I, Nogales Jr F. (1979). The pathology of female genital tuberculosis: A 32 year study of 1436 cases. *Obst. Gynecol.* **53**: 428–442.
39. Parsons C.L., Schmidt J.D., (1982). Control of recurrent lower urinary tract infection in the post menopausal women. *J. Urol.* **128**: 1224–1226.
40. Patton D.L., Cosgrove-Sweeney Y.T., Kuo C.C. (1994). Oral contraceptives do not alter the course of experimentally induced chlamydial salpingitis in monkeys. Sex. *Transm. Dis.* **21**: 89–92.
41. Saht, I.E., (1982). The differential susceptibility of gonococcal pathogenicity variants to sex hormones. *Can. J. Microbiol.* **28**: 301–306.
42. Shaefer G. (1972). Female genital tuberculosis. *Clin. Obst. Gynecol.* **19**: 223–239.
43. Smith M, Zahnley J., Pfeifer D., Goff, D., (1993). Growth and oxidation by Mycobacterium species in tween 80 medium. *Appl. Environ. Microbiol* **59**: 1425–1429.
44. Sugarman B., Mummaw N. (1990). Estrogen binding and effect of estrogen on trichomonas and bacteria. *J. Med. Microbiol.* **32**: 227–232.
45. Sweet R.L., Blankfort-Doyle M., Robie M.O., Schachter J. (1986). The occurrence of chlamydial and gonococcal salpingitis during the menstrual cycle. *JAMA*. **255**: 2062–2064.
46. Zachariae F. (1958). Autoradiographic (35S) and histochemical studies of sulphomucopolysaccharides in the rabbit uterus, oviducts and vagina. *Acta Endocrinol.* **29**: 118–122.

Follicular Growth, Ovulation and Fertilization: Molecular and Clinical Basis
Anand Kumar and Amal K. Mukhopadhyay (Eds.)
Narosa Publishing House, New Delhi, India, 2001

16

Genetic Analysis of Candidate Genes for Premature Ovarian Failure

Clare M. Daniel, Karen A. Burton, Cynthia C. van Ee, John T. France, Ingrid M. Winship* and Andrew N. Shelling

*Research Centre in Reproductive Medicine, Department of Obstetrics and Gynaecology, National Women's Hospital, Auckland, New Zealand
Department of Molecular Medicine, School of Medicine, University of Auckland, Auckland, New Zealand

Introduction

POF is characterised by secondary amenorrhoea, hypoestrogenism, and elevated gonadotrophins in women under the age of 40 years. It is a relatively common condition occurring in 1% of women under the age of 40 years, and will affect 0.1% of women before the age of 30 years (Coulam *et al.*, 1986). Two significant consequences of POF are the loss of fertility, and the clinical effects of hypoestrogenism. Low levels of oestrogen from a young age appear to increase the risk of osteoporosis and coronary heart disease. A significant number of patients with POF have ovarian follicles visible on ultrasound, and approximately 5–10% of these patients will subsequently ovulate and become pregnant (Kalantaridou *et al.*, 1998). This raises the possibility of treatment and restoration of fertility in some women with POF.

POF is clearly a heterogeneous disorder, and only a few causes can be identified (for recent reviews see Conway, 1997; Kalantaridou *et al.*, 1998; Anasti, 1998). It may be due to chromosomal, genetic enzymatic, iatrogenic, autoimmune or infectious aetiology. There are a number of candidate genes for POF. Some POF mutations may be X-linked, as the X chromosome is involved in some types of female infertility (Shelling, 2000). 45, X Turners syndrome which confers primary infertility (Anasti, 1998), and 45, X/46, XX mosaics (Devi *et al.*, 1998) and fragile X carriers with a fragile X premutation (Partington *et al.*, 1996; Conway *et al.*, 1998), all experience early menopause. Autosomal mutations are also seen in POF, and genes coding for the gonadotropins and their receptors have received much attention because of their central role in the female reproductive cycle (Conway, 1996). In particular, defects in FSHR and LHβ have been linked to amenorrhoea. Iatrogenic agents, such as chemotherapy or radiotherapy, are known to reduce follicle numbers and cause POF. Although autoimmune diseases are seen in 10–20% of women with of POF, the role of autoimmunity remains controversial. Infections, such as mumps have also been suggested to cause oophoritis resulting in ovarian failure. Other minor causes include galactosaemia and enzyme deficiencies. For most of these known causes, the pathway of molecular events leading to the development of ovarian failure is still unclear. At a physiological level, POF is most likely to have arisen from either a decreased number of oocytes being formed during development, or an increased rate of oocyte atresia during reproductive life. It is likely that the various identified causes will affect one of these physiological processes, for example, to increase the rate of oocyte atresia.

Normal female reproductive function involves the interaction between a number of hormones. The pituitary gonadotropins, FSH and LH, are two key hormones of human reproduction. They consist of an alpha-subunit common to all glycoprotein hormones, and a specific beta-subunit. FSH is involved in the promotion of follicular growth, selection of preovulatory follicles and the stimulation of oestrogen synthesis. The FSH receptor (FSHR) is expressed solely in the granulosa cells, while the LH receptor (LHR) is expressed in theca and granulosa cells. LH stimulates androgen production by the thecal cells, and progesterone synthesis in luteinised granulosa cells. It triggers ovulation by inducing the rupture of the follicle wall. FSH stimulates aromatase activity in the granulosa cells facilitating conversion of androstenedione and testosterone from the theca cells into oestrogens.

Ovulation requires the coordination of FSH and LH action on the ovary. Defects in these hormones or their receptors could disrupt this process and therefore cause POF. Patients with POF have elevated levels of the gonadotropic hormones FSH and LH and decreased levels of oestrogen, as seen in normally menopausal women where negative feedback has ceased. It has been thought that resistance to the action of gonadotrophins might manifest itself as POF. The term "gonadotrophin resistance" is now referred to as "hormone resistance" and is characterised by mutations that block the action of the hormone receptors (Conway, 1997).

The FSHR was always considered to be a good candidate for mutations in women with POF. Studies in a group of Finnish families identified a loss of function mutation C566T in the FSH receptor (Aittomaki *et al.*, 1995) causing POF with primary amenorrhoea in the affected women. The histological appearance of the ovaries of women with the FSH receptor mutation showed the presence of primordial follicles, and it was concluded that this was a distinct form of POF (Aittomaki *et al.*, 1996). The effects of this mutation are much more dramatic for females than for males. Females have completely unresponsive ovaries whereas men who are homozygous for this mutation have variable and sometimes quite satisfactory gonadal function and fertility. However, it seems that the FSH receptor mutation is quite rare, as it has not been seen in any other populations (Layman *et al.*, 1993, Whitney *et al.*, 1995; Conway, 1997, Kohek *et al.*, 1998; Layman *et al.*, 1998). Mutations in the FSH gene are rare. A patient has been described with primary amenorrhoea and a mutation in her FSHβ gene (Matthews *et al.*, 1993). She was homozygous for the mutation. It was interesting to note that her mother was heterozygous for the mutation and suffered amenorrhoea and infertility. Mutations in FSHβ and FSHR cause mostly primary amenorrhoea, and demonstrate the necessity of normal FSH action for follicular maturation and ovarian function. Therefore genes involved in the FSH pathway are likely candidates for causing POF.

Hormone resistance has also been seen associated with defects in luteinising hormone (LH) and its receptor. Abnormal LH secretion can cause anovulation, premature oocyte maturation, and luteal insufficiency and can be seen in disorders such as polycystic ovary syndrome, recurrent miscarriage, and infertility (Liao *et al.*, 1998). A homozygous mutation in exon 3 of LHβ was demonstrated in a man with hypergonadotrophic hypogonadism (Weiss *et al.*, 1992) causing a substitution of Arg54Gln. However, this mutation does not affect any of the female carriers of the family, although no homozygotes for the mutation have been identified. Recently, another mutation in exon 3 was identified that caused the substitution Gly102Ser (Roy *et al.*, 1996). A subsequent study of 52 infertile women identified the mutation in two women, while no mutations were identified in the normal controls (Liao *et al.*, 1998). While the women in this study were infertile, none were identified as having POF. Other mutations in LHβ have been reported by other groups and found to be associated with primary or secondary infertility, however many of these were also present in normal women and are likely to be polymorphisms. Mutations have also been identified in the LHR in women with ovarian failure (Latronico *et al.*, 1996; Toledo *et al.*, 1996; Arnhold *et al.*, 1999).

This study is concerned with searching for mutations in two candidate genes, FSHR and LHβ, which have been suggested to cause premature ovarian failure. In particular we want to analyse specific mutations; the C566T mutation in FSHR, and G1502A mutation in LHβ.

Methods and Materials

Patient Samples

Patients with POF have been referred to our research group from the Auckland area. On referral, patients were interviewed by Northern Regional Genetic Services. Blood samples were taken after informed consent had been given, and genomic DNA was extracted. The DNA from familial POF patients from five different families, and from 14 other sporadic POF patients, was supplied to our laboratory.

DNA Extraction

Genomic DNA was extracted from 10 ml samples of blood. Lymphocytes were isolated from blood samples using the NYCOMED Lymphoprep™ Kit. Cells were incubated at 65°C for 1 hour with 3.5 ml 6 M $GuHCl_2$, 250 μl 7.5 M NH_4Ac, 50 μl 10 $mgml^{-1}$ Proteinase K and 250 μl 20% Na Sarcosyl.

Cells were added to 2 ml of cold $CHC1_3$, and then spun at 2000 rpm for 3 min. The top layer was collected and added to 10 ml of cold absolute ethanol to precipitate the DNA. DNA was stored in 200 µl TE buffer at 4°C.

Polymerase Chain Reaction (PCR)

The genes concerned were amplified from genomic DNA by PCR under sterile conditions. Sequence-specific primers were those used by Aittomaki *et al.*, (1995) and Roy *et al.* (1996), for the FSHR and LHβ sequences respectively. FSHR primers amplified a 78-base pair fragment from exon 7 of the FSHR gene, which codes for part of the extracellular ligand-binding domain of the receptor protein. LHβ primers gave a 395-base pair fragment from exon 3 of the LHβ-subunit. Genomic DNA (100 ng) was amplified in a 25 µl volume reaction containing 2.5 µl of PCR buffer (1x), 25 nmol of each dNTP, 5 nmol of forward and reverse primers (20 mM), and 0.125 µl Taq DNA polymerase. β-globin was used as positive control and a nil DNA reaction was used as a negative control for all PCR reactions, with a standard PCR program of 30 cycles and an annealing temperature of 55°C.

Restriction Fragment Length Polymorphism for Mutation Detection

Restriction enzymes were used for restriction fragment length polymorphism detection, to identify the two known point mutations in our amplified DNA fragments. In the normal (wild-type) allele, *Bsm* I enzyme digests the 78 bp FSHR fragment into 51 bp and 27 bp fragments, but the digestion site is abolished with the C566T mutation. Similarly, *Eco*0109 I enzyme digests the 395 bp LHβ fragment into 269 bp and 126 bp fragments in the wildtype allele, but this digestion site is abolished with the G1502A mutation. Five microlitres of amplified FSHR DNA was digested for one hour at 65°C with 0.5 µl *Bsm* 1, 2 µl 10x restriction buffer, and 12.5 µl sterile water. Five microlitres of amplified LHβ DNA was digested for one hour at 37°C with 0.5 µl *Eco*0109 I, 0.2 µl BSA, 2 µl 10x restriction buffer, and 12.5 µl sterile water. A negative control reaction (all reagents except restriction enzyme) was also performed, and this provided an undigested marker when digested products were viewed after electrophoresis.

The digestion products were electrophoresed to analyse DNA fragments. The LHβ fragments were run on a 2% agarose gel containing 8 µl of 10 mg/ml ethidium bromide, and photographed with a Kodak Digital camera under UV light. The FSHR fragments were too small for visualisation on an agarose gel, and were run on an 8% polyacrylamide gel, and visualised by silver-staining.

Results

Among our five familial POF patients and fourteen sporadic POF patients, none were found to have either the FSHR exon 7 C566T mutation or the LHβ exon 3 G1502A mutation. Optimisation experiments were performed to ensure that we had complete digestion of the PCR products. Restriction enzyme

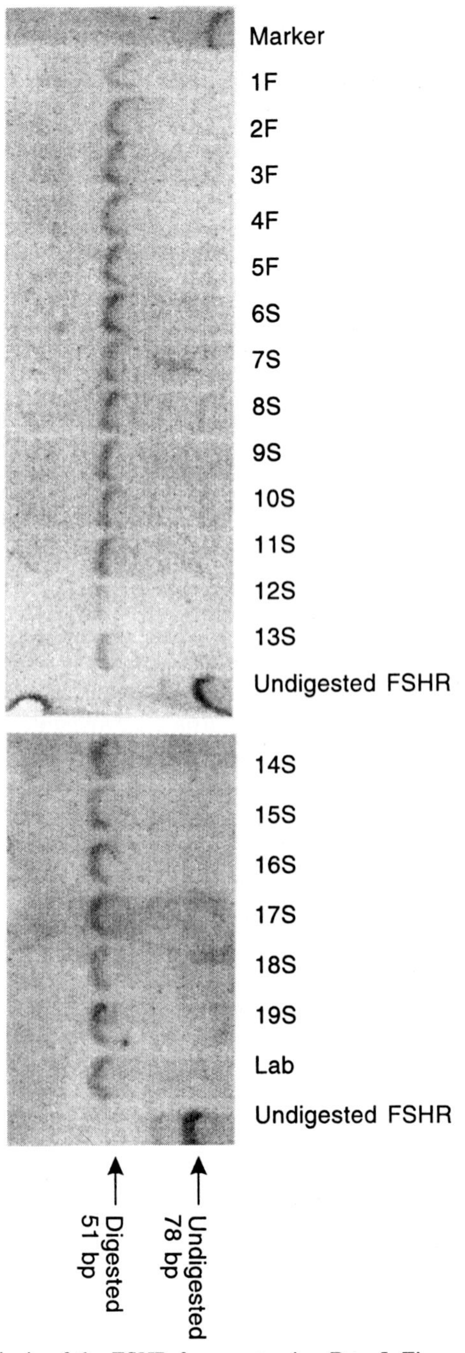

Fig. 1 RFLP analysis of the FSHR fragment using Bsm I. Figure shows the digestion products of the FSHR fragment using Bsml. Undigested DNA gives a single band at 78 bp. Wild-type (WT) yields two bands of 51 bp and 27 bp. Homozygosity for the variant yields a single fragment of 78 bp. A heterozygote carrier will have all three fragments (78 bp, 51 bp and 27 bp). The fragments were run on an 8% polyacrylamide gel, and visualised by silver staining. Familial patients are indicated (F), sporadic patients are indicated (S). Lab refers to a normal control sample.

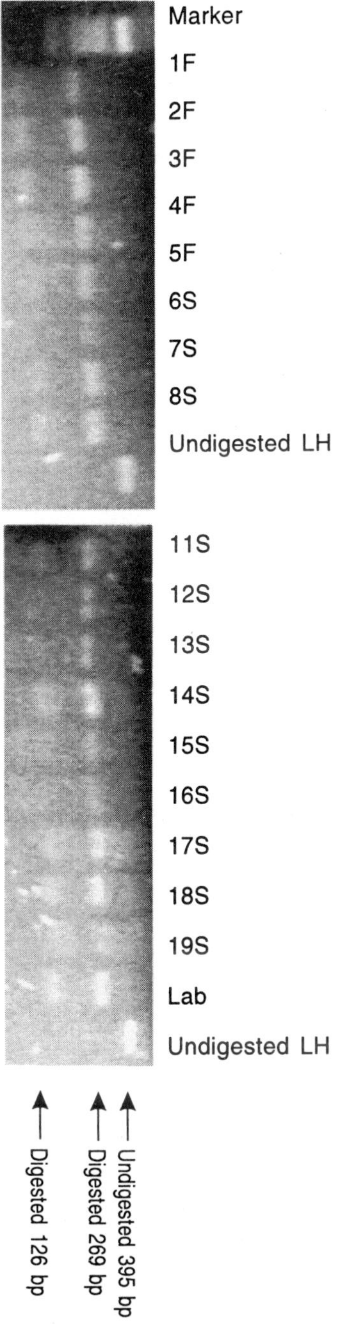

Fig. 2 RFLP analysis of the LHβ fragment using Eco0109I. Figure shows the digestion products of the LHβ fragment using Eco0109I. Undigested DNA gives a single band at 395 bp. Wild-type (WT) yields two bands of 269 bp and 126 bp. Homozygosity for the variant yields a single fragment of 395 bp. A heterozygote carrier will have all three fragments (395 bp, 269 bp and 126 bp). The smaller 126 bp fragment has run off the bottom of this gel, and is therefore not visible. The fragments were run on a 2% agarose gel, and visualised by ethidium bromide staining. Familial patients are indicated (F), sporadic patients are indicated (S). Lab refers to a normal control sample.

digestion in all cases gave two bands smaller than the single uncut DNA band. A shift from the level of the uncut DNA band indicated digestion and thus absence of a mutation. Digested FSHR gave two bands of 51bp and 27bp (uncut size 78bp), and digested LHβ gave two bands of 269bp and 126bp (uncut size 395bp) (Figs. 1 and 2). With complete digestion we did not see any patients with three bands (indicating heterozygosity for a mutation) or a single uncut DNA band (indicating homozygosity for a mutation).

Discussion

POF affects 1% of women and as such, is the topic of much research worldwide. Polymorphisms and mutations linked to POF have been found at several different loci, but there remains debate over the relevance of these mutations versus other exogenous causes of POF.

A mutation in the FSHR gene has been found by Aittomaki *et al.*, (1995) in a population of Finish women with primary ovarian failure. These women have hereditary hypergonadotropic primary amenorrhoea, and their ovaries are streak or hypoplastic type with primordial follicles visible on ultrasound (Aittomaki *et al.*, 1996). This phenotype is caused by a single point mutation in exon 7 of FSHR on chromosome 2p. The C566T mutation changes an Alanine to a valine at codon 189 of the FSHR extracellular ligand-binding domain (Aittomaki *et al.*, 1995). This mutation results in significantly reduced binding capacity and signal transduction of the receptor. While this mutation affects approximately 40% of patients with hypergonadotrophic ovarian failure in Finland (Huhtaniemi and Aittomaki, 1998), to date, it has not been found in other POF patients (Layman *et al.*, 1993, Whitney *et al.*, 1995; Conway, 1997; Kohek *et al.*, 1998; Layman *et al.*, 1998). Our results show no evidence of this mutation in a group of New Zealand women with POF. These results indicate that this point mutation is due to a founder effect in the genetically isolated Finnish population (Aittomaki *et al.*, 1996). Several other studies have identified the presence of FSHR polymorphisms in both POF patients and normal controls, but these appear to have no effect on phenotype (Whitney *et al.*, 1995; Liu *et al.*, 1998).

Examination of the LHβ-subunit gene led to the identification of several new mutations in this gene (Roy *et al.*, 1996; Liao *et al.*, 1998). Many of these are polymorphisms that either have no effect on coding or were not associated with female infertility (Roy *et al.*, 1996). However, a G1502A point mutation in exon 3 was found to be significant in its effect, and was found in two women with endometriosis associated infertility (Liao *et al.*, 1998). The women had differing endocrinological profiles, and one experienced primary infertility while the other experienced secondary infertility. The mutation changes amino acid 102 from glycine to serine, at an amino acid which is normally highly conserved in both humans and other animals. It creates a hydrophilic amino acid where there is normally a hydrophobic residue, and thus may affect molecular conformation and activity (Liao *et al.*, 1998). The exact role of LH in endometriosis is undetermined, but given this hormone's key role in the female reproductive system, it seems a good candidate for involvement in infertility and POF. However, our screen of 19 POF patients failed to find any women with this mutation.

In summary, many putative genetic mutations are under investigation as the search for causes of premature ovarian failure continues. Our study did not find known mutations of the FSHR gene and the LHβ gene in any of our 5 familial and 14 sporadic POF patients. This suggests that these particular mutations may not be commonly involved in POF. As these genes are central in reproduction, it is possible that mutations are not well tolerated, as they will affect fertility. They are likely to remain as sporadic mutations, and will be effectively lost from the gene pool in single generation. Premature ovarian failure is a heterogenous condition and it is likely that mutations in a number of different genes will eventually be found.

Acknowlegements

We would like to thank staff at Fertility PLUS and clinicians for providing patients for this study. Funding was provided by Allied Foods, the University of Auckland Research Committee, the Health Research Council of New Zealand, and the Auckland Medical Research Foundation.

References

Aittomaki, K., Lucena, J. L., Pakarinen, P. *et al.* (1995) Mutation in the follicle-stimulating hormone receptor gene causes hereditary hypergonadotropic ovarian failure. *Cell.* **82**, 959–68.

Aittomaki, K., Herva, R., Stenman, U.H. *et al.* (1996) Clinical features of primary ovarian failure caused by a point mutation in the follicle-stimulating hormone receptor gene. *Journal of Clinical Endocrinology & Metabolism.* **81**, 3722–6.

Anasti, J.N. (1998) Premature ovarian failure-an update. *Fertility & Sterility.* **70**, 1–15.

Arnhold, I.J.P., Latronico, A.C., Batista, M.C. and Mendonca, B.B. (1999) Menstrual disorders and infertility caused by inactivating mutations of the luteinising hormone receptor gene. *Fertility & Sterility.* **71**, 597–601.

Conway, G.S. (1996) Clinical manifestations of genetic defects affecting gonadotrophins and their receptors. *Clinical Endocrinology.* **45**, 657–63.

Conway, G.S. (1997) Premature ovarian failure. *Current Opinion in Obstetrics & Gynaecology*, **9**, 202–206.

Conway, G.S., Payne, N.N., Webb, J. *et al.* (1998) Fragile X premutation screening in women with premature ovarian failure. *Human Reproduction.* **13**, 1184–1187.

Coulam, C.B., Adamson, S.C. and Annegers, J.F. (1986) Incidence of premature ovarian failure. *Obstetrics & Gynaecology.* **67**, 604–6.

Devi, A.S., Metzger, D.A., Luciano, A.A. *et al.* (1998) 45,X/46,XX mosaicism in patients with idiopathic premature ovarian failure. *Fertility & Sterility.* **70**, 89–93.

Kalantaridou, S.N., Davis, S.R. and Nelson, L.M. (1998) Premature ovarian failure. *Endocrinology & Metabolism Clinics of North America.* **27**, 989–1006.

Kohek, M.B.D., Batista, M.C., Russell, A.J. *et al.* (1998) No evidence of the inactivating mutation (C566T) in the follicle-stimulating hormone receptor gene in Brazilian women with premature ovarian failure. *Fertility & Sterility.* **70**, 565–567.

Latronico, A.C., Anasti, J., Arnohold, I.J. *et al.* (1996) Brief report: testicular and ovarian resistance to luteinizing hormone caused by inactivating mutations of the luteinizing hormone-receptor gene. *New England Journal of Medicine* **334**, 507–12.

Layman, L.C., Shelley, M.E., Huey, L.O. *et al.* (1993) Follicle-stimulating hormone beta gene structure in premature ovarian failure. *Fertility & Sterility.* **60**, 852–7.

Layman, L.C., Amde, S., Cohen, D.P. *et al.* (1998) The finish follicle-stimulating hormone receptor gene mutation is rare in North American women with 46, XX ovarian failure. *Fertuity & Sterility.* **69**, 300–302.

Liao, W.X., Roy, A.C., Chan, C. *et al.* (1998) A new molecular variant of luteinizing hormone associated with female infertility. *Fertility & Sterility.* **69**, 102–6.

Liu, J.Y., Gromoll, J., Cedars, M.I. *et al.* (1998) Identification of Allelic Variants in the Follicle-Stimulating Hormone Receptor Genes of Females With or Without Hypergonadotropic Amenorrhea. *Fertility & Sterility.* **70**, 326–331.

Matthews, C.H., Borgato, S., Beck-Peccoz, P. *et al.* (1993) Primary amenorrhoea and infertility due to a mutation in the beta-subunit of follicle-stimulating hormone. *Nature Genetics.* **5**, 83–6

Partington, M.W., Moore, D.Y. and Turner, G.M. (1996) Confirmation of early menopause in fragile X carriers. *American Journal of Medical Genetics.* **64**, 370–2.

Roy, A.C., Liao, W.X., Chen, Y. *et al.* (1996) Identification of seven novel mutations in LH β-subunit gene by SSCP. *Mol Cell Biochem.* **165**, 151–153.

Toledo, S.P., Brumner, H.G., Kraaij, R. *et al.* (1996) An inactivating mutation of the luteinizing hormone receptor causes amenorrhea in a 46,XX female. *Journal of Clinical Endocrinology & Metabolism.* **81**, 3850–4.

Weiss, J., Axelrod, L., Whitcomb, R.W. *et al.* (1992) Hypogonadism caused by a single amino acid substitution in the β subunit of luteinizing hormone. *The New England Journal of Medicine.* **326**, 179–83.

Whitney, E.A., Layman, L.C., Chan, P.J. *et al.* (1995) The follicle-stimulating hormone receptor gene is polymorphic in premature ovarian failure and normal controls. *Fertility & Sterility.* **64**, 518–24.

Follicular Growth, Ovulation and Fertilization: Molecular and Clinical Basis
Anand Kumar and Amal K. Mukhopadhyay (Eds.)
Narosa Publishing House, New Delhi, India, 2001

17

Clinical and Endocrine Changes Following Medical (GnRH Analogue) and Surgical (Laparoscopic Electrocoagulation or Laser Vaporisation) Treatment in Patients with Polycystic Ovary Syndrome (PCOS)

A. Szilágyi, J. Homoki*, W.G. Rossmanith and I. Szabo**
Department of Obstetrics and Gynecology, University of Pécs, Hungary
*Department of Pediatrics and **Department of Obstetrics and Gynecology, University of Ulm, Germany

Introduction

Polycystic ovary syndrome (PCOS) is probably the most prevalent endocrinopathy in women and the most common cause of anovulatory infertility (Adams *et al.*, 1986). The pathophysiology of the disorder has been thoroughly investigated, but its etiology is still unsettled. There are theories supporting a primary hypothalamic-pituitary defect, a primary ovarian steroidogenic defect, a primary adrenal steroidogenic defect and a primary defect of insulin resistance (Dunaif, 1993, Ehrmann *et al.*, 1995, Homburg, 1996, Szilagyi and Rossmanith, 1991). Whatever the pathogenesis of PCOS, the endpoint is an ovary secreting excessive amount of androgens.

PCOS has been defined clinically, biochemically, and by ultrasound. The diagnosis of PCOS is based on proving hyperandrogenemic anovulation and established by transvaginal ultrasound examination of the ovaries characterizing the ovarian morphology. According to the criteria of Adams *et al.*, (Adams *et al.*, 1985), polycystic ovaries (PCO) should be diagnosed when more than eight discrete follicles of <10 mm diameter are seen in the ovary, usually peripherally arrayed around an enlarged, hyperechogenic, central stroma. Most of the patients with PCO have a clinical or biochemical feature consistent with the ultrasound diagnosis and they are likely to face the problems of hyperandrogenism, subfertility and recurrent miscarriage.

There is no single effective treatment regimen for PCOS. The treatment depends on what principal complaint the patient has. It may be infertility, recurrent pregnancy loss or just excessive hair growth with or without menstrual cycle disorders.

The present study consists of two main parts. On one hand the, clinical and hormonal effects of long term gonadotropin-releasing hormone (GnRH) analogue treatment was studied in hirsute PCOS patients and on the other hand infertile, clomiphene citrate resistent PCOS patients were surgically treated to achieve resumption of the menstrual cycle and fertility. Hormonal changes before and after surgery were evaluated, too.

GnRH Analogue Treatment of Hyperandrogenemic PCOS Patients

GnRH analogues exert a profound and prolonged suppression of pituitary gonadotropin secretion which accounts for the marked suppression of ovarian function. Short term studies performed on PCOS patients utilizing a short-acting GnRH agonist for 4 weeks proved marked suppression of ovarian steroidogenesis (Chang *et al.*, 1983). Later on, it became evident that long term GnRH analogue treatment should be considered among the various therapeutic modalities offered in the context of hirsutism (Adashi, 1990). Several studies have aimed to measure serum gonadotropin and androgen levels during GnRH analogue treatment in PCOS (Rittmaster, 1993), but the efficacy of gas chromatographic profiling of urinary steroids in monitoring GnRH analogue therapy has not been studied yet, although the effect of GnRH analogue stimulation test on urinary steroid excretion was reported in a study (Luppa *et al.*, 1995). Furthermore, gas chromatography is an effective method to assess increased 5α-reductase enzyme activity characteristic for PCOS (Steward *et al.*, 1990). Capillary gas chromatographic profiling of urinary steroids may also be used to detect late onset steroid 21-hydroxylase deficiency, the most frequent enzyme deficiency that causes hirsutism, and that should be differentiated from PCOS (Homoki *et al.*, 1988).

Patients and Methods

A long acting GnRH analogue (Decapeptyl Depot® -Ferring, Germany) was administered as monthly intramuscular injection for 6 months in 8 PCOS patients. Clinical and hormonal effects were measured. Serum LH, FSH, prolactin, testosterone, estradiol levels were determined monthly and profiling urinary steroids by column gas chromatography of twenty four hour urine samples was performed before and in the third and sixth month of treatment according to Shackleton and Homoki (Shackleton, 1986, Homoki *et al.*, 1992). To evaluate 5α-reductase enzyme activity in the liver and skin of the patients, the ratios of androsterone (An) to etiocholanolone (Et) and 5α-tetrahydrocortisol (a-THF) to tetrahydrocortisol (THF) were calculated in urine samples before and after the GnRH analogue therapy. Changes in the ratio of androgen metabolites (AM) to cortisol metabolites (CM) during treatment were evaluated, too. Degree of hirsutism was assessed before and after treatment by Ferriman Gallwey score (Ferriman and Gallwey, 1961). Bone mass density before and after therapy was measured by DEXA (dual energy X-ray absorptiometry).

Results

LH has decreased significantly following the first injection of the GnRH analogue (from 11.9 ± 2.9 to 1.05 ± 0.5 U/1) (Fig. 1). FSH and prolactin levels have not changed during treatment. Testosterone levels were normalized (<3 nmol/1), estradiol was suppressed near to postmenopausal levels (50–100 pmol/1).

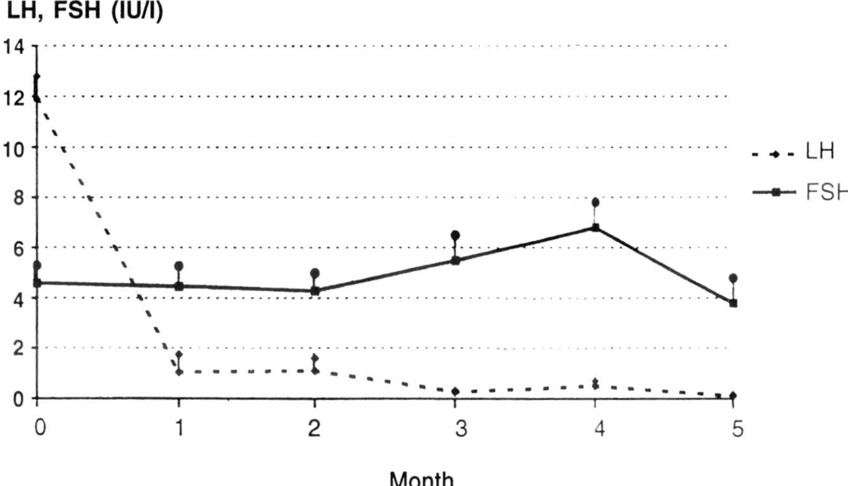

Fig. 1 Changes in circulating LH and FSH levels during 6 months of GnRH analogue (Decapeptyl Depot) treatment (n = 8).

An/Et and a-THF/THF ratios in the urine samples have decreased significantly during therapy (An/Et from 2.1±0.25 to 1.6±0.2, a-THF/THF from 1.1±0.1 to 0.8±0.1), but they were still higher than that of in healthy females (normal ratios: An/Et 1, a-THF/THF 0.6). The elevated ratio of androgen metabolites to cortisol metabolites characteristic of PCOS has decreased during treatment to normal values (AM/CM from 0.8±0.2 to 0.5±0.1) (Fig. 2).

Clinically, all the PCOS patients became amenorrheic during GnRH agonist treatment, and 6 out of the 8 patients complained of mild hot flushes. Hirsutism assessed by Ferriman Gallwey score was diminished, but not significantly. Decrease in bone mass density was within normal ranges, and it has not reached pathological osteopenia.

Conclusions

Long acting GnRH analogue treatment in PCOS is effective in reducing serum LH, estradiol and androgen levels. Gas chromatographic profiling of urinary steroids by using specific ratios is a sensitive tool for monitoring changes in steroid hormone production during GnRH analogue administration. Use of specific ratios may help to avoid errors of urine collection. Increased 5α-reductase enzyme activity characteristic for PCOS (Steward *et al.*, 1990) is diminished during treatment, but it remains still higher than that of in healthy females. Production of androgen metabolites is reduced, too.

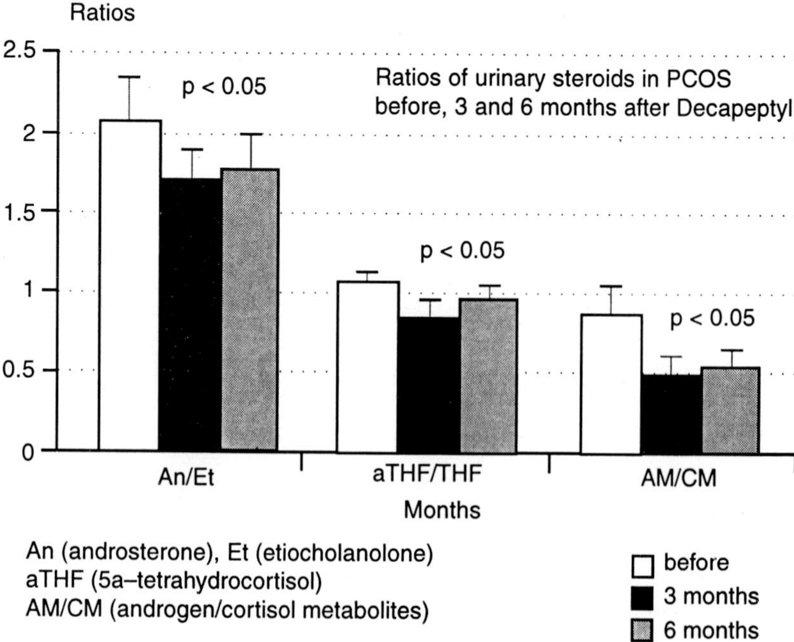

Fig. 2 Changes in the ratios of urinary steroids (androsterone/etiocholanolone, 5α-tetrahydrocortisol/tetrahydrocortisol, androgen metabolites/cortisol metabolites) in PCOS patients during GnRH analogue (Decapeptyl Depot) treatment (*n*=8).

Degree of hirsutism improves during GnRH analogue administration, but for a significant improvement probably a longer or combined treatment would be required. In a long term GnRH analogue treatment, DEXA may help to select those patients with decreased bone mass density, who may need low dose estrogen or oral contraceptive add back therapy to prevent side effects of estrogen deficiency. Addition of estrogens and progestins to GnRH analogue therapy may result in reduction of bone loss and further improvement of hirsltism, too (Carr *et al.*, 1995, Lemay and Faure, 1994).

Surgical Treatment of PCOS Associated Infertility

For the treatment of infertility associated with PCOS ovulation induction is the most appropriate treatment. Clomiphene citrate in doses of 50–200 mg/day from day 4 or 5 of the cycle for 5 days may be the simplest method to achieve ovulation in 80% of the cases, however, the overall pregnancy rate is only 30–40%. This discrepancy may partly be explained by the antiestrogenic

Gjönnaess, H. (1984) Polycys
through the laparoscope. *F*
Gjönnaess, H. (1990) A simple
Forum, **11**, 214–217.
Homburg, R. (1996) Polycys
multisystem endocrinopath
Homoki, J., Solyom, J., Teller, V
deficiency by capillary gas
and adolescents. *Eur. J. Pe*
Homoki, J., Solyom, J., Wach
pregnenolones in patients w
to steroid 21-hydroxylase c
Keckstein, G., Rossmanith, W.
treatment of polycystic ovar
4, 103–107.
Kovacs, G., Buckler, H., Ban
polycystic ovarian syndrom
Gynecol., **98**, 30–35.
Lemay, A., Faure, N. (1994) S
releasing hormone agonis
hyperandrogenism: A pilot
Luppa, P., Muller, B., Jacob, K
in serum and urine in polycy
for an altered corticoid exc
Rittmaster, R.S. (1993) Use of
of hyperandrogenism. *Clin.*
Rossmanith, W.G., Keckstein,
surgery on the gonadotropi
Endocrinol., 34, 223–230.
Shackleton, C.H.L. (1986)
Chromatography, **379**, 91–
Shoham, Z., Borenstein, R.,
following clomiphene citrat
Endocrinol., **33**, 271–278.
Stein, I.F., Cohen, M.R. (1939
sterility. *Am J. Obstet. Gyn*
Steward, P.M., shackleton, C.H
polycystic ovary syndrome.
Szilágyi, A., Rossmanith, W.
hormonal levels after ovar
syndrome. *Arch. Gynecol.*
Szilágyi, A., Rossmanith, W.G
periphere Regulations-storu
Szilágyi, A., Hole, R., Keckst
dopaminergic and opioider
women with polycystic ova

effect of clomiphene citrate on the cervical mucus and endometrium, and the increased LH concentrations during clomiphene citrate action that may seriously compromise pregnancy rates in these patients (Homburg, 1996; Shoham *et al.*, 1990). If clomiphene treatment fails to induce ovulation ("clomiphene failures") the next step may be stimulation of the ovaries with exogenous gonadotropins or surgical manipulation of the ovaries.

Surgical treatment, originally by bilateral wedge resection of the ovaries (Stein and Cohen, 1939), or later by laparoscopic approaches (electrocautery, laser photodiathermy, laser drilling) is found to be effective in restoring ovulation for at least 6 months, but the benefit of monopolar current may last for several years (Gjönnaess, 1984; Daniell and Miller, 1989, Keckstein *et al.*, 1990). As a consequence of profound changes in the gonadotropin secretion in response to ovarian surgery, resumption of normal menstrual cyclicity with subsequent ovulation has been observed (Szilágyi *et al.*, 1990, Rossmanith *et al.*, 1991). Although wedge resection by laparotomy has not been performed for 15–20 years, our data are available on the hormonal effects of this procedure. The aim of the present study was to compare the hormonal and clinical effects of ovarian wedge resection, electrocoagulation and laser vaporisation in patients with polycystic ovary syndrome.

Patients and Methods

The diagnosis of PCOS was based on the clinical manifestations of chronic hyperandrogenemia, such as menstrual disturbances (oligo- to amenorrhea), hirsutism and infertility. Endocrinological features comprised elevated serum LH with normal or subnormal FSH concentrations and the finding of elevated serum testosterone, androstenedione and/or dehydroepiandrosterone sulfate (DHEAS) levels. The presence of enlarged ovaries with multiple subcapsular cysts was sonographically confirmed in all PCOS women. All patients had failed to ovulate in response to clomiphene citrate regimens (100 mg/day for 5 days) administered for at least 3 months.

Nine patients underwent classical wedge resection of the ovaries, 11 patients were treated by laparoscopic Nd: YAG laser vaporisation (contact approach for the laser beam, continuous wave mode), and in 56 patients laparoscopic multiple point electrocoagulation of the ovaries by monopolar current was performed.

Results

Changes in serum LH and FSH before and after the different surgical modalities are summarised in Figs. 3 and 4. A significant decrease in serum LH five days after surgery could be achieved only after the laparoscopic approaches (laser or electrocoagulation). On the other hand, laparoscopic procedures induced an increase in serum FSH that may be favourable in initiating menstrual cyclicity.

Serum testosterone levels before and after surgery are shown in Figure 5. Serum testosterone has decreased five days after surgery in some wedge resected patients, but a pronounced decrease could be achieved only by the laparoscopic methods. Changes in the pulsatile pattern of LH and FSH were

secretion (Szilágy
abnormalities in
A further adv
stimulation with
ovarian response c
et al., 1991).

Summary anc

Finally, we may
successfully be r
requirements of t
be treated by long
be combined wit
to reduce the risk
Genazzani *et al.*,
combined therapy
of the ovaries are
The laparoscopic
conventional ovu

References

Adams, J., Franks,
endocrine feature
ii, 1375–1378.
Adams, J., Polson,
with anovulation
Adashi, E.Y. (1990)
management of
Carr, B.R., Breslau,
releasing hormo
clinical research
Chang, R.J., Laufer
ovarian disease
hormone agonis
Daniell, J.F., Miller,
Fertil. Steril., **51**
Dunaif, A. (1993) In
687, 60–64.
Ehrmann, D.A., Ba
a form of functi
secretion. *Endoc*
Ferriman, D., Gallw
J. Clin. Endocri
Genazzani, A.D., Ga
improves the rec
Endocrinol., 10,

Follicular Growth, Ovulation and Fertilization: Molecular and Clinical Basis
Anand Kumar and Amal K. Mukhopadhyay (Eds.)
Narosa Publishing House, New Delhi, India, 2001

18

Induction of Ovulation in Hypergonadotropic Amenorrhoea

P.K. Meherji, M. Desai and S. Kadam
Institute for Research in Reproduction (ICMR)
Parel, Mumbai-400 02, India

Introduction

Ovarian failure either primary or premature is considered an irreversible entity and attempts to induce ovulation in these women is a frustrating task. Though, there have been number of isolated case reports documenting ovulation or pregnancies in women with hypergonadotropic amenorrhoea either during or immediately after estrogen replacement therapy (Shangold *et al.*, 1977), our attempts in the past at inducing ovulation or achieving pregnancies by use of exogenous estrogens has yielded poor results.

Recently, successful ovulation and pregnancies have been reported following ovulation induction with human menopausal gonadotropins during or after serum gonadotropin suppression by exogenous estrogens or GnRH analogues in women with apparent menopause (Szlachter *et al.*, 1979; Check *et al.*, 1989; Check *et al.*, 1984; Surrey and Cedar, 1989). We report here our attempts at ovulation in some of these women with elevated gonadotropin levels using estrogens to suppress endogenous gonadotropins and clomiphene citrate (CC) to induce ovulation.

Materials and Methods

Twelve women with elevated gonadotropins and under 35 years of age were studied. All except one were married and desired pregnancy. Serum FSH in all were beyond 2000 ng/ml and endogenous estradiol deficiency was evident by their failure to have withdrawal bleeding with medroxy progesterone acetate (MPA) (10 mg × 5 days). Routine investigations including semen analysis in males and laparoscopic examination of tubes in females were normal in eleven infertile couples. Ovaries were either small or normal in size in all the cases. Ovarian biopsies were not performed.

Two suppression regimes were used. Seven patients assigned to Group I received cyclic therapy with ethynyl estradiol (EE) given orally in doses of

0.05 mg/day for 20 days and MPA 10mg/day during last 5 days of estrogen administration. The therapy was continued till adequate pituitary suppression was achieved. This was defined as a level of serum FSH < 20% of the basal level obtained on 14th day of estrogen therapy prior to progestin administration. Steroids were stopped and clomiphene citrate was administered following withdrawal bleeding. Group II (n = 5) patients received estrogen therapy in similar manner. But once the FSH suppression was achieved EE was continued and CC was administered simultaneously to avoid rebound of gonadotropins to presuppressional levels. Follicular development was monitored by serial pelvic ultrasonography and ovulation documented by estimation of pregnanediol glucuronide (PdG) in urine (Khatkhatay *et al.*, 1987).

Results

Characteristics of patients studied and clinical responses to their therapy are shown in Tables 1 and 2. Their ages ranged from 21–35 years with a mean age of 29.8 years. Following estrogen therapy elevated baseline FSH levels were reduced in each group to within follicular range; the suppression being achieved after 2 weeks of drug therapy in the majority. The patients in both the groups were given clomiphene citrate (CC) in doses upto 150 mg/day.

None of the patients in Group I showed evidence of growing follicles on sonography nor an ovulatory level of urinary PdG. However, small follicles of sizes 13–16 mm probably resting follicles were seen only in one individual.

Amongst the patients of ovarian failure in Group II, follicular development and rupture was observed on sonography in four out of five individuals following continuous gonadotropin suppression and clomiphene stimulation while the urinary PdG levels suggested ovulation in two with corpus luteum deficiency (CLD) in one of them. Of the two, one of them showed follicular growth only when stimulated with a higher (150 mg/day) dose of the drug. In the other three individuals though the follicular growth was observed, the PdG levels were anovulatory and the menses had to be induced.

Discussion

Successful ovulation and pregnancies following endogenous suppression and ovulatory induction have been reported by (Check *et al.*, 1990). The possible explanation given for such a mode of therapy is that clinically elevated gonadotropin levels in women with premature ovarian failure down regulate granulosa cell FSH receptors rendering resting follicles refractory to exogenous stimulation therefore pretreatment with estrogens allowed regeneration of ovarian FSH receptors and results in a positive response to ovulation inducers. Based on this assumption, ovulation induction was attempted in women with hypergonadotropic amenorrhoea using clomiphene citrate instead of human menopausal gonadotropins as the drug is cheaper, more readily available and if successful would serve as an economical alternative.

Table 1 Characteristics of patients : Group 1

Subjects	*Age*	*Duration of Amenorrhoea*	*Serum FSH ng/ml*		*Ovarian size at*	*Therapy offered*	*U.S. Findings*	*Response to C.C.*
			Pre–E	*Post–E*				
1.	32	1 Year	2620	304	Normal	C.C. 50 mg/d	No follic.growth	Anov.
2.	30	11 Years	2689	127	Small	C.C. 50 mg/d	No follic.growth	Anov.
3.	27	1 Year	2258	406	Small	C.C. 50 mg/d	No follic.growth	Anov.
4.	35	8 Years	3200	340	Small	C.C. 50 mg/d	No follic.growth	Anov.
5.	31	2 Years	2133	204	Small	C.C. 50 mg/d	Follicles of 13 to 16 mm which regressed	Anov.
6.	29	6 Years	2100	403	Small	C.C. 50 mg/d	No follic. growth	Anov.
7.	27	1 Year	2000	186	Small	C.C. 50 mg/d	No follic. growth	Anov.

Anov = Anovulation; CLD = Corpus luteum deficiency; CLA = Corpus luteum adequacy.

Table 2 Characteristics of patients: Group II

Subjects	*Age*	*Duration of Amenorrhoea*	*Serum FSH ng/ml*		*Ovarian size at*	*Therapy offered*	*U.S. Findings*	*Response to C.C.*
			Pre–E	*Post–E*				
1.	35	5 Months	2765	223	Small	E.E. + C.C. 50 mg/d	Follicular growth and rupture	Anov.
2.	34	2 Years	2592	225	Small	E.E + C.C. 50 mg/d	Follicular growth and rupture	Anov.
3.	22	2 years	2839	236	Normal	E.E. + C.C. 50 mg/d	Follicular growth and rupture	Ovul. with CLD
						E.E. + C.C. 100 mg/d	No follicular development	Anov.
4.	21	6 Years	3200	140	Small	E.E. 50 mg/d E.E. 100 mg/d	No follicular development	Anov.
						E.E. 150 mg/d	Follicular growth and rupture	Ovul. with CLA
5.	34	1 Year	>2000	343	Normal	E.E. + C.C. 50 mg/d	No follicular development	Anov.

Anov = Anovulation; CLD = Corpus luteum deficiency; CLA = Corpus luteum adequacy.

In a group of hypergonadotropic women pretreated with estrogens (Group I) no ovulation was observed following clomiphene citrate administration. Hence, it was decided to continue estrogen administration alongwith clomiphene citrate to prevent a rebound of gonadotropins to baseline levels and subsequent receptor down regulation. With this mode of therapy, follicular growth and rupture was observed in four of the 5 individuals. However, true ovulation as determined by estimation of pregnanediol glucuronide levels was observed in two only i.e. 40%. This rate is comparable to 12-60% as reported by others (Tanaka *et al.*, 1982; Check *et al.*, 1990). To our knowledge, CC has not been tried so far and though ovulation rate is low, we believe ovulation with this drug should be tried instead of more expensive HMG therapy in women with normal or smaller looking ovaries but not with streak ovaries, since they may have some follicular stock deep in the parenchyma that would respond to such a mode of therapy and raise some hopes in the otherwise hopeless cases.

References

Check, J.H. and Chase, J.S. (1984). Ovulation induction in hypergonadotropic amenorrhoea with estrogen and human menopausal gonadotropin therapy. *Fertil. Steril.* **42**: 919–922.

Check J.H., Chase, J.S. and Spence, M. (1989). Pregnancy in premature ovarian failure after therapy with oral contraceptives despite resistance to previous human menopaussal gonadotropin therapy. *Am. J. Obstet. Gynaec.* **160**: 114–115.

Check, J.H., Nowroozi, K., Chase, J.S. *et al.*, (1990). Ovulation induction and pregnancy in 100 consecutive women with hypergonadotropic amenorrhoea. *Fertil. Steril.* **53**: 811–816.

Khatkhatay, M.I., Sankolli, G.M., Meherji, P.K. *et al.*, (1987). Application of penicillinase linked ELISA of pregnanediol glucuronide for detection of ovulation and assessment of corpus luteum function. *Endocrinol. Japonica* **34(3)**: 465–472.

Shangold, M.M., Turksey, R.N., Bashford, R.A. *et al.* (1977). Pregnancy following the "insensitive ovary syndrome". *Fertil. Steril.* **28**, No. 11: 1179–1181.

Surrey, E.S. and Cedar, M.I. (1989). The effect of gonadotropin supression on the induction of ovulation in premature ovarian failure patients. *Fertil. Steril.* **52**: 36–41.

Szlachter, B.N., Nachtigall, L.E., Epstein, J. *et al.*, (1979). Premature menopause: A reversible entity. *Obstet. Gynecol.* **54**: 396–398.

Tanaka, T., Sakuragi, N., Fujimoto, S. *et al.*, (1982). HMG-HCG therapy in patients with hypergonadotropic ovarian anovulation: One pregnancy case report and ovulation and pregnancy rate. *Int. J. Fertil.* **27**: 100–102.

Assisted Reproduction

Follicular Growth, Ovulation and Fertilization: Molecular and Clinical Basis
Anand Kumar and Amal K. Mukhopadhyay (Eds.)
Narosa Publishing House, New Delhi, India, 2001

19

Tools for Quality Assessment and Quality Management in Reproductive Techniques in Germany

M.S. Kupka, O. Richter and H. van der Ven
Department of Gynecological Endocrinology and Reproductive Medicine
University-Clinic Bonn, Germany

Introduction

Tools for quality assessment and quality management are integrated into German public health system. Minimized budgets for national health services force to lower costs even in reproductive medicine (Hamberger and Sjoqvist, 1994).

For medical students, subjects like health-management and medical informatics became a normal subject in their studies (Siekmann, 1989).

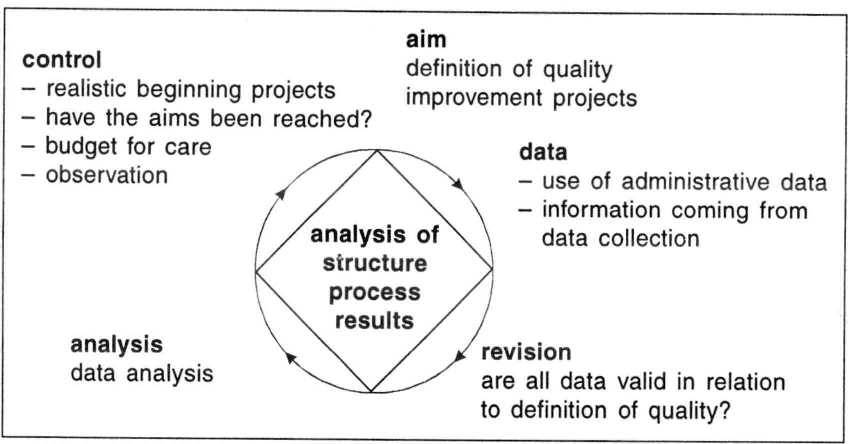

Fig. 1 Mechanisms of quality management.

Managed care and the rise of consumer groups has forced health care providers and organization to begin analysing how and why they deliver care to patients. The incorporation of quality improvement tools such as evidence-based practice guidelines have been shown to reduce practice variability, improving the process and outcomes of care. So the correct development and implementation of these guidelines can be an effective tool to assist the gynaecologist in improving the care delivered to the patients.

Quality of care and costs are getting closer together. Whereas costs and quality management did not play a substantial role 30 years ago, the consumption of resources nowadays is part of the outcome of quality of care. The definition of quality must be seen in the dimensions of structure, process and result. Resulting from newly developed clinical practice, guidelines are planned as instruments for cost containment in near future. Those guidelines may end up in a quality management system. The most well known basic of such a quality management system are the DIN EN ISO 9000 ff. and the EFQM. The ISO 9001 and the European Quality Award became the most common base of evaluation for certification of quality management systems in Europe-wide organisations. Whereas the ISO 9001 dose not give any information about the real achieved quality, the European Quality Award reflects the process. Guidelines are necessary to prove the cost effectiveness of measures of quality control and quality assurance since too much quality control and assurance may result in increased overall consumption of resources, leading to a reduction in the quality of care when ensuring that the overall budget is covered.

The management of analytical quality also depends on the careful evaluation of the imprecision and inaccuracy of laboratory methods and the application of statistical quality control procedures to detect medically important analytical errors that may occur during routine analysis. A system of quality standards is recommended to incorporate different types of requirements, such as clinical outcome criteria, analytical outcome criteria, and analytical performance criteria. For practical applications, all need to be translated into operating specifications for the imprecision, inaccuracy, control rules, and number of control measurements that are necessary to assure analytical quality during routine production of test results.

In America another system will be established based on total quality management. The American patient and consumer groups are increasing pressure on physicians and health care organisations to improve and document the quality of care that they deliver. A leading cause of inadequate care is the significant variations in medical practice that cannot be explained by patient differences, which persist through out the country.

Numerous projects for quality assessment have been established in obstetrics and gynecology (surgery for out-patients, mammography, pap-smear, perinatology, neonatology). Guidelines and quality assurance programs for cancer-treatment have been conceived. Information technologies have been integrated in this fields (Valet and Brockhaus, 1997). Therefore, new software has been established in hospitals and private clinics. A German law for hospital tumor registration has attempted to unify different strategies. Items like analysis of structure-process-result, planning-control-guidance, total quality management and evidence based medicine were integrated into clinical practice.

Some hospitals already have ISO 9001 certification. In January 1991 the guidelines for embryo-protection became law in Germany. In assisted reproductive technologies (ART) it is well known, that quality management can increase the

Gynecology	*Obstetrics*	*Reproductive medicine*
– "Qualitatssicherung in der operativen Gynakologie" project 1992–1996 > 50 hospitals in Germany – ambulatory surgery – "Deutsche Mammographie-Studie" – Diagnostic Cytology – "Leitlinien fur chronische Unterbauchschmerzen" – guidelines for cancer-treatment – standard recommendations of the Association of the Scientific Medical Soceities in Germany (AWMF = Arbeitsgemeinschaft der wissenschaftlichen medizinischen Fachgesellsch).	– quality assessment programme of prenatal and perinatal care Perinatal-Erhebung since 16 years in (West)-Germany "Munchner Erhebung" > 50% computerised – quality assurance programme of neonatal care "Neonatal-Erhebung" since 11 years – guidlines for neonatal care management "Leitlinien zum rationalen arztlichen Handeln in der Neonatologie und padiatrischen Intensivmedizin" – guidlines and standard recommendations of the Work Group for medical and scientific societies (AWMF)	– "Deutsches IVF-Register" since 1989 for Germany – guidlines and standard recommendations of the AWMF

Fig. 2 Projects for quality assessment in Germany.

success rates of treatment. Since 1982 treatment-cycles for in-vitro-fertilisation (IVF) have reached 35,000 per year in Germany. In 1999 100 centres offered ART-treatment.

Material and Methods

Since 1990 the German IVF Record (DIR, Deutsches IVF-Register) collects data for ART-treatment. This was computer-based and started with a simple database-software (d-base©). Because of rapid improvement and integration of new techniques (PZD, SUZI, ICSI) this software has been updated 18 times.

Since 1997 a new software was established (FileMakerPro 4.1c) and updated several times. There are also two internet information platforms which offer information about all relevant guidelines (for hormonal treatment, laboratory working procedures, IVF and ICSI-treatment, and genetic diagnostics): http://www.uni-duesseldorf.de/WWW/AWMF/awmf-frp.htm and http://www.dir-online.de/

Results

The board of the German IVF Record tried to establish a helpful tool for data-collection. It is necessary to reduce the items for one IVF-cycle, otherwise the compliance of the participants decrease. At the moment 150 items can be registered, the number of compulsory data fields depend on the treatment.

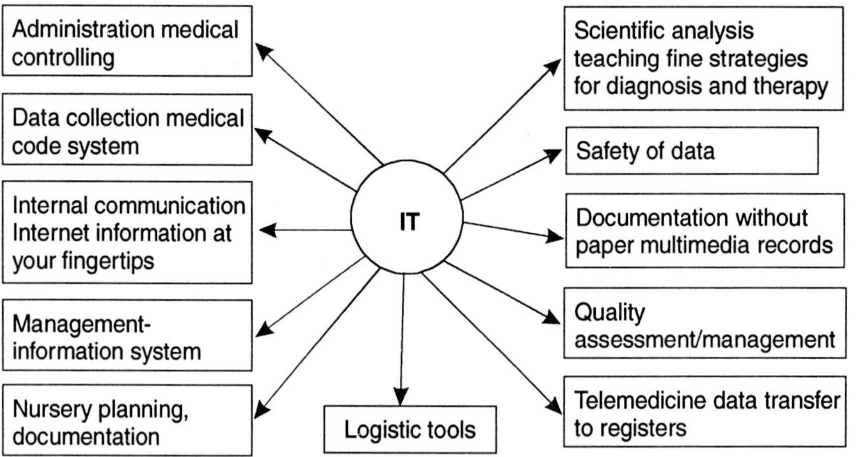

Fig. 3 Requirements for medical information technology.

Analysis of 84.042 cycles registrated with the old software show good compliance in IVF-centres with more than 500 follicular punctures per year.

The formal evaluation of 329 German guidelines (Helou and Ollenschlager, 1998) presented on the internet showed that most of them do not fit the international experience of clinical practice guidelines (information for whom the guidelines are, relevant literature, names of the experts etc.).

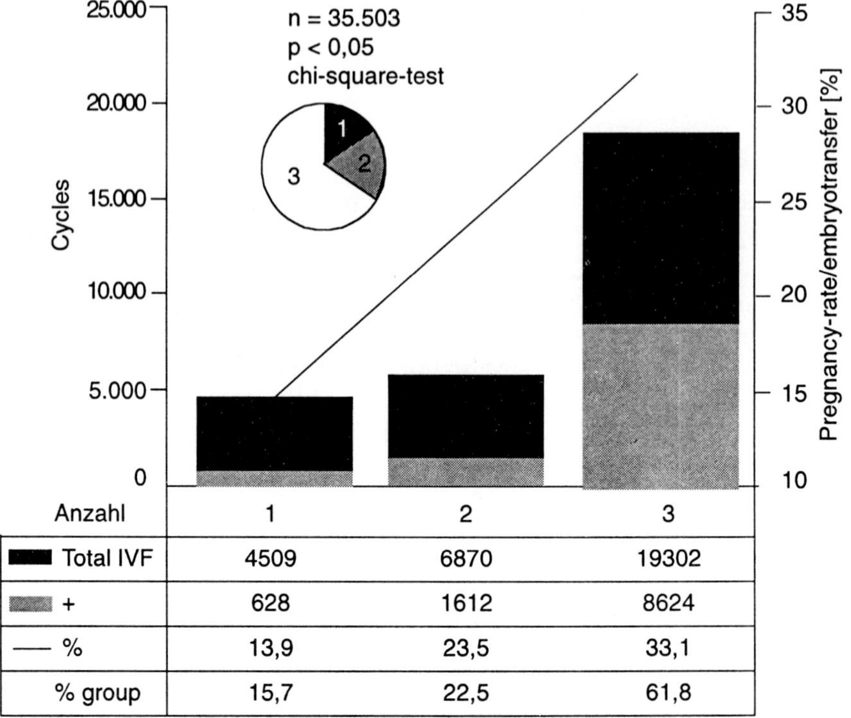

Anzahl	1	2	3
Total IVF	4509	6870	19302
+	628	1612	8624
— %	13,9	23,5	33,1
% group	15,7	22,5	61,8

Fig. 4 Number of transferred embryos–IVF.

Conclusion

We are convinced that the foundation of the German IVF Record has been the most important tool for quality assessment in reproductive medicine. The cooperation is voluntary. In the US the success-rates of each IVF-centre is not published at the end of the year. Development of new guidelines must follow fixed rules.

References

Hamberger, L., Sjoqvist, B. (1994). The information technology revolution-how it may affect gynecology and obstetrics, *Int. J. Gynaecol. Obstet.*; **47**, 211–213.

Helou, A., Ollenschlager, G. (1998). Methodological quality of clinical practice guidelines in Germany: results of a systemic assessment of guidelines presented on the Internet, *Z. Arztl. Fortbild. qualitatssich*; **92(6)**, 421–428.

Siekmann, U. (1989). Seminar: EDV in Gynäkologie und Geburtshilfe, *Arch. Gynecol. Obstet*; **245**, 1104–1107.

Valet, A., Brockhaus, M. (1997). Computers in general practice of the established gynecologist: market status-electronic data processing solutions-electronic data processing problems-future perspectives, *Zentralbl. Gynakol*: **119**, 435–438.

Follicular Growth, Ovulation and Fertilization: Molecular and Clinical Basis
Anand Kumar and Amal K. Mukhopadhyay (Eds.)
Narosa Publishing House, New Delhi, India, 2001

20

Environmental Evaluation of Assisted Reproduction Techniques Laboratories in Germany and the United States of America

F. Geisthövel, A. Ochsner and A.V. Gilligan*
Inst. Gyn. Endocrinol. & Reprod. Medicine, Gemeinschaftspraxis WTG, Freiburg i.Br., Germany
*Alpha Environmental Inc., Jersey City, NJ, USA

1. Introduction

Environmental factors which are known to have adverse effects on cells and sensitive tissues are microparticles, heavy metals, organic compounds, bacteria, viruses, fungi, temperature, UV-light and possibly electromagnetic fields. In consequence, these factors could harm sperm, oocytes, zygotes and embryos. The adverse effect could affect fertilization, cleavage, implantation, and ongoing pregnancy rates.

There are few data concerning this complex issue available in literature. In typical animal studies, agents are administered to animals directly, at relatively high doses with observation of the toxic effects. Some examples of such studies are cited below. Male rats given 50,000 ppm acetone in the drinking water for 13 weeks had depressed sperm motility and increased incidence of abnormal sperm (Dietz *et al.*, 1991). Reduced pregnancy rates occurred in untreated female rats mated with ethanol treated males (Cicero *et al.*, 1994). Formaldehyde has been found to affect sperm mobility, count and viability in male rats given 10 mg/kg/d for 30 days (Majumder & Kumar, 1995). In addition, acetaldehyde has been shown to be 100% embryolethal to explanted rat embryos when added to culture medium at concentrations of 20 μg/dl; and to cause growth retardation and teratogenic effects at concentrations of 10 μg/dl (Giavini *et al.*, 1991). These data found in laboratory rodents are suggestive of potential effects on human reproductive success.

Considering that gametes and embryos in the Assisted Reproductive Techniques (ART) laboratory can also be directly exposed, it is advisable to inspect the ambient conditions of the ART lab (Cohen *et al.*, 1997; Sharara *et al.*, 1998; Geisthövel *et al.*, 2000). The present review discusses environmental issues concerning microparticles, volatile organic compounds (VOCs) and aldehydes. Own data, represented here are the results of a collaboration between our institute and Antonia V. Gilligan. The first important review on the impact of environmental factors in an Assisted Reproductive Techniques (ART) lab (Cohen *et al.*, 1997) was done using analytical data developed by her. She also has advised leading ART labs in the US. The results of several studies performed in US clinics are compared with ours.

2. Sources and Evaluations

2.1 Microparticles

Microparticles or particulates are small fragments that may be either suspended in air or media. Sources of microparticles are dust [outdoor, e.g. traffic, construction work; indor, e.g. debris from the staff, shoes, unfiltered air conditioning, interior construction work], soot (diesel engines, smoke), fibers (clothes, paper towels, isolation by both asbestos and fiberglass, wipers, packages), powder (talcum on clinical gloves), pollen as well as cellular debris (e.g. dead skin cells). The diameters of these particles ranges between 0.01 (e.g. tobacco smoke: contains > 4,000 chemical compounds!) to 100 microns (e.g. pollen). One should monitor the level of microparticles in the air because this may serve as carrier of bacteria, virus, fungi as well as VOCs. Additionally, they can be used as a convenient monitoring tool for overall environmental quality (Boone *et al.* 1999).

2.2 Volatile Organic Compounds (VOCs)

2.2.1 Introduction

There are numerous VOCs consider to be aggressive compounds mainly because of their allergenic, carcinogenic and mutagenic effects such as: chlorinated hydrocarbons (trichloroethylene), aromatic compounds [benzene and benz(a)pyrene], ethers (dimethyl ether), aldehydes (formaldehyde) as well as highly reactive species, such as epoxides. There are a wide variety of sources for VOCs: exhaust from internal combustion engines and industrial chimneys, indoor surface covering of furniture, floor covering, floor polish, plasticware (dishes, tubes, pipettes, needles, catheters, bottles), refrigerators, incubators, CO_2 gas, tubing, anesthetic gases, heating ventilation air conditioning (HVAC) device, staff (clothes, shoes, breath, cosmetics, hairsprays, deodorants, perfumes, tabacco smoking), cleaning material, disinfecting material alcohols, building material, lacquers, varnishes, tinctures. It has to keep in mind that there are thousands of potential pollutants, albeit at mg/m^3 to $\mu g/m^3$ levels, and even more combinations of those are present in the air. Our knowledge of these materials, and their relevant levels and interactions is very limited.

2.2.2 Basic concentrations in our center

Antonia V. Gilligan (A.V.G.) performed in March/April 1999 a series of environmental measurements of VOCs and aldehydes in and around our facility. The center is located on one of the busiest traffic locations in the city of Freiburg i.Br.. Volatile organic compounds were measured by the US EPA to-14 sampling followed by gas chromatography/mass spectroscopy (Compendium of Methods, 1984/1988). Samples were taken over 24 hours using a flow controller and six-litre summa canister. Analyses were done on the exterior environment, in the ART lab, in the incubators and from the CO_2 cylinders supplying incubators.

Our outdoor results (Table 1) were comparable with those found in exteriors surrounding several US clinics; the sampling showed a significant number of VOCs (Table 2). An interesting finding was that our CO_2 cylinder sampling (Table 3) showed remarkably less concentrations of VOCs comparable to the results measured in US clinics (Table 4) aside from dimethyl ether that was present in high concentrations (1.000 $\mu g/m^3$) in our center. Our VOCs levels were found to be much higher in the ART lab and the incubator (Inc #3 and Inc #4) than in the outdoor conditions Table 5); the samples had extremely high concentrations of 1,1-difluoroethane (1,1-DFE in Inc #4 (30.000 $\mu g/m^3$); elevated levels of isopropyl alcohol (IPA) were detected both in the ART lab (5.300 $\mu g/m^3$) and in the incubators (Inc #3: 2.000 $\mu g/m^3$, Inc#4: 1.500 $\mu g/m^3$). In general, the atmospheres inside the incubator (Table 6) and in the ART lab are higher in VOCs than the outside air.

2.2.3 Concentrations following filtering

In addition, the application of the Coda filter system (<genX> international, Inc, US) was examined. This system consists of the Coda CO_2 Tri-Gas Inline Filter™ (ILF), the Coda Incubators Filtration Unit™ (IncFU) and the Coda Tower™. These units use activated carbon filter (filers VOCs and aldehydes) and High Efficiency Particulate (HEPA) filter (filers microparticles); since the carbon filter may release microparticles by itself the HEPA filter is placed behind the carbon filter. The Coda IncFU system uses a self-contained pump to circulate incubator air. The Coda TowerTM put in the ART lab is a separate air cleaner; it uses activated carbon, a HEPA filter and a fan to circulate the laboratory's air through the purifying media. The tests performed in our facility evaluated the ILF and IncFU only.

Surprisingly, inconsistent data were collected in our study after using the Coda ILF and IncFU systems (Table 5). A marked increase of IPA was determined in Inc #3 (5.300 $\mu g/m^3$) containing the IncFU only, and Inc #4 (5.000 $\mu g/m^3$) containing both the IncFU and the ILF. A dramatic rise of IPA (Table 5) was observed in Inc #4. On the other hand, there was a lost of numerous VOCs ranging between 10 and 1.000 $\mu g/m^3$ in Inc #3 (Fig. 1), and between 10 and 300 $\mu g/m^3$ in Inc#4 (Fig. 2), respectively.

2.2.4 Summary

Concluding the data, it has been shown in the present study that: (i) VOCs are observed in higher levels indoors and in the incubators vs. outdoors

Table 1 Volatile organic compounds: Outside results of our center

Material	*$\mu g/m^3$*
Ethanol	35.0
Toluene	24.0
2-Methylbutane	20.0
1,4-Dichlorobenzene	17.0
Acetone	16.0
2-Butanone	10.0
Isobutane + Acetaldehyde	10.0
Butane	10.0
Pentane	10.0
2-Methylpentane	9.0
Isopropyl Alcohol	8.5
Chlorobenzene	7.0
C4 Alkene	6.0
Hexamethylcyclotrisiloxane	5.0
Propene	4.8
m&p-Xylenes	4.6
Benzene	4.4
Benzaldehyde	4.0
a-Methylstyrene + Unidentified Silane or Siloxane	4.0
n-Hexane	3.7
Tetrachloroethene	3.7
Dichlorodifluoromethane	3.5
Hexanal	3.0
3-Methylpentane	2.2
Ethyl Acetate	2.2
Methylcyclopentane	2.0
Pentanal	2.0
Isooctane	2.0
Chloromethane	1.8
Trichlorofluoromethane	1.7
tert-Butyl Methyl Ether	1.4
Ethylbenzene	1.4
o-Xylene	1.4
Acetonitrile	1.3
Butyl Acetate	1.2
Heptane	1.0
Methylene Chloride	0.9
Carbon Disulfide	0.9
Styrene	0.9
Acrolein	0.8
Trichlorotrifluoroethane	0.7

Table 2 Volatile organic compounds: Outside results of 14 US clinics*

Material	*Mean* ($\mu g/m^3$)	*SDM*
Isopropyl Alcohol	43.4	55.0
Ethanol	32.2	30.7
Acetone	27.2	17.0
Toluene	18.1	27.2
m&p-Xylenes	9.8	19.9
2-Butanone	8.8	14.5
Tetrachloroethene	5.9	9.2
Trichlorofluoromethane	5.8	7.9
Dichlorodifluoromethane	3.6	1.9
Chloromethane	3.5	6.1
o-Xylene	2.7	5.4
Ethylbenzene	2.7	6.0
1,4-Dichlorobenzene	2.4	5.7
Benzene	2.4	1.4
Propene	2.0	1.4
tert-Butyl Methyl Ether	1.9	3.0
Trichlorotrifluoroethane	0.9	0.6
Ethyl Acetate	0.8	1.5
Acrolein	0.7	1.1
Butyl Acetate	0.7	0.9
Acetonitrile	0.4	0.8
Styrene	0.1	0.2

*measured by Antonia V. Gilligan

Table 3 Volatile organic compounds: Our CO_2 grab results

Material	$\mu g/m^3$
Dimethyl Ether	1,000.0
Isopropanol	73.0
Propane	30.0
Ethyl Methyl Ether	20.0
Acetone	16.0
Ethanol	12.0
n-Butane	10.0
2,2,4,6,6-Pentamethyl-3-heptene	10.0
n-Decane	7.7
n-Undecane	6.5
Toluene	5.7
1,2,4-Trimethylbenzene	4.8
m- & p-Xylene	2.8
n-Dodecane	2.7

Table 4 Volatile organic compounds: CO_2 grab result of US clinics*

Volatile organic compounds	*μg/m³*
Benzene	100
Unknown Freon	100
Isopropanol	80
n-Pentane	50
Actaldehyde	50
n-Butane	30
Isohexane+Acetic Acid	30
Acetone	24
Ethanol	20
Toluene	12
n-Heptane	10
C9H12 Alkyl Benzene	10
n-Undecane	10
C7H16 Alkane	9
C12 H26 Alkane	7
Trichloroethene	4.7
m- & p-Xylenes	3.8
Ethylbenzene	1.6

*measured by Antonia V. Gilligan

Table 5 Volatile organic compounds and total aldehydes: Our overall results

Location	*Compounds*	*Concentrations (μg/m³)*			
		0–40 hrs	*41–80 hrs*		
		Basal	*Basal*	*Filtration*	
		Inc #3	*Inc #4*	*IncFU*	*IncFU+ILF*
Outdoor					
	IPA	8			
	TAs	17	15		
ARTs lab			–		
	IPA	5.300	10.000		
	TAs	70	89		
Inc#3				1.000	
	IPA	2.000		5.200	
	TAs	56		45	
Inc #4					
	IPA	1.500			5.000
	TAs	52			31

1,1-DEF: 1,1-difluoroethane; IPA: isopropyl alcohol.

Table 6 Volatile organic compounds: Results of incubators in ART labs of 19 US clinics*

Material	*Mean* ($\mu g/m^3$)	*SDM*
Isopropyl Alcohol	1086.5	1937.9
Ethanol	697.4	1518.1
Acetone	295.7	659.5
Ethyl Acetate	34.2	141.9
Toluene	29.8	48.7
1,1,1-Trichloroethane	28.3	92.5
n-Decane	23.1	35.6
n-Hexane	19.6	44.0
Dichlorotetrafluoroethane	18.4	55.4
2-Butanone	18.3	35.7
Trichloroethene	16.7	64.0
n-Undecane	15.5	24.8
Methylene Chloride	14.3	25.5
1,2,4-Trimethylbenzene	12.6	27.3
Chloroethane	12.1	32.3
1,4-Dichlorobenzene	11.6	30.5
Acetonitrile	11.2	22.3
m&p-Xylenes	10.9	12.2
Nonane	10.8	15.0

*Measured by Antonia V. Gilligan.

although the outside area is strongly loaded; (ii) VOCs are present in the CO_2 supplying the incubators; (iii) high levels of IPA were detected and have increased during the working procedure indicating an accumulating presence of IPA in the entire system in spite of limited use; (iv) the distribution of VOCs varies on different days, in the different compartments, and even in the different incubators; (v) few material that did increase in the incubators were affected by the increase in the air of the lab; (vi) levels of single substances might increase inspite of the installation of a filter system; on the other hand, it is able to diminish many other substances, and (vii) the use of the ILF may not reduce significantly VOCs levels.

There might be several reasons why these data are inconsistent: (i) the capacity of the filters may have been exceeded by the laboratory's level of VOCs; (ii) the incubators were in active use and repeatedly opened during tests, (iii) the IncFu may not have been capable of dealing with swings in VOCs concentration, (iv) although the measurements have been carried out over two successive 48 hour periods prolonged determinations may evaluate better the effectiveness of the filters. The tests indicate the need to manage VOCs by an integrated strategy of ART laboratory design, laboratory practices (judicious use of antiseptic agents), and localized improvement of the air particularly in the incubator.

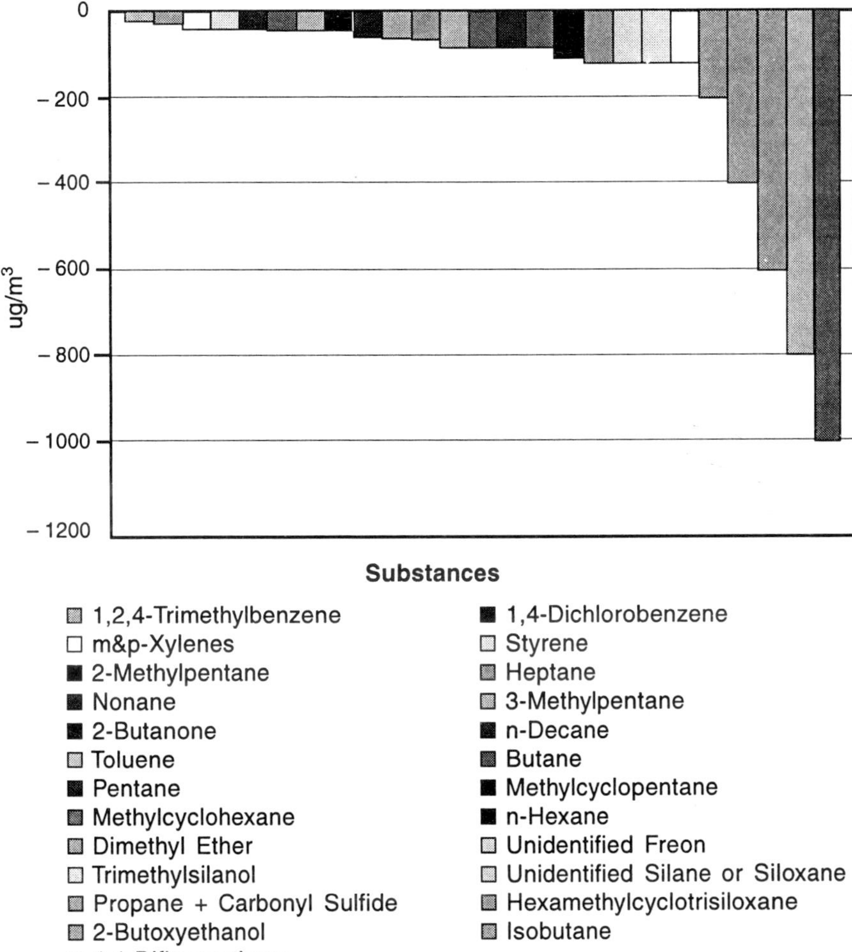

Fig. 1 Volatile organic comounds lost using Coda IncFU (Inc #3).

It has to be mentioned, that several VOCs has been found by A.V.G. in a variety of plastic dishes. Chamber studies have shown the release of 920 ng/kg of styrene from one brand of petri dish. Styrene is routinely found in ART incubators and occasionally found in ART laboratory atmospheres. Mineral oil used in ART embryo culture can act as a sink and/or source for organic (Miller *et al.*, 1994) depending on the solubility of the material. These findings suggest that VOCs may pass the oil phase and enter the media phase if they are hydrophilic. Conversely, VOCs may be held in the mineral oil phase if the VOC is hydrophobic. It has been suggested before that plastic dishes may release cytotoxic substances (De Haan, 1971).

In general, it has to be considered, however, that absolute concentrations of VOCs are not equivalent to the chemically aggressive potential of these compounds.

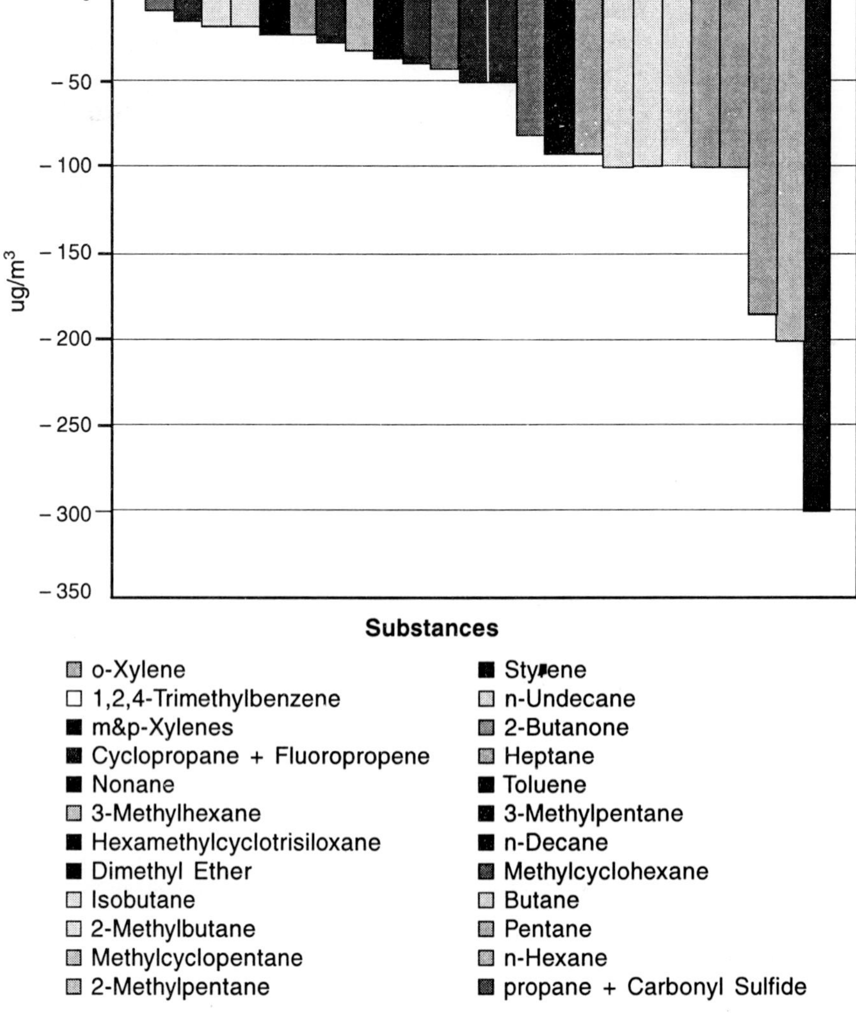

Fig. 2 Volatile organic compounds lost using Coda IncFU and Coda ILF (Inc #4).

2.3 Aldehydes

The sources of aldehydes are similar to those described for VOCs. Aldehydes were determined by US EPA to-11 using absorptive tubes followed by High Performance Liquid Chromatography (HPLC) (Compendium of Methods, 1984/1988). As seen for VOCs, it became evident by measurements of aldehydes performed by A.V.G. in US ART labs that aldehydes are present in significantly higher concentrations indoor (Table 7) and in the incubator (Table 8) vs outdoor (Table 9). These results are supported by the investigation in our institute (total aldehydes outside: 17 $\mu g/m^3$, ART lab: 70 $\mu g/m^3$, Inc #3: 56 $\mu g/m^3$), Inc#4: 52 $\mu g/m^3$ (Table 5). A clear decline was achieved by the filter system: Inc#3: 45 $\mu g/m^3$, Inc#4: 31 $\mu g/m^3$ (Table 5) indicating that both the Inc FU and the ILF are of benefit regarding total aldehydes.

Table 7 Aldehydes: Results of ART lab levels of 14 US clinics*

Material	*Mean ($\mu g/m^3$)*	*SDM*
Formaldehyde	14.9	12.4
Acetaldehyde	9.0	7.6
Propionaldehyde	2.9	0.9
Hexaldehyde	2.6	3.7
Crotonaldehyde	0.7	0.3
Butyraldehyde	1.3	1.2
Benzaldehyde	1.2	0.2
o-Tolualdehyde	0.1	0.2
m-Tolualdehyde	0.6	1.6
Isovaleraldehyde	0.1	0.1
Valeraldehyde	1.0	1.3
Acrolein	0.0	0.1

*Measured by Antonia V. Gilligan.

The levels include three locations with exceptionally high aldehyde levels. The levels in well designed and productive IVF laboratories ($N = 10$) are generally in the following ranges: formaldehyde 5.9 $\mu g/m^3$; acetaldehyde 3.4 $\mu g/m^3$: propionaldehyde 0.6 $\mu g/m^3$.

Table 8 Aldehydes: Results in incubators of 7 US clinics*

Material	*Ave. $\mu g/m^3$*	*SDM*
Formaldehyde	11.3	5.1
Acetaldehyde	7.8	4.3
Propionaldehyde	12.9	9.9
Hexaldehyde	3.2	2
Crotonaldehyde	0.1	0.1
Butyraldehyde	0.9	0.5
Benzaldehyde	1.4	0.3
o-Tolualdehyde	0	0
m-Tolualdehyde	0.2	0.2
Valeraldehyde	1.7	1
Acrolein	0	0

*Measured by Antonia V. Gilligan.

In evaluating the data above, it is important to consider that many of the locations were experiencing difficulty during the testing. For facilities that are highly productive, the formaldehyde levels are well below 1.0 $\mu g/m^3$. This is a small sample, primarily taken from facilities with certain unique environmental concerns. Therefore, there may be a positive basis in the data. The strikingly high level of acetaldehyde is troublesome.

Table 9 Aldehydes: Outside levels of 12 US-clinics*

Material	*μg/m³*	*SDM*
Formaldehyde	3.1	2.0
Acetaldehyde	1.9	0.9
Propionaldehyde	1.9	3.3
Hexaldehyde	0.3	0.5
Butyraldehyde	0.5	0.7
Valeraldehyde	0.2	0.3
Acrolein	0.4	0.9
Crotonaldehyde	0.0	0.1
Benzaldehyde	0.1	0.2
Isovaleraldehyde	0.0	0.0
o-Tolualdehyde	0.0	0.0
m-Tolualdehyde	0.0	0.0
p-Tolualdehyde	0.0	0.0
2,5-Dimethyl Benzaldehyde	0.0	0.0

*Measured by Antonia V. Gilligan.

2.4 Summary Concerning Air Conditions in the Incubators

Measures for optimizing the intra-incubator air conditions should take into consideration the following three factors: (i) CO_2 supplying system, (ii) ART lab air, and (iii) release from VOCs by plastic materials.

3 Measures

3.1 General recommendations

In most clinics, the measures taken to reduce VOCs will be a compromise between the cost of building alterations, personnel practices and potential gain.

Changing procedures will require an educational process which is time consuming and needs patience. Staff practices that may require attention are:

- Scrub discipline in the procedure room and IVF laboratory. Clean scrubs are crucial to eliminate microbiological contamination being introduced into the ART laboratory.
- Personal use of perfumes, colognes, etc. should be eliminated.
- Cleaning in the ART laboratory with alcohol should be avoided.
- Opening packages of plastic ware prior to use to allow for the offgasing of VOCs. This can be done by placing the plastic ware in a laminar flowhood overnight before use.
- Maintaining the ART laboratory's ventilation system for optimal performance. This will require filter changes and periodic inspection.
- Evaluation the purity of CO_2 used in the ART laboratory.

- Instituting an ongoing program of ART process control. This requires that ART success statistic be compiled along with data on patient age, endocrinology work up, fertilization rate, number of embryo transferred, etc., as well as environmental quality. Activities that can have a significant adverse effect on the ART laboratory are: painting, new construction using adhesives, tar and carpets, and air pollution crisis.

There are measures on three financial levels, in order to improve the conditions in the lab.

3.2 Low Budget

3.2.1 Microparticles

The aim might be to reduce class 5 grade ($10^5/m^3$) to class 4 grade ($10^4/m^3$) contamination [class 2 grade ($10^2/m^3$) or even class 1 grade ($10^1/m^3$) contamination are required in the chip producing industry]. Several measures are necessary to reduce the concentration of microparticles. The amount of horizontal levels of tables, furniture and shelves should be reduced. Crucial is a correct documentation and controlling system (e.g. regular changing of the filters in the HVAC system). Special shoes (Tech stales, Inc. USA) and coats (CRTM GmbH, Munic, Germany) consisting of 100% polyester (including antistatic fibers) decrease the amount of microparticles (e.g. skin cells, dust and fibers) derived from both the skin and underwear, and from the coat itself; the disadvantage of these coats is the fast disposition to perspire which can be reduced by stable room temperature conditions [< 24 °C; humidity 40–60% (this number appears to avoid mold issues); s. below]. The use of wipers manufactured by non-woven polyester/cellulose blend (Contec, USA) and powder-free gloves may attenuate the amount of fibers and microparticles in the air.

3.2.2 Volatile organic compounds, aldehydes

Laboratory staff people should be entreated to eliminate the use of hair spray, perfume, deodorants and tobacco smoking. Patients should be instructed in the same manner. Laboratory staff is required to wear caps constantly and mouth masks when dealing with oocytes and embryos. Cleaning with alcohol should be minimized and supplemented by pure sterile water. The floors might be treated with neutral cleaning material. Heavy floor cleaning should be done when the ART laboratory is not actively handling materials. As mentioned above it is advisable to have an ongoing maintenance plan for the HVAC. The use of pure "5.4" CO_2 gas (purity: vol. % 9.99994) is recommended. The application of an almost "closed" oocyte retrieval/flushing technique and working under oil might lower the contamination with adverse substances (including microparticles).

It is also recommended allowing the emission of gases from new laboratory products prior use (Cohen *et al.*, 1997). Anaesthetic gases generated e.g. in the adjacent surgery room during oocyte retrieval should be shunt off by a tubing system.

3.3 Medium Budget

3.3.1 Microparticles

Permanent adhesive mats (Dycem limited, Bristol, England) on the floors are able to capture dust from soles of the shoes. Those mats may emit gases by itself and, therefore, should be used not before a sufficient time of emission has passed, or if a high ventilation plant (s.below) has been installed.

3.3.2 Volatile organic compounds, aldehydes

The CO_2 supplying system can be upgraded by using high purity materials. The use of very pure "5.5" CO_2 gas (purity: vol. %: 9.99995) might be better as the "5.4" CO_2 gas. The "5.5" CO_2 gas is supercritical fluid extraction (SFE) grade gas used in analytical chromatography analysis; this gas contains the following impurities: $O_2 \leq 1.0$ vpm, $N_2 \leq 2.0$ vpm, $CH_4 \leq 0.1$ vpm, $CO \leq 0.5$ vpm. Gas cylinders should be certificated as being cleaned before refilled. High vacuum cleaning is the preferred cleaning method. Due to the exterior contaminants on the cylinder, it is best if they are located outside the ART laboratory. High purity nickel gas regulators with stainless steel membranes, stainless steel (Finetron) piping and expansion stations should be used in constructing the CO_2 distribution system. Alternately, Teflon lined tubing can be used in low-pressure portions of the system. When installing such a system, it is of paramount concern that the system of all construction-derived materials is purged before connecting it to the incubator; furthermore, the pH of the media should be checked.

Many companies who sell CO_2 (and the cylinders) will ignore the specific demands, which are required for an ART lab. The ART lab must educate its supplier on their high purity requirements. Concerning our data, the Coda ILF might reduce aldehydes; however, it does not appear to significantly reduce VOCs. The filtration of the air in the ART lab by the Coda tower (not evaluated in our study!) and in the incubators by Coda IncFU may improve partly the environmental conditions.

3.4 High Budget

An ideal laboratory would use stainless steel work benches, stainless steel utility carts with wire shelves or perforated metal surfaces as well as wire shelves which may reduce both the accumulation of microparticles on plane surfaces and the release of VOCs; alternatively, work surfaces of Corian, a semisynthetic material with zero VOCs release potential, can be used. In addition, surfaces of walls and ceilings should be free of holes to the exterior environment; be "hard", made of plaster or sheetrock and be painted with low VOCs release potential paint. Suspended ceiling tiles often allow a contamination of the ART laboratory by exterior environmental influences. Flooring should be of sheet vinyl which has been formulated for medical use (Medintech).

The most effective and also costly tool appears be an enhanced HVAC system including a heating and cooling system (room temperature < 24 C°, humidity 40–60%) as well as activated carbon and HEPA filter stations; the air volume of the *related space* should be changed 10 to 15 times/h with fresh air at a very low pressure of 0.05-0.10 inches of water which would be less than one would experiences in going from ground level to a pressurized aircraft; likewise, the intra-incubator milieu that might be checked by pH determinations will not be affected. The wall of the air supplying pipe line, at least the post-filter final part, should be composed of zero VOCs releasing material. Such an installation would be optimal to stabilize the ambient conditions and reduce significantly the contamination with microparticles, VOCs and aldehydes particularly of the incubator air; resulting in less VOCs contamination than the outside air because the fresh incoming air has been passed through purifying tools. The use of laminar flow benches, which reduce the number of bacteria and mold becomes less critical. A proper equipment should be individually adapted on the surrounding, ambient and adjacent conditions.

In cases of unexplainably low pregnancy and/or high miscarriage rates it might make sense to perform mouse embryo testing of e.g. plastic batches, media or mineral oil and/or to seek for sources of VOCs contamination e.g. by the CO_2 supplying system using an appropriate equipment of determinations. A simultaneous mouse embryo testing, however, is not allowed in incubators where human material is placed.

4 Summary

The review including own and US data presents some ideas concerning environmental conditions in the ART lab. Sources of pollution are many, and concentrations of the pollutants are higher in the air of the ART lab and of the incubators than in the outside air. Steps for improving the conditions in the ART lab and in the incubators require both procedural changes and potentially expensive facility upgrades. The Coda incubator system might reduce both VOCs and aldehydes in the incubator but the results are preliminary and inconsistent so far. Changes in proper equipment require profound and individualized balancing of benefits, disadvantages, and costs. Further work by our group is underway to explore the ambient conditions in the ART lab and in the incubator, and to develop cost-effective technologies and work practices that might result in improved environmental conditions concerning particularly both the working space and the incubators in order to be consequently followed by an increased ongoing pregnancy rate.

5. Acknowledgement

Prof. M. Wikland's (Sweden) pioneering work in the field of quality control in ART labs gave us a lot of new ideas. The assistance of U. Knecht, Repromed

GmbH, Munich, Germany, in technical questions, and the generous financial support by Serono Pharma GmbH and Organon GmbH, Germany, on performing our studies are greatly appreciated.

References

Boone WR, Johnson JE, Locke AJ, Crane IV MM, Pricė TM. Control of air quality in an assisted reproduction technology laboratory. *Fertil Steril 1999*; **71**: 150–157.

Cicero TJ, Nock B, Oconnor LH *et al.* Acute paternal alcohol exposure impairs fertility and fetal outcome. *Life Sci 1994*; **55**: PL33–36.

Cohen J, Gilligan A, Esposito W, Schimmel T, Dale B. Ambient air and its potential effects on conception. *Hum Reprod 1997*; **12**: 1742–1749.

Compendium of Methods for the Determination of Toxic Organic Compounds on Ambient Air—United State Environmental Protection Agency Document 600/4-84-041, April 1984/1988. Available from the United States Superdentent of Document, Washington, DC De Haan RL. Toxicity of tissue culture media exposed to polyvinyl chloride plastic. *Nature New Biol 1971*; **231**: 85–86.

Dietz DD, Leininger, JR, Rauckmann EJ *et al.* Toxicity studies of acetone administered in the drinking water of rodents. *Fundam Appl Toxicol 1991*; **17**: 347–360.

Geisthövel F., Ochsner A, Gilligan AV. Environmental evaluation and corresponding measures in an Assisted Reproduction Techniques laboratory. Submitted, 2000,

Giavini E, Broccia ML Prati M *et al.* Effects of ethanol and acetaldehyde on rat embryos developing in vitro. *In Vitro Cell Dev Biol 1992*; **23A**: 205-210

Majumder PK and Kuma VL. Inhibitors effects of formaldehyde on the reproductive system of male rats. *Indian J Physiol Pharmacol 1995*; **39**: 80-82

Miller KF, Goldberg JM, Collins RL. Covering embryo culture with mineral oil alters embryo growth by acting as a sink for an embryotoxic substance. *J Ass Reprod Genetics 199*; **11**: 342–345.

Sharara FI, Seifer DB, Flaws JA. Environmental toxicants and female reproduction. *Fertil Steril 1998*; **70**: 613–622.

Follicular Growth, Ovulation and Fertilization: Molecular and Clinical Basis
Anand Kumar and Amal K. Mukhopadhyay (Eds.)
Narosa Publishing House, New Delhi, India, 2001

21

The New Importance of Testicular Biopsy in the Era of Assisted Reproduction

W. Schulze
Department of Andrology, University of Hamburg (UKE)
Martinistr. 52, 20246 Hamburg, Germany

Testicular biopsy has been used to evaluate disturbances of male gamete production for more than 50 years. Opinion about its value in clinical management of male infertility, however, has varied considerably. During the last decades, the development and perfection of non-invasive methods such as chromosome analysis, sonography and endocrinological analysis has reduced the indication for testicular biopsy. Reasonable indications left for this procedure were seen by many specialists only in two conditions: azoospermia in the presence of normal serum FSH concentration to differentiate post testicular obstruction from spermatogenic arrest and confirmation or exclusion of suspected testicular malignancies consequent to pathological sonographic findings. A major breakthrough was made during the mid-nineties, when it was shown that intracytoplasmic injection of surgically recovered testicular spermatozoa (TESE = testicular sperm extraction) into the oocytes (ICSI) resulted in fertilization followed by successful pregnancies similar to that obtained with ejaculated spermatozoa [1]. With this, a possibility of a new treatment regime opened for couples desiring a child where the man suffered from azoospermia because of inborn or acquired defects of the testicular parenchyma. It is known that many men do have localized areas of spermatogenic activity in the testis.

An optimized treatment strategy to identify and characterize such areas has been introduced as so-called "cryo-TESE concept" [2]. This concept combines the removal of testicular tissue via biopsies for the diagnostic evaluation of the infertility with the possibility of a TESE/ICSI option [3,4,5]. It involves the subdivision of a biopsy sample into at least four portions. One portion, in which fragments of three testicular lobules are included, is subjected to a histological examination by means of the semi-thin sectioning technique. A second portion, which spatially overlaps the first, is immediately used for a mild enzymatic sperm extraction procedure (test-TESE) using collagenase type Ia. The remaining two biopsy portions are cryopreserved. This tissue can be used even many months later for the definitive TESE/ICSI treatment, when sperm had been detected by the test-TESE or during histological examination. Testicular samples taken for the diagnostic procedure (histology, test-TESE) and storage overlap and cover several layers of tissue ("sandwich pattern"). This guarantees high predictive value as to the suitability of the cryopreserved portions for TESE. The protocols for freezing/thawing and subsequent preparation of the testicular tissue are listed in Tables 1 and 2.

Table 1 Freezing of human testicular biopsies

- SpermFreeze (MediCult/Denmark) as ready-to-use medium: Earle's Balanced Salt Solution + 15% glycerol as cryoprotectant, HEPES-buffered (18mM).
- Submerge biopsies into 0.5 ml SpermFreeze within Nunc Vials, equilibrate for a maximum of 15 minutes.
- In Nicoolbag 10 System (Air Liquide/Germany): cool to – 60°C within 5 minutes.
- Start exponential cooling to –120°C over 55 minutes.
- Submerge samples into liquid nitrogen until use.

Table 2 Enzymatic sperm extraction from frozen testicular tissue

1. Thaw sample in 37° C water bath for approx. 3 minutes.
2. Transfer to 1 ml culture medium (MediCult) and dissect gently.
3. Incubate sample for 30 minutes (37° C; 5% CO_2).
4. Add collagenase (Sigma type Ia) at a final concentration of 400 U/ml.
5. Incubate for 30 minutes, dissect further without squeezing the tubules.
6. Incubate for another 60 minutes with collagenase (total incubation time can be modified according to the previous finding in the test-TESE, in severe cases overnight).
7. Squeeze remaining tubules, remove debris mechanically
8. Suspend the sediment, transfer to test tube and centrifuge (500–800 g, 10 minutes).
9. Remove collagenase supernatant, resuspend pellet in small volume of culture medium.
10. Transfer suspension into medium droplets under oil and incubate for at least 30 minutes.
11. Collect sperm from the edges of the droplets and transfer into PVP.
12. Perform ICSI as usual.

The application of cryoprotectant can lead—presumably as a consequence of local hyperosmolarity—to a shrinkage of cell nuclei and cytoplasm in some immature germ cells. Primary spermatocytes are the most likely cells to be affected. Spermatids, on the other hand, hardly ever exhibit such changes. Occasionally the acrosome appears swollen, although both the acrosome membrane and the outer cell membrane remain intact. These changes show that the cryoprotectant itself causes impairment of cell function, although it prevents even greater damage from the freezing process. The best preservation of germ cell structures is observed when testicular tissue is incubated in the cryoprotectant for 3 to 15 min. During this short time period no equilibrium is reached and a gradient of cryoprotectant with lower concentrations towards the center of the tissue sample is formed. Accordingly cells at the outside shrink to a higher degree than in the middle of the sample. Upon freezing this protects peripheral tissue from cell damage, since the outer layers are dehydrated and vitrify without formation of ice crystals. Consequently, no heat is generated. Towards the centre cells are less protected by the cryoprotectant, but this is compensated for by two mechanisms: less osmotic damage by the lower concentrations of cryoprotectant and an almost optimal cooling rate. The adaption of the cooling rate is caused by the poor heat conductivity of the amorphously solidified outer tissue layer, which slows the cooling process towards the centre of the biopsy sample. Thus the cells of the whole tissue samples are frozen in an ideal way. Upon thawing the sample, no recrystallisation heat is set free in the outer regions of the biopsies (because little water is present). Therefore only little heat is transferred to the inner part of the sample, and the cells here, which are probably less cryoprotected, can thaw without formation of larger ice crystals.

Until March 1999 complete data sets were available from 1113 patients who underwent testicular biopsies according to the cryo-TESE concept. This group comprised men with azoospermia during previous semen analyses. In addition, men were included whose previous semen analyses had shown such low numbers of ejaculated sperm, that there was a high risk of a futile ICSI attempt without any back up sperm from testicular biopsies. The indications for the application of the cryo-TESE concept are listed in Table 3.

Table 3 Indications for testicular biopsy with cryo-TESE option

- During surgery for post-testicular obstruction.
- Inoperable post-testicular obstruction.
- Bilateral aplasia of the vasa deferentia.
- Idiopathic normogonadotropic azoospermia.
- Hypergonadotropic azoospermia.
- Anejaculation (resistent to treatment).
- Cryptozoospermia.
- Necrozoospermia.

The postsurgical testicular sperm extraction (test-TESE) was possible in 847 (76.1%) of all cases. In the remaining 266 (23.9%) cases no sperm could be extracted although simultaneously performed histology showed sperm in 15 of these men, the TESE was negative.

69 patients showed only "round" spermatids as the most advanced cellular differentiation during testicular histology. But in 55 of them mature spermatids could be isolated after TESE. Thus only 14 (1.2%) of all men investigated showed a spermatogenetic arrest at the early spermatid stage.

772 patients suffered from azoospermia. A positive TESE result was obtained in 530 (68.7%) of all cases. If the analysis is limited to azoospermic men with FSH serum concentrations below 8 IU/L, which represent the majority of obstructive azoospermia, 259 of 289 patients (89.6%) had a positive TESE result. But even when the FSH concentrations were elevated (> 8 IU/L) in

azoospermia (483 patients), there was still a 55% change of isolating sperm from the test material.

In a series of 655 patients with bilateral testicular biopsies 35.7% of these cases revealed differences of the status of spermatogenesis between left and right testis. Such a difference becomes clinically important, when no mature spermatids are seen on one side, whereas contralaterally at least few mature spermatids can be found. This situation was present in 17.1% of all bilaterally biopsied patients. Interestingly, in general results of histological evaluation and test-TESE were significantly better on the right side. Based on these data a bilateral biopsy is recommended whenever possible.

The importance of a careful histological analysis is impressively supported by the fact, that in 11 patients testicular biopsies revealed an early stage testicular tumour (carcinoma in situ = CIS).

The experience to date has shown that microinjection of frozen/thawed testicular sperm yields very similar fertilization and pregnancy rates to those obtained with spermatozoa from the ejaculate. The results depend on the age of the female partner, the quality of the testicular spermatozoa as well as on the number of transferred embryos (Fig. 1).

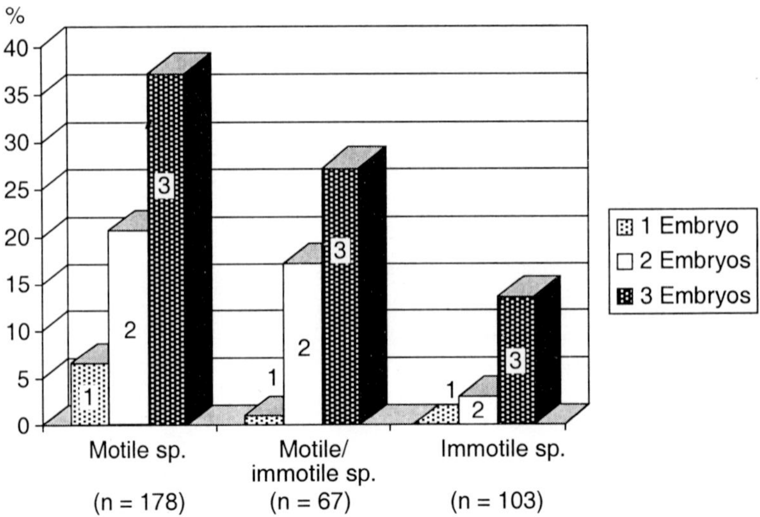

Fig. 1 Pregnancy rates in dependency from number of transferred embryos and quality of sperm retrieved by cryo-TESE.

In conclusion, the cryo-TESE/ICSI concept offers optimal conditions for a successful clinical treatment of severe male fertility disturbances. Based upon an extensive andrological diagnostic approach in the initial stage (histology; test-TESE) one can fairly well predict the chance of a successful TESE/ICSI treatment and minimize the risk of costly and futile preparations of the female partner for an IVF procedure. Moreover, if the preliminary diagnosis indicates the presence of sperm, the cryopreserved samples can be used for repeated TESE/ICSI cycles without the need for further surgery of the male. Also in terms of preventive medicine (e.g. potential CIS) the concept described here is superior to other methods which are directed solely towards retrieval of sperm.

References

1. Silber, S.J., Van Steirteghem, A.C., Liu, J. et al. (1995) High fertilization and pregnancy rate after intracytoplasmic sperm injection with spermatozoa obtained from testicle biopsy. *Hum. Reprod.*, **10**, 148–152.
2. Salzbrunn, A., Benson, D.M., Holstein, A.F., Schulze, W. (1996) A new concept for the extraction of testicular spermatozoa as a tool for assisted fertilization (ICSI). *Hum. Reprod.* **11**, 752–755.
3. Fischer, R., Baukloh, V., Naether, O.G.J., Schulze, W., Salzbrunn, A., Benson, D. M. (1996) Pregnancy after intracytoplasmic sperm injection of spermatozoa extracted from frozen-thawed testicular biopsy. *Hum, Reprod.* **11**, 2197–2199.
4. Jezek, D., Knuth, U.A., Schulze, W. (1998) Successful testicular sperm extraction (TESE) in spite of high serum follicle stimulating hormone and azoospermia: correlation between testicular morphology, TESE results, semen analysis and serum hormone values in 103 infertile men. Human *Reprod.* **13**, 1230–1234.
5. Schulze, W., Thomas, F., Knuth, U.A. (1999) Testicular sperm extraction: comprehensive analysis with simultaneously performed histology in 1418 biopsies from 766 subfertile men. *Hum. Reprod.* 14 (Suppl. 1), 82–96.

Follicular Growth, Ovulation and Fertilization: Molecular and Clinical Basis
Anand Kumar and Amal K. Mukhopadhyay (Eds.)
Narosa Publishing House, New Delhi, India, 2001

22

Hormonal Influence on the Fertilization Potential of Sperm

N.R. Moudgal[1], H. Krishnamurthy[1,2] and M.R. Sairam[2]

[1]Jawaharlal Nehru Center for Advanced Scientific Research & Department of Molecular Endocrinology, Reproduction, Development & Genetics, Indian Institute of Science, Bangalore, India

[2]Molecular Reproduction Research Laboratory, Clinical Research Institute of Montreal, Montreal, (H2W 1R7) Canada

Introduction

The causative factors of human male infertility, which appears to be on the rise, are varied. It is only in a small percentage of cases that lack of hormonal support has been recognized as a deficiency. Although, it is accepted that as little as three million sperms/ml in the ejaculate is adequate to confer fertility in man, investigators are realizing that *in situ* fertilization is difficult to achieve with this number if the sperm quality is poor. The hormonal factors that regulate human spermatogenesis are reasonably well understood. Though it is held that LH/testosterone is the major regulator, FSH is invariably used as part of hormonal therapy to initiate and maintain quantitative spermatogenesis (Matsumoto *et al*, 1986; Tapanainen *et al*, 1997; Simoni *et al*, 1999). Effective blockade of pituitary LH secretion and consequent testicular testosterone (T) production leads to azoospermia but the role of FSH, if any, in regulating spermatogenesis was until recently poorly understood.

Using the bonnet monkey (Macaca radiata) as a human surrogate model we have over the last several years succeeded in creating specific hormonal deficiencies in the adult male of this species. This has allowed us to study the effect on spermatogenesis as well as fertility. These studies were undertaken with the hope that the results would permit one to a better understanding of the regulation of spermatogenesis in the primate and as such the causes of infertility in the human male. In this review we have made an attempt to correlate this with our present knowledge of how lack of specific hormone(s) could be the underlying cause of infertility in men.

Effect of LH/Testosterone Deficiency on Spermatogenesis in the Bonnet Monkey

Adult male bonnet monkeys were rendered specifically LH/testosterone deficient by either immunizing them with a heterologous LH preparation like ovine LH (Suresh *et al*, 1995) or by dosing them with central inhibitors like Norethisterone (NET-En) alone (Shetty *et al*, 1997). These treatments that did not lead to any significant change in serum FSH levels induced azoospermia. Testicular biopsy samples taken following induction of LH/testosterone deficiency were analyzed by DNA flow cytometry for quantitative germ cell transformations. These studies clearly established that the lack of LH blocks spermatogonial proliferation and meiosis. The overall effect of T deficiency thus was azoospermia leading to infertility. Androgen loss also caused behavioral changes.

Effect of FSH and Estrogen Deficiency on Spermatogenesis in the Bonnet Monkey

FSH function in adult male bonnet monkeys was specifically blocked by immunizing them with either oFSH, its beta subunit (Moudgal *et al*, 1992, Aravindan *et al*, 1993) or a recombinant FSH receptor protein fragment (Moudgal *et al*, 1997a). While the antibodies to FSH/its beta subunit neutralize circulating hormone, the antibodies against the receptor protein effectively block the receptor from interacting with the hormone ligand. The resultant effect in both treatments is essentially similar. Such treatment did not lead to any change in serum T levels confirming the specificity of the deficiency obtained. DNA flow cytometry of testicular biopsy samples revealed that blockade of FSH function by either of the above immunization procedures led to quantitative as well as qualitative decline in spermatogenesis. A significant reduction in spermatogonial proliferation as well as transformation of spermatogonia to primary spermatocytes was observed, indicating its effect on spermatocytogenesis (Aravindan *et al*, 1993 & Suresh *et al*, 1995). That it affects spermiogenesis became apparent from the significant reduction in round and elongating/elongated spermatids (Aravindan *et al*, 1993 and 1997). Thus, the overall effect of blockade in FSH function was significant inhibition in sperm production (>70%) leading to oligospermia.

We recently produced genetic mutant (FORKO) mice that lack FSH receptor to examine the effects on testicular function and fertility (Sairam *et al*, 1998 and Dierich *et al*, 1998) in a rodent species. Mutant males exhibit weight loss of testis, epididymis and seminal vesicle as well as low levels of testosterone. Except for reduced seminiferous tubular diameter and evidence of accelerated apoptosis, no gross changes were apparent upon histological examination. Analysis of testicular germ cells by flow cytometry revealed a significant increase in the percentage of 2C cells (spermatogonia and non-germ cells) and a significant increase in the percentage of HC cells (elongated spermatids) of FORKO males. The absolute number of homogenization-resistant elongated spermatids was also significantly reduced in the mutant males. A two-fold

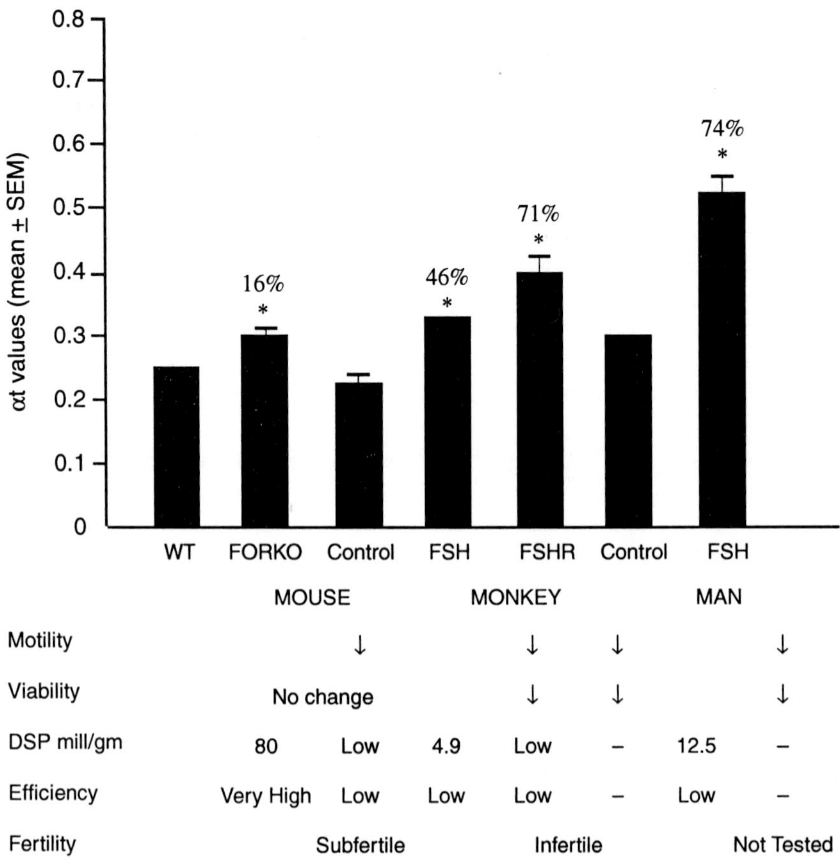

Fig. 1 Deficiency of hormone (FSH) or its receptor compromises sperm quality. Figure summarizes flow cytometric analysis of sperm from three different species. αt values indicated on the Y-axis are determined by the sperm chromatin structure assay (SCSA, see text) as an index of chromatin packaging. SCSA provides an assessment of the degree of DNA denaturation. αt values can range from 0 to 1 with upward shift indicating the greater susceptibility of sperm DNA to acid-induced denaturation which signifies poor chromatin compaction. Monkeys showing sperm of high αt values were infertile while FORKO males exhibit reduced fertility. Fertility testing was not part of the study in the FSH immunized volunteer men. DSP = Daily sperm production rate. Note that in rodents, spermatogenic efficiency is considered very high.

increase in the c-kit positive 2C cells was recorded in the mutant males indicating the perturbation of Sertoli-germ cell interactions. Elongated spermatids of FORKO males showed a dramatic increase in propidium iodide binding suggesting reduced nuclear compaction (Krishnamurthy *et al*, 2000a).

The role of estrogen, for long considered a classic female hormone, in regulation of testicular function remains unclear. Recent reports of studies in a variety of animal species indicate variable effects ranging from no effect to an essential requirement for normal testicular function and epididymal maturation of sperm (Robertson *et al*, 1999). Recent studies carried out in adult male proven fertile bonnet monkeys at our laboratory using a specific and highly potent aromatase inhibitor (CGS 16949A; Shetty *et al*, 1998) showed that continuous inhibition of estrogen synthesis in the testis does not have any effect on spermatocytogeneses and meiosis. Blockade of estrogen synthesis, however, revealed a marked effect on spermiogenesis, the transformation of round to elongating/elongated spermatids being significantly inhibited (Shetty *et al.*, 1998). It is interesting that Robertson *et al*, (1999) also observed in aromatase knock-out mice a marked inhibition in spermatid formation and maturation. The estrogen receptor α-mutant mice are also infertile (Eddy *et al*, 1996).

Effect of Specific FSH and Estrogen Deficiency on Fertility Status of Monkeys

Blockade of either FSH or estrogen function in the adult proven fertile male bonnet monkey led to oligospermia but this was not accompanied by any change in serum T levels or libido. The fertility status of FSH/FSH receptor immunized as well as estrogen synthesis blocked animals was tested in controlled mating studies with proven fertile female bonnet monkeys. The treated animals became infertile when either of these two hormones were lacking (Moudgal *et al*, 1992, 1997b & Shetty *et al*, 1998). Fertility, however, was restored when the hormonal status returned to normalcy following withdrawal of treatment and recuperation.

In order to explain the basis of infertility even without achieving azoospermia, the quality of sperm ejaculated by these monkeys was carefully assessed. The fertilizing ability of sperm in situ is dependent upon several factors. The ability to penetrate the cervical mucus in sufficient numbers and reach the site of fertilization is dependent not only on sperm counts but also on viability and motility. Marked reduction (~70%) in motility, viability and gel penetrability was noticed in ejaculated sperm of FSH immunized monkeys. Significant decrease in markers like acrosin and hyaluronidase activities indicated that the sperm might be incapable of dispersing the cumulus and penetrating the zona (Moudgal *et al.* 1992). Sperm cyclic AMP levels (Raj *et al*, 1991) and ATPase activity (Srivastava & Das 1992) are also significantly reduced following immunization with oFSH. More interestingly, attempts to obtain in vitro fertilization using both zona-intact homologous monkey eggs (Raj *et al.* 1991) and zona-denuded hamster eggs (Sharma & Das 1992) failed indicating that fertilizing ability of sperms was impaired. In monkeys lacking either FSH or estrogen, Fertility Index, (a product of viable sperm counts and motility score) correlated with the actual fertility status as assessed by mating studies. A 10-fold reduction in fertility index generally indicated induction of infertility status.

The studies of Ballachey *et al.* (1987) and Engh (1992) using both bull and human sperm have established a good correlation between chromatin status of sperm and ability to fertilize. Sperm nuclear chromatin stabilizes as they undergo maturation. The process however, starts with the transformation of round to elongating/elongated spermatids and replacement of histones by protamines. Inadequate and/or poor protamination as well as deficient disulfide bridge formation during the transit of the sperm through the epididymis has been observed to be one of the main causes of defective chromatin packaging. Such sperms exhibit poor forward motility and are unable to penetrate the egg and achieve fertilization. Sensitive flow-cytometric methods were therefore used to assess the chromatin status of ejaculated sperm of monkeys in which FSH function was blocked. Denaturation of sperm DNA followed by binding of fluorescent dyes like acridine orange showed that the ratio of single to double strandedness had significantly increased (αt values) in sperm of FSH/FSH receptor immunized monkeys (Krishnamurthy *et al*, 2000a). This clearly indicated that lack of FSH support produces defective chromatin packaging. Further tests indicated that the sperm DNA of FSH blocked monkeys were indeed more vulnearable to chemical denaturation, suggesting poor protamination. Labeling studies with radioactive iodoacetamide, which also gives a better idea of the degree of protamination, indicated that it was significantly inhibited in the absence of FSH actions. Since protamination occurs in the testis and the formation of disulphide bridging is promoted in the epididymis as part of the maturation process both testicular and epididymal sperms were examined for such defects in chromatin compaction. Similar changes observed in FORKO mice also suggest the inadequate condensation of sperm chromatin. These data allow us to conclude that genetic disruption of FSH receptor signaling in the rodent induces major changes that might contribute to reduced fertility (Krishnamurthy *et al*, 2000b). The results clearly indicate that defective spermiogenesis must be the root cause of the maturational defects seen during the transit of sperm in the epididymis. This leads to ejaculation of sperm incapable of impregnating eggs.

Examination of chromatin status in the ejaculated sperm of monkeys, in which estrogen synthesis was blocked by using an aromatase inhibitor, indicated the sperm DNA to be hypercondensed, quite the opposite of what was observed during FSH deficiency. These investigations lend experimental support to the conclusion of hypo as well as hypercondensation of sperm DNA as a cause of infertility in men (Engh *et al*, 1992).

Relevance of the Above Studies to the Human

It is now generally well recognized that patients with secondary hypogonadism due either to Kallman syndrome or hypopituitarism can be effectively treated with either pulsatile GnRH or a combination of both FSH and LH to induce spermatogenesis and achieve fertility (Simoni *et al*, 1999). Experimental evidence (detailed above) supports the concept of irreplaceable role for FSH in promoting quantitative and qualitative spermatogenesis in the primate. Compelling clinical evidence also suggests that in some category of infertile/subfertile men, exogenous FSH therapy helps restore production of quality sperms that can fertilize (Phillip *et al*, 1998; Baccetti *et al*, 1997; Gromoll *et al*, 1996; Bartoov *et al*, 1994; Acosta *et al*, 1992; Matsumoto *et al*, 1986).

Recently we have had the opportunity to examine the chromatin status of ejaculated sperm of a group of five human male volunteers who were immunized with oFSH vaccine as part of a phase I clinical study (Moudgal *et al*, 1997b). Like in the monkey study, the washed sperm samples were subjected to both acid and reductant-induced DNA denaturation followed by binding to fluorescent dyes and flow cytometry. The results have clearly indicated that blockade of FSH function even for a limited period equivalent to about one spermatogenic cycle in the human male, leads to production of sperm having defective/inadequate chromatin packaging (Krishnamurthy *et al*, 2000). We infer that as in the case of the monkey this must be a consequence of impaired spermiogenesis. Alterations in the binding to sperm DNA reflecting poor chromatin packaging has been correlated to human infertility (Engh *et al*, 1992 & Sakkas *et al*, 1996). An inverse relationship between dye binding and *in vitro* fertilization has also been noted (Bianchi *et al*, 1996). In the light of varied evidences available, it may not be unreasonable to predict that continuous blockade of FSH function in the human male by immunizing with either FSH/its beta sub-unit or an FSH receptor protein fragment may lead to infertility due to production of sperm incapable of effecting fertilization.

Conclusions

It appears from the above that there is sufficient experimental data currently available to support the use of FSH therapy in infertile/subfertile hormone deficient men to not only augment the quality and quantum of sperm output but also its fertilizing potential. Infertility due to testosterone deficiency alone can be corrected by providing exogenous testosterone therapy. The importance of FSH as a major determinant of sperm quality in primates is supported by experimental and clinical evidence. The summary presented in figure 1 clearly shows that FSH deprivation achieved through various methods affects sperm quality in rodents as well as in primates. At present it is not clear if natural estrogen deficiency/lack does occur in the human and if so whether it is accompanied by testicular dysfunction and infertility. The widely discussed implication of environmental considerations including xenoestrogens affecting fertility in men require further experimental verification along the lines we have discussed in this article (Turner and Sharpe 1997).

Acknowledgements

The investigations mentioned in this article were supported by the Department of Biotechnology (India), Indian National Science Academy, FRSQ (Canada)-INSERM (France) and MRC of Canada.

References

Acosta, A.A., Khalifa, E., and Oehninger, S. (1992) Pure human follicle stimulating hormone has a role in the treatment of severe male infertility by assisted reproduction. Norfolk's total experience. *Hum Reprod.*, **7**, 1067–1072.

Aravindan, G.R., Gopalakrishnan, K., Ravindranath, N. *et al.* (1993) Effect of altering endogenous gonadotropin concentrations on the kinetics of testicular germ cell turnover in the bonnet monkey (Macaca raidata). *J. Endocrinol.*, **137**, 485–495.

Aravindan, G.R., Krishnamurthy, H. and Moudgal, N.R. (1997). Enhanced susceptibility of follicle-stimulating hormone deprived infertile bonnet monkey (M. radiata) spermatozoa to dithiothreitol induced DNA decondensation *in situ. J Androl.*, **18**, 688–697.

Baccetti, B., Strehler, E. and Capitani, S. *et al* (1997) The effect of follicle stimulating hormone therapy on human sperm structure. *Hum Reprod.* **12**, 1955–1968.

Ballachey, B.E., Hohenboken, W.D. and Evenson, D.P. (1987) Heterogeneity of sperm nuclear chromatin structure and its relationship to bull fertility. *Biol Reprod.*, **36,** 915–925.

Bartoov, B., Dov Har-Even. and Eltes, F. (1994) Sperm quality of subfertile males before and after treatment with human follicle-stimulating hormone. *Fertil Steril.*, **61**, 727–734.

Bianchi, P.G., Manicardi, G.C. and Urner, F. (1996) Chromatin packaging and morphology in ejaculated human spermatozoa: evidence of hidden anomalies in normal spermatozoa. *Mol Hum Reprod.*, **2**, 139–144.

Dierich, A., Sairam, M.R., Monaco, L. *et al.* (1998) Impairing follicle-stimulating hormone (FSH) signaling in vivo: targeted disruption of the FSH receptor leads to aberrant gametogenesis and hormonal imbalance. *Proc. Natl Acad Sci USA*, **95**, 13612–13617.

Eddy, E.M., Washburn, T.F., Bunch, D.O. *et al.* (1996) Targeted disruption of the estrogen receptor gene in male mice causes alteration of spermatogenesis and infertility. *Endocrinol*, **137**, 4796–4805.

Engh, E., Clausen, O.P.F. and Scholberg. A. *et al.* (1992) Relationship between sperm quality and chromatin condensation measured by sperm DNA fluorescence using flow cytometry. *Int. J Androl.*, **15**, 407–415.

Gromoll, J., Simoni, M. and Neishlag E. (1996) An activating mutation of the follicle-stimulating hormone receptor autonomously sustains spermatogenesis in a hypophysectomized man. *J. Clin Endocrinol Metab.*, **81**, 1367–1370.

Krishnamurthy, H., Prasanna Kumar, K.M., and Joshi, C.V. *et al.* (2000a) Alternations in the sperm characteristics of FSH immunized men are similar to those of FSH deprived infertile male bonnet monkeys. *J. Androl.*, **21**, 316–327.

Krishnamurthy, H., Danilovich, N, Morales, C.R. and Sairam, M.R. (2000b) Qualitative and Quantitative Decline in Spermatogenesis of the Follicle Stimulating Hormone Receptor Knockout (FORKO) Mouse. *Biol Reprod.*, **62**, 000–000 (in press).

Matsumoto, A.M., Karpas, A.E. and Bremner, W.J. (1986) Chronic human chorionic gonadotropin administration in normal men: Evidence that follicle stimulating hormone is necessary for the maintenance of quantitative normal spermatogenesis in men. *J. Clin Endocrino Metab,* **62**, 1184–1190.

Moudgal, N.R., Ravindranath, N., Murthy, G.S. *et al.* (1992) Long term contraceptive efficacy of vaccine of ovine follicle stimulating hormone in male bonnet monkeys (*Macaca radiata*). *J Reprod. Fertil*, **96**, 91–102.

Moudgal, N.R., Sairam, M.R. and Krishnamurthy H.N. *et al.* (1997a) Immunization of male bonnet monkeys (M. radiata) with a recombinant FSH receptor preparation affects testicular function and fertility. *Endocrinol*, **138**, 3065–3068.

Moudgal, N.R., Murthy, G.S., Prasanna Kumar, K.M. *et al.* (1997b) Responsiveness of human male volunteers to immunization with ovine follicle stimulating hormone (oFSH) Vaccine-Results of a Pilot Study. *Hum Reprod.* **12**, 457–463.

Phillip, M., Arbelle, J.C., Seger, Y, *et al.* (1998) Male hypogonadism due to a mutation in the gene for the β-subunit of follicle stimulating hormone. *New Engl J. Med.*, **338**, 1729–1732.

Raj, H.G.M., Kotagi, S.G., Letellier, R., *et al.* (1991) Active immunization with gonadotropins in the crab eating monkey (*M. fascicularis*): Evaluation for male contraception. In Moudgal, N.R., Yoshinaga, K., Rao, A.J. and Adiga, P.R. (eds), *Perspectives in Primate Reproductive Biology*. Willey Eastern, New Delhi, p. 307–316.

Robertson, K.M., O'Donnell, L., Jones, M.E., *et al.* (1999) Impairment of spermatogenesis in mice lacking a functional aromatase (cyp 19) gene. *Proc Natl Acad Sci USA,* **96**, 7986–91.

Sairam, M.R., Dierich, A., Monaco., *et al.* (1998) Targeted disruption of the FSH receptor leads to aberrant gametogenesis, hormonal imbalances causing infertility/reduced fertility. Proceedings of the 80th annual meeting of the Endocrine Society, USA, Abstract No OR46-3.

Sakkas, D., Urner, F., Bianchi PG, *et al.* (1996) Sperm chromatin anomalies can influence decondensation after intracytoplasmic sperm injection. *Hum Reprod.,* **11**, 837–843.

Sharma R.K. and Das, R.P. (1992) Effect of fertility regulating agents on motility and zona free hamster egg penetration by spermatozoa of bonnet monkey. *Indian J. Exp. Biol.* **30**, 977–981.

Shetty, G., Krishnamurthy, H., Krishnamurthy, H.N. *et al.* (1997) Use of norethisterone and estradiol in mini doses as a contraceptive in the male. Efficacy studies in the adult male bonnet monkey (Macaca radiata) *Contraception*, **56**, 257–265.

Shetty, G., Krishnamurthy, H., Krishnamurthy, H.N., *et al.* (1998) Effect of long-term treatment with aromatase inhibitor on testicular function of adult male bonnet monkeys (M. radiata) *Steroids*, **63**, 414–420.

Simoni, M., Weinbauer, G.F., Gromoll, J., and Nieschlag E. (1999) Role of FSH in male gonadal function. *Ann Endocrinol* (Paris). **60**, 102–106.

Srivastava, A., and Das, R.P. (1992) Sperm production and fertility of bonnet monkeys (M. Radiata) following immunization with oFSH. *Ind J Exp Biol,* **30**, 574–577.

Suresh, R., Medhamurthy, R., and Moudgal, N.R. (1995) Comparative studies on the effects of specific immunoneutralization of endogenous FSH or LH on testicular germ cell transformations in the adult bonnet monkey (Macaca radiata). *Am J Reprod Immunol,* **34**, 35–43.

Tapananien, J.S., Aittomaki, K., Min, J., *et al.*, (1997) Men homozygous for an inactivating mutation of the follicle-stimulating hormone (FSH) receptor gene present variable suppression of spermatogenesis and fertility. *Nat Genet,* **15**, 205–206.

Turner, K.J. and Sharpe, R.M. (1997) Environmental oestrogens-present understanding. *Rev Reprod.,* **2**, 69–73.

Follicular Growth, Ovulation and Fertilization: Molecular and Clinical Basis
Anand Kumar and Amal K. Mukhopadhyay (Eds.)
Narosa Publishing House, New Delhi, India, 2001

23

Hormonal Regulation of Acrosomal Exocytosis

Michael Schaefer[1], Ursula F. Habenicht[2], Günter Schultz[1] and Thomas Gudermann[3]

[1]Institue für Pharmakologie, Universitätsklinikum Benjamin Franklin, Freie Universität Berlin, 14195 Berlin, Germany

[2]Forschungslaboratorien der Schering AG, 13342 Berlin, Germany

[3]Institut für Pharmakologie und Toxikologie, Fachbereich Humanmedizin, Philipps-Universität Marburg, 35033 Marburg, Germany

Capacitation

Capacitation is a prerequisite of sperm to undergo the acrosome reaction, an exocytotic event leading to release of hydrolytic enzymes and substantial reorganization of the plasma membrane (Yanagimachi, 1994). In vitro, capacitation can be achieved in culture media of defined composition including energy substrates, albumin, $NaHCO_3$, and Ca^{2+}. The time-dependent acquisition of fertilizing ability is correlated with changes in sperm metabolism, serine/ threonine and tyrosine phosphorylation, lipid and protein composition of the sperm cell membrane, intracellular ion concentrations, and the development of hyperactivated sperm motility (Yanagimachi, 1994; Cross, 1998; Kopf *et al.*, 1999).

While albumin serves to remove cholesterol from the sperm plasma membrane to bring about membrane fluidity changes, entry of HCO_3^- into the cell is thought to evoke the increase in intracellular pH noted during capacitation. HCO_3^- influx in conjunction wth a rise of $[Ca^{2+}]_i$ modulates cAMP concentrations via effects on sperm adenylyl cyclase (see: Kopf *et al.*, 1999). At least 9 membrane-associated adenylyl cyclases have been cloned that respond to extracellular stimuli via G proteins (Sunahara *et al.*, 1996). The sperm adenylyl cyclase, however, is unique and fundamentally differnt from that of somatic cells. Sperm adenylyl cyclase activity depends on the divalent cation Mn^{2+}, is activated by HCO_3^- and is insensitive to regulation by G proteins (Braun *et al.* 1977; Hildebrandt *et al.*, 1985; Okamura *et al.*, 1991). A soluble adenylyl cyclase insensitive to G protein and forskolin regulation has recently been cloned after purification of enzyme activity from rat testis cytosol (Buck *et al.*, 1999). Subsequent Northern analysis and in situ hybridization indicate that high-level mRNA expression is confined to male germ cells with accumulation of message in round spermatids (Sinclair *et al.*, 2000). Bicarbonate directly stimulates mammalian soluble adenylyl cyclase activity in a pH-dependent manner. Future studies will have to show whether the newly cloned soluble adenylyl cyclase accounts for all functional characeristics of the long sought sperm enzyme or whether additional gene products are to be expected. Immunological studies suggested the expression of the olfactory Ca^{2+}-regulated adenylyl cyclase III on the acrosomal membrane (Gautier-Courteille *et al.*, 1998).

Capacitation of spermatozoa is accompained by tyrosine phosphorylation of a subset of proteins ranging from Mr 40,000 to 120,000 (Visconti and Kopf, 1998). Interestingly, protein tyrosine phosphorylation during capacitation is enhanced by cAMP-dependent serine/threonine protein kinase (PKA) activity downstream of albumin, Ca^{2+} and HCO_3^- exposure. The systematic search for protein substrates which are tyrosine phosphorylated in a cAMP-dependent manner in capacitating sperm initially led to the identification of A-kinase anchoring protein 82 (AKAP82) associated with the fibrous sheath of the flagellum (Turner *et al.*, 1998). AKAPs serve as scaffolding proteins to organize signaling complexes in discrete locations within the cell, and posttranslational modification of AKAPs may help establish the specific flagellar bending pattern conducive to hyperactivated motility required to penetrate the zona pellucida.

Acrosome Reaction

To date, three physiological inducers of the acrosome reaction are known: Subsequent to species-specific bindng of spermatozoa to the egg's extracellular matrix, one of the three major proteins forming the mouse zona pellucida, ZP3, elicits the acrosome reaction (Wassarman, 1999). The steroid hormone progesterone secreted by ovarian follicular cells surrounding the oocyte also triggers acrosomal exocytosis (Osman *et al.*, 1989) as does prostaglandin E produced locally in the fallopian tube and by granulosa cells in the vicinity of the oocyte (Schaefer *et al.*, 1998). ZP3 is generally regarded as the main inducer of acrosomal exocytosis while progesterone and E prostaglandins may serve to prime spermatozoa to effectively respond to ZP3 (Roldan *et al.*, 1994).

The male gametes are among the most highly differentiated cells in mammalian organisms. Due to the high degree of specialization and the resulting morphological and functional differences as compared to somatic cells, signal transduction pathways in spermatozoa leading to acrosome reation are understood poorly. However, it is tempting to speculate that fundamental cellular processes like transmembrane signaling and exocytosis in spermatozoa, involve a similar set of proteins and are subject to regulation by the same basic principles that are also operative in somatic cells. One of the essential featurs of acrosomal exocytosis is an influx of Ca^{2+} from the extracellular space required to promote the fusion between the outer acrosomal membrane and the overlying sperm plasma membrane (Yanagimachi, 1994). The lack of knowledge about signaling steps leading to Ca^{2+} influx and acrosome reaction contrasts with detailed information on the expression and even subcellular localization of signal transduction components (Fig. 1) like receptor tyrosine kinases (see: McLeskey *et al.*, 1998), G-protein-coupled receptors, receptor kinases, G proteins, and effectors such as enzymes and ion channels (summarized in: Schaefer *et al.*, 1998). The delineating of signaling cascades in spermatozoa is an essential first step to understand the physiology of fertilization at the molecular level.

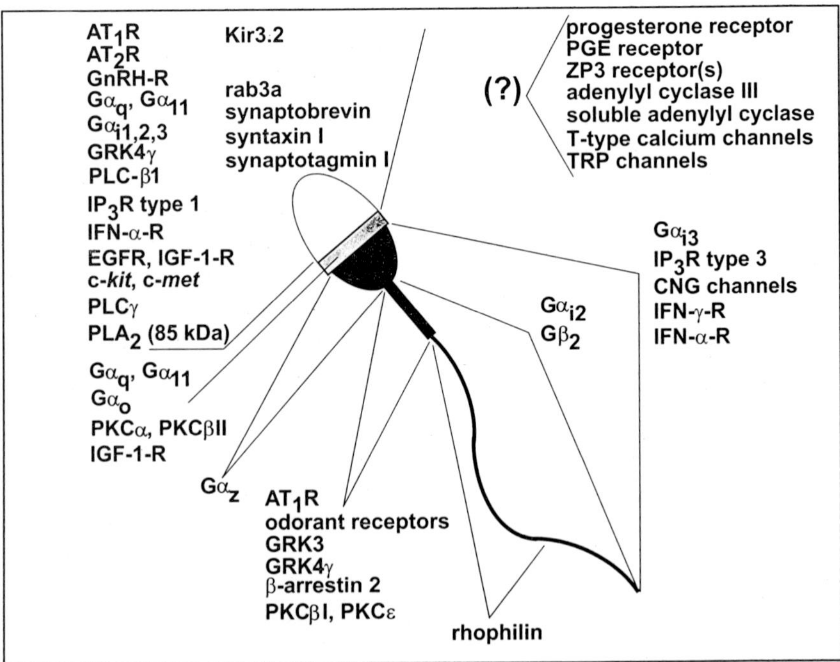

Fig. 1 Examples of signaling proteins expressed in spermatozoa. A variety of proteins known to be involved in signal transduction processes in somatic cells are also expressed in spermatozoa. Their differential expression in the acrosomal compartment, the equatorial region, the postacrosomal region, the middle, principle and end piece of the sperm tail is shown. Those signaling components whose expression is postulated but not proven or whose exact localization in spermatozoa is not known are labeled by a question mark. *Abbreviations used:* AT_1R, AT_1 angiotensin receptor; AT_2R, AT_2 angiotensin receptor; CNG channels, cyclic nucleotide-gated channels; EGFR, epidermal growth receptor; Gα, a subunit of a heterotrimeric guanine nucleotide-binding protein; Gβ, β subunit of a heterotrimeric guanine nucleotide-binding protein; GnRH-R, gonadotropin-releasing hormone receptor; GRK, G-protein-coupled receptor kinase; IFN-γ-R, interferon-γ-receptor; IGF-1-R, insulin-like growth factor-1-receptor; IP_3R, inositol trisphosphate receptor; Kir3.2, a G-protein-gated inwardly rectifying K^+ channel subunit; PGE, prostaglandin E; PKC, protein kinase C; PLA_2, phospholipase A_2; PLC, phospholipase C; TRP channels, transient receptor potential channels; ZP3, zona pellucida protein 3.

Steroid Hormone Signaling in Spermatozoa

In human spermatozoa progesterone and 17α-hydroxyprogesterone induce a rapid influx of Ca^{2+} from the extracellular space and trigger acrosomal exocytosis (Meizel, 1995; Baldi *et al.*, 1999; Blackmore, 1999). A rise of $[Ca^{2+}]_i$ is followed by an increase in phospholipase C activity resulting in the breakdown of phosphatidylinositol-4,5-bisphosphate (Thomas and Meizel, 1989). So far, the identity of phospholipase C isoforms involved in this process still remains obscure, yet phospholipase C-β and -δ species are likely to play a role. Although all phospholipase C isozymes are activated by Ca^{2+} *in vitro*, phospholipase C-δ appears to be more sensitive to Ca^{2+} than the other isoenzymes. Thapsigargin, an inhibitor of Ca^{2+}-ATPases in the endoplasmatic reticulum, is also able to induce acrosomal exocytosis (Meizel and Turner, 1993). As spermatozoa do not contain an endoplasmic reticulum, Ca^{2+} stores which could be emptied by thapsigargin treatment might be located in the acrosome, the mitochondria or in a perinuclear compartment (Blackmore, 1993).

The progesterone effect on spermatozoa cannot be abrogated by potent progesterone receptor antagonists like RU486 or ZK98299. Other synthetic progestins such as norgestrel, medroxyprogesterone acetate, norethindrone, and R5020, which display agonist properties at the nuclear receptor, have no effect on spermatozoa (Blackmore *et al.*, 1991). The steroid hormone exerts a rapid, non-genomic effect on the sperm plasma membrane. Comparative analysis of the effects of progesterone derivatives on Ca^{2+} fluxes through the sperm membrane revealed that progesterone binds to the membranous steroid receptor via its β-face, whereas it binds to the genomic receptor via its α-face. It was further concluded that the active progestins establish a close contact to the sperm progesterone receptor across the C/D-ring upper edge (C-11, C-12, C-17), and that proper orientation of the C-21 methyl group is important for biological activity (Blackmore *et al.*, 1996).

The signal transduction cascade which links binding of progesterone to Ca^{2+} influx still remains elusive. As the addition of AlF_4^-, an activator of heterotrimeric G proteins, does not induce Ca^{2+} entry (Blackmore *et al.*, 1990) and as the cellular response to progesterone is unaffected by pretreatment of spermatozoa with pertussis toxin which uncouples $G_{i/o}$ proteins from G-protein-coupled receptors (Foresta *et al.*, 1993), experimental evidence supporting the notion of G protein involvement in progesterone signaling is lacking. Although incubation of human spermatozoa with progesterone does not lead to increased cAMP production (Schaefer *et al.*, 1998) a cAMP-dependent process has been suggested to be involved in the progesterone action on spermatozoa (Parinaud and Milhet, 1996; Kobori *et al.*, 2000). Activation and aggregation of membranous sperm progesterone receptors stimulates the tyrosine phosphorylation of a 94-kDa phosphoprotein (Tesarik *et al.*, 1993), thus fostering speculations that the sperm progesterone receptor might have tyrosine kinase activity. In addition, the existence of a unique steroid receptor/Cl^- channel complex pertaining to progesterone-induced acrosomal exocytosis has been inferred from studies with antagonists and blockers of $GABA_A$ receptor/Cl^- channels (summarized in: Revelli *et al.*, 1998). However, the pharmacological profile of the extragenomic sperm progesterone receptor is not compatible with that of the neuronal $GABA_A$ receptor/Cl^- channel (Blackmore *et al.*, 1994). The multiple responses of spermatozoa to progesterone have nourished speculations that progesterone might even act via at least three types of cell surface receptors: (1) a Ca^{2+}-permeable channel, (2) a receptor tyrosine kinase, and (3) a Cl^- channel (Revelli *et al.*, 1998).

Recently, a putative functional estrogen receptor was described on human spermatozoa (Luconi *et al.*, 1999). The membranous receptor is recognized by an antibody directed against the ligand-binding domain of the genomic estrogen receptor and has an apparent molecular mass of 29 kDa. 17β-Estradiol induces a modest increase in $[Ca^{2+}]_i$, yet mitigates progesterone-dependent Ca^{2+} fluxes. Thus, it is speculated that estrogens may play a role in the modulation of non-genomic actions of progesterone during the fertilization process (Luconi *et al.*, 1999; Baldi *et al*, 2000).

The molecular identity of progesterone receptors in sperm membranes is still a controversial and utterly unresolved issue. Inhibition of progesterone-induced Ca^{2+} influx and acrosome reaction has been observed after incubating sperm with a monoclonal antibody (C-262) raised against the C-terminal steroid binding domain of the genomic progesterone receptor (Sabeur *et al.*, 1996), and the presence of trace amounts of progesterone receptor mRNA in mature human spermatozoa has recently been reported (Sachdeva *et al.*, 2000). Therefore, the contribution of a membrane-associated genomic progesterone receptor to the steroid effects observed in spermatozoa cannot be excluded with certainty. In addition, antisera raised against a porcine liver endomembrane progesterone-binding protein inhibit progesterone-induced acrosomal exocytosis by approximately 30% (Buddhikot *et al.*, 1999) and have minor effects on progesterone-dependent Ca^{2+} transients (Falkenstein *et al.*, 1999). Immunofluorescence studies recognized antigen in the posterior region of human spermatozoa (Buddhikot *et al.*, 1999). These results were interpreted to mean that the porcine liver progesterone binding protein might be part of a larger progesterone binding complex on the sperm plasma membrane.

Progesterone may represent a physiological endogenous ligand of sigma binding sites (Su *et al.*, 1988) which are abundant in testis (Wolfe *et al.*, 1989). The purified and cloned receptor binds progestrone with an affinity in the upper nM range (Hanner *et al.*, 1996), which is well within the concentration range of progesterone acting on human spermatozoa. Thus, it is tempting to speculate that progesterone interacts with sigma receptors on the sperm plasma membrane. However, Ca^{2+} fluxes in human spermatozoa mediated by progesterone and the steroidal sigma agonist RU 3117 are not affected by the high-affinity sigma ligands haloperidol and DTG [1,3-di-(2-tolyl) guanidine] (Schaefer *et al.*, 2000a). Furthermore, RU 3117 desensitizes spermatozoa to further progesterone challenge while leaving the response to prostaglandin E_1 untouched. Collectively these results would corroborate the notion of RU 3117 interacting with a membranous progesterone receptor rather than progesterone binding to sigma receptors. Thus sigma receptors appear not to be involved in the action of progesterone on human spermatozoa (Schaefer *et al.*, 2000a), and for the time being identification of a sperm membrane progesterone receptor still remains the holy grail for all researchers interested in rapid, non-genomic steroid effects.

Prostaglandin Signaling in Spermatozoa

Early reports on the hamster and guinea pig sperm indicated a role for E prostaglandins and $PGF_{2\alpha}$ for the sperm acrosome reaction (Meizel and Turner, 1984; Joyce *et al.*, 1987). With the help of the Ca^{2+} indicator quin-2, Aitken *et al.*, (1986) observed a rise in $[Ca^{2+}]_i$ in human sperm subsequent to stimulation with PGE_1 and PGE_2. However, it was recently shown that PGE_1 binds to human sperm membranes with high affinity and induces a pertussis toxin-insensitive, transient rise of $[Ca^{2+}]_i$ (Schaefer *et al.*, 1998; Shimizu *et al.*, 1998). In a population of immobilized fura-2 loaded human spermatozoa, every single sperm responds to PGE_1 challenge with a Ca^{2+} transient which relies on the presence of extracellular Ca^{2+} (Fig. 2A). PGE_1-induced Mn^{2+} influx and subsequent quenching of fura-2 fluorescence demonstrates that the increased $[Ca^{2+}]_i$ originates from an influx of cations from the extracellular space (Fig. 2B). The kinetic profile of PGE-dependent Ca^{2+} transients is similar to that observed after progesterone addition. Interestingly, sequential addition of both agonists does not lead to cross-desensitization, and a co-stimulation of sperm with PGE_1 and progesterone has an additive effect (Schaefer *et al.*, 1998). These observations suggest that distinct signaling mechanisms are responsible for progesterone- and PGE_1-mediated Ca^{2+}-influx in human sperm.

E prostaglandins were found to be the only prostanoids with agonistic properties (EC_{50} value for PGE_1: < 10 nM). Pharmacological characteristics of the PGE effect on spermatozoa are not compatible with those of any cloned prostanoid receptor indicating the expression of a distinct membrane receptor. Stimulation of human sperm membranes with PGE results in GTP loading of $G_{q/11}$ proteins (Schaefer *et al.*, 1998). Thus, in human sperm PGE induces Ca^{2+} influx and acrosome reaction via a $G_{q/11}$-coupled E prostanoid receptor that so far has eluded isolation and molecular cloning.

The characteristics of the cation permeable channel(s) activated by PGE and by progesterone are identical: In addition to Ca^{2+} and Mn^{2+} other divalent cations like Sr^{2+} and Ba^{2+} as well as Na^+ (Foresta *et al.*, 1993) also enter human sperm upon progesterone and PGE_1 stimulation (Schaefer *et al.*, 1998). Blockers of voltage-gated L-type and T-type Ca^{2+} channels as well as L-*cis*-diltiazem, an effective blocker of cGMP-gated cation channels, do not impair progesterone- and PGE_1-induced Ca^{2+} transients (Schaefer *et al.*, 1998). However, agonist-induced Ca^{2+} influx can be attenuated effectively by lanthanids (La^{3+}, Gd^{3+}) and Zn^{2+} with IC_{50} values of 2, 10, and 30 μM, while Ni^{2+}, Cd^{2+}, and Co^{2+} are ineffective up to a concentration of 1 mM. Thus, it appears that the distinct signaling pathways initiated by progesterone and PGE_1 may converge on a common non-selective cation channel whose molecular identity is unknown.

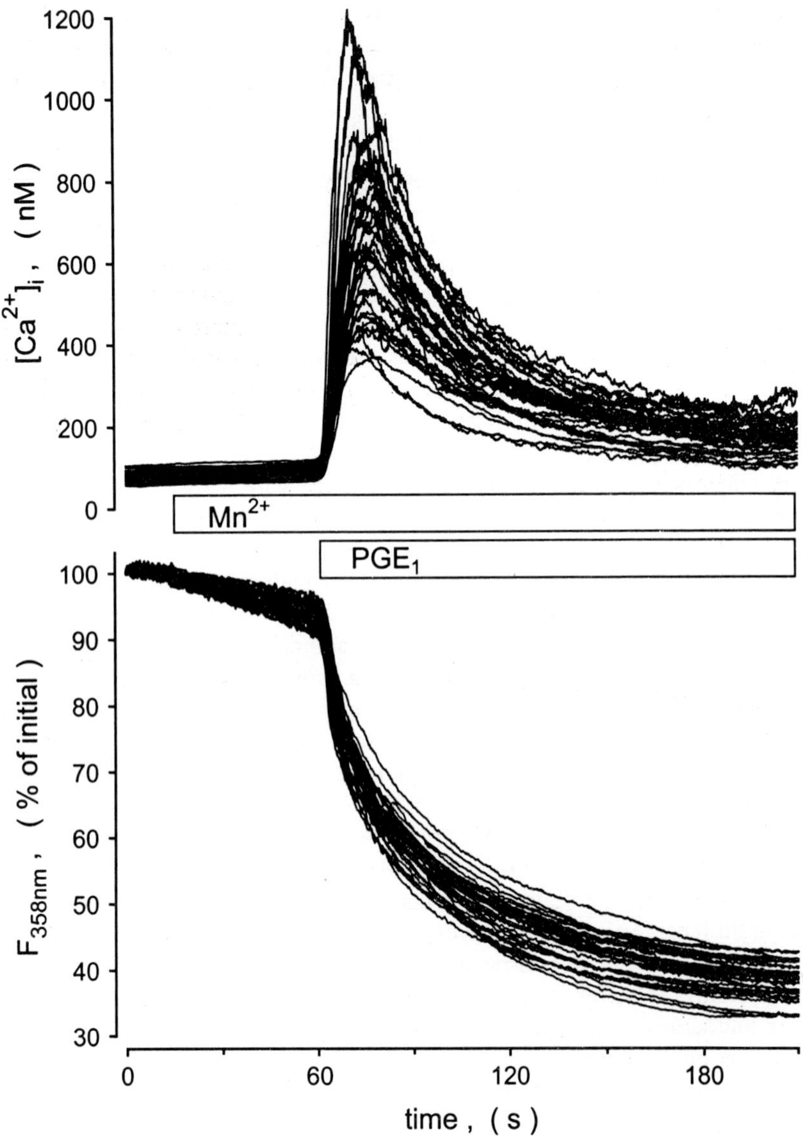

Fig. 2 Prostaglandin E_1 induces a transient elevation of $[Ca^{2+}]_i$ and Mn^{2+} influx in single sperm heads. Capacitated human spermatozoa were immobilized on poly-L-lysine-coated glass coverslips and loaded with the fluorescent $[Ca^{2+}]_i$-indicator fura-2. Fura-2 was alternately excited at 340 nm, 358 nm, and 380 nm. The bath solution was supplemented with Mn^{2+} (500 μM) as indicated by the bar. Mn^{2+}-influx was monitored at the isosbestic wavelength of 358 nm by observing the formation of a nonfluorescent fura-2-Mn^{2+}-complex. Spermatozoa were subsequently stimulated by adding 1 μM prostaglandin E_1 to the bath solution. Mean fluorescence over the regions of interest was used to calculate $[Ca^{2+}]_i$. Regions of interest were defined over head regions of 35 individual spematozoa. Background signals were assessed after lysing cells with reduced 0.1% Triton X-100 and substracted from the data.

In human semen, sperm are exposed to high PGE_1 concentrations (approx. 80 μM) contributed by the seminal vesicles (Mann and Mann, 1981). Prostatic secretions which are added to the ejaculate prior to those of the seminal vesicles, are responsible for mM Zn^{2+} concentrations in human semen, thus protecting spermatozoa from a massive Ca^{2+} influx which would profoundly impair the fertilizing potential of sperm (Aitken *et al.*, 1996) (Fig. 3). Zn^{2+} ions may stabilize sperm membranes during storage and ejaculation, and removal of Zn^{2+} appears to be required to prepare sperm for fertilization (Andrews *et al.*, 1994). The latter view is supported by the observation that incubation of hamster sperm with Zn^{2+} chelators accelerates capacitation (Andrews and Bavister, 1989) and that Zn^{2+} inhibits human sperm motility and acrosome reaction (Riffo *et al.*, 1992). Therefore, it is likely that one important role of Zn^{2+} in the male reproductive tract is that of an endogenous cation channel blocker to protect and maintain spermatozoa in a transitory quiescent state. During their ascent in the female reproductive tract, sperm escape mM Zn^{2+} concentrations and are exposed only to serum concentrations of Zn^{2+} (approx. 20 μM) nearly all of which is bound to albumin (see Fig. 3).

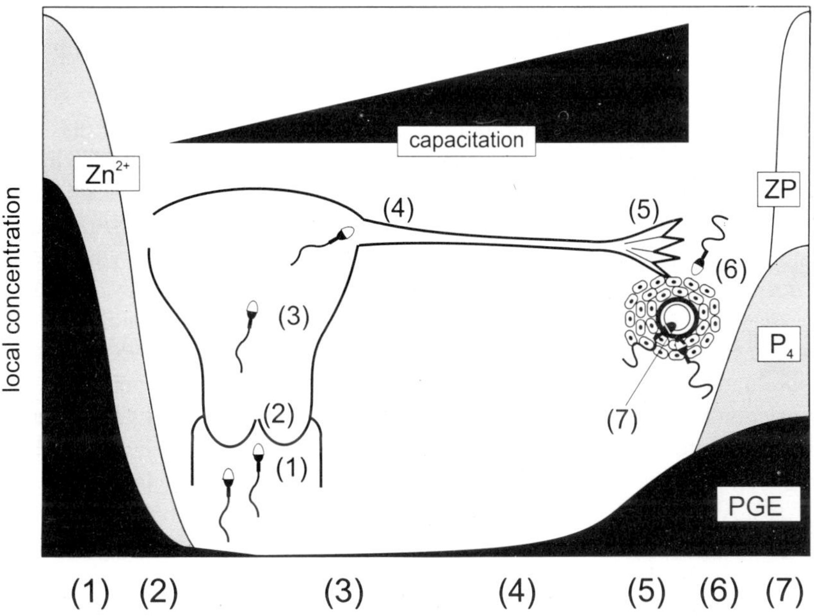

Fig. 3 Schematic overview of local factors influencing sperm capacitation and acrosome reaction. In human semen spermatozoa are exposed to high concentration of prostaglandin E (PGE) [(1)] but are protected from massive Ca^{2+} influx by mM concentrations of Zn^{2+}. Spermatozoa capacitate during their ascent in the female reproductive tract [(2) cervix, (3) corpus uteri, (4) oviduct] and are further on exposed to serum concentrations of Zn^{2+}. In the vicinity of the ovulated egg sperm are confronted with rising concentrations of E prostaglandins produced within the fallopian tube [(5) ampulla] and progesterone (P_4) produced locally by granulosa cells surrounding the oocyte [(6)]. These local factors prime spermatozoa to effectively undergo acrosomal excytosis upon contact with the zona pellucida [ZP,(7)].

When approaching the ovulated egg, concentrations of E prostaglandins produced locally rise, and the combined stimulation of sperm by PGE and μM concentrations of progesterone in the cumulus oophorus may then help prime spermatozoa for effective acrosomal exocytosis after binding to the zona pellucida (Fig. 3).

Zona Pellucida-Mediated Signal Transduction

The plasma membrane of all mammalian eggs is completely surrounded by a zona pellucida, composed of only three glycoproteins, ZP1, ZP2, and ZP3 (Wassarman, 1999). Each of these proteins is heterogeneously glycosylated with both complex-type N-linked and O-linked oligosaccharides. In the mouse, the sperm-binding and acrosome reaction-inducing activities of the intact zona pellucida are conferred by ZP3. While ZP3 preferentially interacts with acrosome-intact spermatozoa, ZP2 is responsible for binding of acrosome-reacted sperm via the inner acrosomal membrane (Kopf *et al.*, 1999; Wassarman, 1999). The molecular identity and function of sperm-associated ZP3-binding proteins/receptors has been discussed controversially for years. Four candidates have received the most attention and will be introduced briefly. For a more thorough overview the reader is referred to recent review articles on this topic (Kopf *et al.*, 1999; Wassarman, 1999).

(1) In the mouse, a membranous β-1-4 galactosyltransferase has been proposed as a receptor of N-acetylglucosamin-containig oligosaccharide ligands of the zona pellucida and as an activator of G_i proteins (Miller *et al.*, 1992; Gong *et al.*, 1995). However, genetic disruption of the galactosyltransferase gene yields null males which are fertile demonstrating that galactosyltransferase is not absolutely critical for sperm-egg interaction (Lu and Shur, 1997).

(2) sp56 is a peripheral sperm membrane protein involved in the species-specific binding of mouse sperm to ZP3 (Bookbinder *et al.*, 1995). However, re-evaluation of the cellular localization of sp56 in mouse sperm revealed association with the acrosomal matrix (Foster *et al.*, 1997) and calls into question its function to mediate sperm binding to the zona pellucida. Furthermore, it is unclear whether sp56 plays any role in cellular signaling processes.

(3) A 95-kDa phosphoprotein has been proposed to correspond to a receptor tyrosine kinase called zona receptor kinase (ZRK) or Hu9 which specifically binds ZP3 and gets activated by this interaction (Burks *et al.*, 1995). Later it was realized that ZRK corresponds to c-mer, a proto-oncogene of the axl family of tyrosine kinases (Bork, 1996) whose role in the fertilization process is obscure.

(4) Spermadhesins are small (12–16 kDa) boar sperm-associated proteins which bind zona pellucida glycoproteins (Sanz *et al.*, 1992; see: McLeskey *et al.*, 1998). a signaling function of these proteins has not been described.

(5) Hardy and Garbers (1995) purified a sperm membrane protein, called zonadhesin, which is expressed in spermatids and in the apical region of the sperm head and which binds to the zona pellucida in a species-specific manner. Mouse zonadhesin has a single transmembrane span and contains

multiple structural motifs (e.g. mucin, EGF, von Willebrand factor-related D domains) thought to be involved in cell adhesion processes within its large extracellular N-terminus (Gao and Garbers, 1998). So far, any cell signaling function of zonadhesin is unknown. Zonadhesin's D domains are homologous to von Willebrand factor type D domains and to α-tectorin, a major non-collagenous protein of the tectorial membrane in the inner ear (Legan *et al.*, 1997). Interestingly, β-tectorin is characterized by a single zona pellucida domain. Missense mutations in the zonadhesin-like domain of α-tectorin as well as in the zona pellucida domain of β-tectorin are associated with autosomal dominant hearing loss (Alloisio *et al.*, 1999) indcating that the two tectorial proteins interact in a way similar to sperm-egg adhesion (Legan *et al.*, 1997).

While there are still many open questions with regard to the identity of ZP3 receptors on spermatozoa, relevant signal transduction pathways in sperm initiated after sperm-zona pellucida interaction have been deciphered. ZP3-dependent signaling in spermatozoa appears to involve G_i-proteins. Incubation of sperm membranes with solubilized zona pellucida results in selective activation of G_{i1} and G_{i2} (Ward *et al.*, 1992, 1994). ADP-ribosylation and thus inactivation of sperm G_i proteins by pertussis toxin has been reported to inhibit zona pellucida-induced acrosomal exocytosis (Lee *et al.*, 1992). Stimulation of sperm with ZP3, however, affects the concentration of several signaling molecules like cAMP, inositol trisphophate, diacylglycerol, arachidonic acid, phosphatidic acid and additionally alters intracellular ion concentrations (Ca^{2+}, Na^+, K^+, H^+) (Kopf *et al.*, 1999). These multiple ZP3 effects on sperm may be explained by envisioning ZP3 as a multivalent ligand engaging various different sperm membrane receptors characterized by distinct signaling properties.

Molecular Ca^{2+} Influx Mechanisms in Spermatozoa

A crucial feature of ZP3-induced acrosomal exocytosis is an elevation of sperm $[Ca^{2+}]_i$. High voltage- and low voltage-activated calcium channels have been implicated in ZP3-dependent Ca^{2+} influx preceeding the acrosome reaction (Darszon *et al.*, 1996). As indicated by reverse transcription polymerase chain reaction experiments using spermatogenic cell RNA, the genes for the α_{1A} and α_{1E} subunits of high voltage-activated calcium channels may be expressed during mouse spermatogenesis (Liévano *et al.*, 1996). However, P-, Q-, or R-type high voltage-activated currents expected to result from the expression of the above mentioned channel α subunits have not been detected during spermatogenesis. On the contrary, T-type low voltage-activated Ca^{2+} currents can be detected in mouse spermatogenic cells by whole cell patch clamp methods (Arnoult *et al.*, 1996; Liévano *et al.*, 1996). T-type Ca^{2+} currents in spermatogenic cells can be blocked by several organic and inorganic substances with the following rank order of potency: PN200–110 > nifedipine > pimozide > Ni^{2+} > verapamil > amiloride > Cd^{2+} (Arnoult *et al.*, 1998). The expression of T-type low voltage-activated Ca^{2+} channels in spermatozoa is hard to prove because patch clamp recordings in spermatozoa pose a daunting technical challenge. High temporal resolution (2–4 msec) fluorescence imaging methodology recently provided further insight into the role of T-type channels in ZP3-induced Ca^{2+} influx and acrosome reaction (Arnoult *et al.*, 1999). Addition of zona pellucida glycoprotein extracts to mouse sperm produced a rapid elevation of $[Ca^{2+}]_i$ which returned to basal levels after approx. 180 msec (Arnoult *et al.*, 1999). Such rapid Ca^{2+} transients which can unequivocally be classified as T-type currents by pharmacological means, cannot solely account for the delayed (several min) and sustained $[Ca^{2+}]_i$ elevations observed in capacitated sperm populations exposed to solubilized zona pellucida (see below).

A characteristic feature of low voltage-activated currents is a low threshold for voltage-dependent inactivation. Therefore, sperm must maintain a hyperpolarized membrane potential prior to interaction with the zona pellucida. Previous assessments of the sperm membrane potential were derived from population-averaged data of sperm suspensions and suggested that sperm were sufficiently depolarized to result in complete inactivation of low voltage-activated currents. Arnoult *et al.*, (1999) correlated the membrane potential of single sperm with the capacitation state and showed that the membrane potential of capacitated single sperm is sufficiently hyperpolarized (–60 to –80 mV) to relieve voltage-dependent inactivation. Hyperpolarization of capacitating sperm (membrane potential of uncapacitated sperm: –20 to –30 mV) may be mediated by an increase in K^+ permeability. The K^+ channels responsible for this affect have not been identified yet. In this regard it is of interest to note that a novel splice varient of a G-protein-gated inwardly rectifying K^+ channel subunit (Kir3.2) is predominantly expressed in the acrosomal vesicles of mouse spermatids (Inanobe *et al.*, 1999).

The ZP3 contact depolarizes the sperm membrane potential (to values of approx. –25 mV) sufficiently enough to activate T-type Ca^{2+} currents. Depolarization is mediated by a voltage-independent cation channel permeable to various mono- and divalent cations (Na^+, Ca^{2+}, Co^{2+}, Mn^{2+}, Ni^{2+}). This non-selective cation channel can be blocked by lanthanids or large organic cations such as N-methyl-*D*-glucamin (see: Florman *et al.*, 1998). Sperm cation channels with similar characteristics are activated by progesterone or PGE (Aitken, 1997; Foresta and Rosato, 1997; Schaefer *et al.*, 1998). Therefore, it is a tantalizing hypothesis to assume that progesterone, PGE and ZP3 all activate the same species of non-selective cation channels in the sperm plasma membrane. So far, the molecular identity of the cation channel(s) involved has not been revealed. Mammalian members of the *Drosphila* transient receptor potential (TRP) channel family are potential candidates, because TRP channels are voltage-independent, fairly non-selective cation channels which are activated in a receptor/phospholipase C-dependent manner (Hofmann *et al.*, 1999, 2000; Schaefer *et al.*, 2000b). In addition, TRP channels might be involved in store-operated calcium entry (Birnbaumer *et al*, 1996; Hofmann *et al*, 2000; Philipp *et al.*, 2000). Mouse TRPC2 was reported to be mainly expressed in testis (Vannier *et al.*, 1999), and the expression of a truncated bovine ortholog is restricted to spermatocytes (Wissenbach *et al.*, 1998). Cyclic nucelotide-gated (CNG) cation channels form a second family of attractive candidates to mediate a depolarizing cation influx, and these channels are the only ones whose expression in sperm has been proven at the molecular level (Weyand *et al.*, 1994). CNG channels are expressed along the flagellum of sperm, and their expression pattern may provide the structural basis for the control of flagellar beating (Weisner *et al.*, 1998).

The current model of ZP3-induced ion fluxes commences with the activation of a non-selective cation channel by a receptor-mediated mechanism (Florman *et al.*, 1994, 1999). The resulting transient increase in $[Ca^{2+}]_i$ is not sufficient to trigger acrosomal exocytosis. However, the accompanying Na^+ influx depolarizes the sperm membrane and thereby activates T-type Ca^{2+} channels. Due to their rapid voltage-dependent inactivation low voltage-activated Ca^{2+} channels can only account for a short-lived Ca^{2+} transient unable to trigger the acrosome reaction. It is further speculated that a ZP3-dependent G_i-mediated transient alkalinization in concert with the afore mentioned Ca^{2+} transients promote a sustained $[Ca^{2+}]_i$ elevation potentially mediated by a store-operated mechanism. This sustained rise of $[Ca^{2+}]_i$ then leads to acrosome reaction (Florman *et al.*, 1999).

Conclusions

Outlining signal transduction cascades in spermatozoa is a first essential step to understand the process of fertilization at the molecular level. Still there are many open questions with regard to the identity of receptor and effectors involved in the initiation of acrosomal exocytosis. The major goal of future studies will be to identify and characterize pertinent signaling proteins which on the one hand may turn out to be instrumental in deciphering idiopathic male infertility and on the other hand may become useful drug targets to modify the fertilizing ability of spermatozoa.

Acknowledgements

The authors' own work reported herein was funded by the Deutsche Forschungsgemeinschaft (DFG), Bundesministerium fur Bildung, Wissenschaft, Forschung und Technologie (BMBF), and Stiftung P.E. Kempkes, Marburg, Germany. Whenever possible, review articles rather than original reports were listed. The authors apologize to all researchers whose work could not be cited due to strict space limitations.

References

Aitken, R.J. (1997) The extragenomic action of progesterone on human spermatozoa. *Hum. Reprod.,* **12**, 38–42.

Aitken, R.J., Buckingham, D.W. and Irvine, D.A. (1996) The extragenomic action of progesterone on human spermatozoa: Evidence for a ubiquitous response that is rapidly down-regulated. *Endocrinology,* **137**, 3999–4009.

Aitken, R.J., Irvine, S. and Kelly, R.W. (1986) Significance of intracellular calcium and cyclic adenosine 3',5'-monophosphate in the mechanisms by which prostaglandins influence human sperm function. *J. Reprod. Fertil.*, **77**, 451–462.

Alloisio, N., Morle, L., Bozon, M., Godet, J., Verhoeven, K., Van Camp, G., Plauchau, H., Muller, P., Collet, L. and Lina-Granade, G. (1999) Mutation in the zonadhesin-like domain of α-tectorin associated with autosomal dorminant non-syndromic hearing loss. *Eur. J. Hum. Genet.,* **7**, 255–258.

Andrews, J.C., Nolan, J.P., Hammerstedt, R.H. and Bavister, B.D. (1994) Role of zinc during hamster sperm capacitation. *Biol. Reprod.*, **51**, 1238–1247.

Andrews, J.C. and Bavister B.D. (1989) capacitation of hamster spermatozoa with the divalent cation chelators D-penicillamine, t-histidine and L-cysterine in a protein free culture medium. Gamete Res. **23**, 159–170.

Arnoult, C., Cardullo, R.A., Lemos, J.R. and Florman, H.M. (1996) Activation of mouse sperm T-type Ca^{2+} channels by adhesion to the egg zona pellucida. *Proc. Natl. Acad. Sci. USA*, **93**, 13004–13009.

Arnoult, C., Villaz, M. and Florman, H.M. (1998) Pharmacological properties of the T-type Ca^{2+} current of mouse spermatogenic cells. *Mol. Pharmacol.,* **53**, 1104–1111.

Arnoult, C., Kazam, I.G., Visconti, P.E., Kopf, G.S., Villaz, M. and Florman, H.M. (1999) Control of the low voltage-activated calcium channel of mouse sperm by egg ZP3 and by membrane hyperpolarization during capacitation. *Proc. Natl. Acad. Sci. USA,* **96**, 6757–6762.

Baldi, E., Luconi, M., Bonaccorsi, L., Maggi, M., Francavilla, S., Gabriele, A., Properzi, G. and Forti, G. (1999) Nongenomic progesterone receptor on human spermatozoa: Biochemical aspects and clinical implications. *Steroids,* **64**, 143–148.

Baldi, E., Luconi, M., Muratori, M., Forti, G. (2000) A novel functional estrogen receptor on human sperm membrane interferes with progesterone effects. *Mol. Cell. Endocrinol.*, **161**, 31–35.

Birnbaumer, L., Zhu, X., jiang, M., Boulay, G., Peyton, M., Vannier, B., Brown, D., Platano, D., Sadeghi, H., Stefani, E. and Birnbaumer, M. (1996) On the molecular basis and regulation of cellular capacitative calcium entry: roles for trp proteins. *Proc. Natl. Acad. Sci. USA*, **93**, 15195–15202.

Blackmore, P.F. (1993) Thapsigargin elevates and potentiates the ability of progesterone to increase intracellular free calcium in human sperm: possible role of perinuclear calcium. *Cell Calcium,* **14**, 53–60.

Blackore, P.F. (1999) Extragenomic actions of progesterone in human sperm and progesterone metabolites in human platelets. *Steroids*, **64**, 149–156.

Blackmore, P.F., Beebe, S.J., Danforth, D.R. and Alexander, N. (1990) Progesterone and 17α-hydroxyprogesterone: novel stimulators of calcium influx in human sperm. *J. Biol. Chem.*, **265**, 1376–1380.

Blackmore, P.F., Fisher, J.F., Spilman, C.H. and Bleasdale, J.E. (1996) Unusual steroid specificity of the cell surface progesterone receptor on human sperm. *Mol. Pharmacol.,* **49,** 727–739.

Blackmore, P.F., Im, W.B. and Bleasdale, J.E. (1994) The cell surface progesterone receptor which stimulates calcium influx in human sperm is unlike the A ring reduced steroid site on the $GABA_A$ receptor/chloride channel. *Mol. Cell. Endocrinol.*, **104,** 237–243.

Blackmore, P.F., Neulen, J., Lattanzio, F. and Beebe, S.J. (1991) Cell surface-binding sites for progesterone mediate calcium uptake in human sperm. *J. Biol. Chem.*, **266,** 18655–18659.

Bookbinder, L.H., Cheng, A. and Bleil, J.D. (1995) Tissue- and species-specific expression of sp56, a mouse sperm fertilization protein. *Science,* **269**, 86–89.

Bork, P. (1996) Sperm-egg binding protein or proto-oncogene? *Science*, **271**, 1431–1432.

Braun, T., Frank, H., Dods, R. and Spepsenwol, S. (1977) Mn^{2+}-sensitive, soluble adenylate cyclase in rat testis. Differentiation from other testicular nucleotide cyclases. *Biochem. Biophys, Acta,* **481**, 227–235.

Buck, J., Sinclair, M.L., Schapal, L., Cann, M.J. and Levin, L.R. (1999) Cytosolic adenylyl cyclase defines a unqiue signaling molecule. *Proc. Natl. Acad. Sci. USA*, **96**, 79–84.

Buddhikot, M., Falkenstein, E., Wehling, M., and Meizel, S. (1999) Recognition of a human sperm surface protein involved in the progesterone-initiated acrosome reaction by antisera against an endomembrane progestrone binding protein from porcine liver. *Mol. Cell, Endocrinol.*, **158**, 187–193.

Burks, D.J., Carballada, R., Moore, H.D.M. and Saling, P.M. (1995) Interaction of a tyrosine kinase from human sperm with the zona pelucida at fertilization. *Science*, **269**, 83–86.

Chen, Y., Cann, M.J., Litvin, T.N., Lowgenko, V., Sinclair, M.L., Levin, L.R. and Buck, J. (2000) Soluble adenynyl cyclase as an evolutionarily conserved bicarbonate sensor. Since, **289**, 625–628.

Cross, N.L. (1998) Role of cholesterol in sperm capacitation. *Biol. Reprod.*, **59**, 7–11.

Darszon, A., Lievåno, A. and Beltran, C. (1996) Ion channels: key elements in gamete signaling. *Curr. Top. Dev. Biol.*, **34**, 117–167.

Falkenstein, E., Heck, M., Gerdes, D., Grube, D., Christ, M., Weigel, M., Buddhikot, M., Meizel, S. and Wehling, M. (1999) Specific progesterone binding to a membrane protein and related nongenomic effects on Ca^{2+}-fluxes in sperm. *Endocrinology*, **140**, 5999–6002.

Florman, H.M. (1994) Sequential focal and global elevations of sperm intracellular Ca^{2+} are initiated by the zona pellucida during acrosomal exocytosis. *Dev. Biol.*, **165**, 152–164.

Florman, H.M., Arnoult, C., Kazam, I.G., Li, C. and O'Toole, C.M.B. (1998) A perspective on the control of mammalian fertilization by egg-activated ion channels in sperm: a tale of two channels. *Biol. Reprod.*, **59**, 12–16.

Florman, H.M., Arnoult, C., Kazam, I.G., Li, C. and O'Toole, C.M.B. (1999) Calcium channels of mammalian sperm: Properties and role in fertilization. In The Male Gamete: From Basic Science to Clinical Application, eds. Gagnon, C. (Cache River Press, Vienna, IL), pp. 187–193.

Foresta, C. Rossato, M. (1997) Calcium influx pathways in human spermatozoa. *Mol. Hum. Reprod.*, **3**, 1–4.

Foresta, C., Rossato, M. and Di Virgilio, F. (1993) Ion fluxes through the progesterone activated channel of the sperm plasma membrane. *Biochem. J.*, **294**, 279–283.

Foster, J.A., Friday, B.B., Maulit, M.T., Blobel, C., Winfrey, V.P., Olson, G.E., Kim, K.S. and Gerton, G.L. (1997) AM67, a secretory component of the guinea pig sperm acrosomal matrix, is related to mouse sperm protein sp56 and the complement component 4-binding proteins. *J. Biol. Chem.*, **272**, 12714–12722.

Gao, Z. and Garbers, D.L. (1998) Species diversity in the structure of zonadhesin, a sperm-specific membrane protein containing multiple cell adhesion molecule-like domains, *J. Biol. Chem.*, **273**, 3415–3421.

Gautier-Courteille, C., Salanova, M. and Conti, M. (1998) The olfactory adenylyl cyclase III is expressed in at germ cells during spermiogenesis. *Endorcrinology*, **139**, 2588–2599.

Gong, X., Dubois, D.H., Miller, D.J. and Shur, B.D. (1995) Activation of a G protein complex by aggregation of β-1,4-galactosyltransferase on the surface of sperm. *Science*, **269**, 1718–1721.

Hanner, M., Moebius, F.F., Flandorfer, A., Knaus, H.G., Striessnig, J., Kempner, E. and Glossmann, H. (1996) Purification, molecular cloning, and expression of the mammalian sigma1-binding site, *Proc. Natl. Acad. Sci. USA*, **93**, 8072–8077.

Hardy, D.M. and Garbers, D.L. (1995) A sperm membrane protein that binds in a species-specific manner to the egg extracellular matrix is homologous to von Willebrand factor. *J. Biol. Chem.*, **270**, 26025–26028.

Hildebrandt, J.D., Codina, J., Tash, J.S., Kirchick, H.J., Lipschultz, L., Sekura, R.D. and Birnbaumer, L. (1985) The membrane-bound spermatozoal adenylyl cyclase system does not share coupling characteristics with somatic cell adenylyl cyclases. *Endocrinology*, **116**, 1357–1366.

Hofmann, T., Obukhov, A.G., Schaefer, M., Harteneck, C., Gudermann, T. and Schultz, G. (1999) Direct activation of human TRPC6 and TRPC3 channels by diacylglycerol. *Nature*, **397**, 259–263.

Hofmann, T., Schaefer, M., Schultz, G. and Gudermann, T. (2000) Transient receptor potential channels as molecular susbtrates of receptor-mediated cation entry. *J. Mol. Med.*, **78**, 14–25.

Inanobe, A., Horio, Y., Fujita, A., Tanemoto, M, Hibino, H, Inageda, K. and Kurachi, Y. (1999) Molecular cloning and characterization of a novel splicing varient of the Kir3.2 subunit predominantly expressed in mouse testis. *J. Physiol.,* **521**, 19–30.

Joyce, C.L., Nuzzo, N.A., Wilson, L. Jr. and Zaneveld, L.J. (1987) Evidence for a role of cyclooxygenase (prostaglandin synthetase) and prostaglandins in the sperm acrosome reaction and fertilization. *J. Androl.,* **8**, 74–82.

Kobori, H., Miyazaki, S. and Kuwabara, Y. (2000) Characterization of intracellular Ca^{2+} increase in response to progesterone and cyclic nucleotides in mouse spermatozoa. *Biol. Reprod.* **63**, 113–120.

Kopf, G.S., Ning, X.P., Visconti, P.E., Purdon, M., Galantino-Homer, H. and Fornés, M. (1999) Signaling mechanisms controlling mammalian sperm fertilization competence and activation. In *The Male Gamete: From Basic Science to Clinical Application*, eds. Gagnon, C. (Cache River Press, Vienna, IL), pp. 105–118.

Lee, M.A., Check, J.H. and Kopf, G.S. (1992) A guanine nucleotide-binding regulatory protein in human sperm mediates acrosomal exocytosis induced by the human zona pellucida. *Mol. Reprod. Dev.*, **31**, 78–86.

Liévano, A., Santi, C.M., Serrano, C.J., Treviño, C.L., Bellvé, A.R., Hernández-Cruz, A. and Darszon, A. (1996) T-type Ca^{2+} channels and α_{1E} expression in spermatogenic cells, and their possible relevance to the sperm acrosome reaction. *FEBS Lett.* **388**, 150–154.

Legan, P.K., Rau, A., Keen, J.N. and Richardson, G.P. (1997) The mouse tectorins. Modular matrix proteins of the inner ear hommologous to components of the sperm-egg adhesion system. *J. Biol. Chem.*, **272**, 879–8801.

Lu, Q. and Shur, B.D. (1997) Sperm from β 1,4-galactosyltransferase-null mice are refractory to ZP3-induced acrosome reactions and penetrate the zona pellucida poorly, *Development*, **124**, 4121–4131.

Luconi, M., Muratori, M., Forti, G. and Baldi, E. (1999) Identification and characterization of a novel functional estrogen receptor on human sperm membrane that interferes with progesterone effects. *J. Clin. Endocrinol. Metab.,* **84**, 1670–1678.

Mann, T. and Lutwak-Mann, C. (1981) In *Male Reproductive Function and Semen: Themes and Trends in Physiology, Biochemistry and Investigative Andrology* (Springer, Berlin), pp. 269–336.

McLeskey, S.B., Dowds, C., Caballada, R., White, R.R. and Saling P.M. (1998) Molecules involved in mammalian sperm-egg interaction. *Int. Rev. Cytol.,* **177**, 57–113.

Meizel, S. and Turner, K.O. (1984) The effects of products and inhibitors of arachidonic acid metabolism on the hamsters sperm acrosome reaction. *J. Exp. Zool.*, **231**, 283–288.

Meizel, S. and Turner, K.O. (1993) Initiation of the human sperm acrosome reaction by thapsigargin, *J. Exp. Zool.,* **267**, 350–355.

Meizel S. (1995) Initiation of human sperm acrosome reaction by progesterone. In *Human Sperm Acrosome Reaction*, eds. Fenichel, P. and Parinaud, J. (John Libbey, London), pp. 151–164.

Miller, D.J., Macek, M.B. and Shur, B.D. (1992) Complementarity between sperm surface β-1,4-galactosyltransferase and egg-coat ZP3 mediates sperm-egg binding. *Nature,* **357**, 589–593.

Okamura, N., Tajima, Y., Onoe, S. and Sugita, Y. (1991) Purification of bicarbonate-sensitive sperm adenylylcyclase by acetamido-4'-isothiocyanostibene-2,2'-disulfonic acid-affinity chromotography. *J. Biol. Chem.,* **266**, 17754–17759.

Osman, R.A., Andria, M.L., Jones, A.D. and Meizel, S. (1989) Steroid induced exocytosis: the human sperm acrosome reaction. *Biochem. Biophys. Res. Commun.*, **160**, 828–833.

Parinaud, J and Milhet, P. (1996) Progesterone induces Ca^{++}-dependent 3',5'-cyclic adenosine monophosphate increase in human sperm. *J. Endocrinol. Metab.,* **81**, 1357–1360.

Philipp, S., Trost, C., Warnat, J., Rautmann, J., Himmerkus, N., Schroth, G., Kretz, O., Nastainczyk, W., Cavalié, A., Hoth, M. and Flockerzi, V. (2000) TRP4 (CCE1) is part of native CRAC-like channels in adrenal cells. *J. Biol. Chem.*, **275**, 23965–23972.

Revelli, A., Massobrio M., and Tesarik, J. (1998) Nongenomic actions of steroid hormones in reproductive tissues. *Endocr. Rev.*, **19**, 3–17.

Riffo, M., Leiva, S. and Astudillo, J. (1992) Effect of zinc on human sperm motility and the acrosome reaction. *Int. J. Androl.,* **15**, 229–237.

Roldan, E.R., Murase, T. and Shi, Q.X. (1994) Exocytosis in spermatozoa in response to progesterone and zona pellucida. *Science*, **266**, 1578–1581.

Sanz, L., Calvete, J.J., Mann, K., Schafer, W., Schmidt, E.R. and Töpfer-Petersen, E. (1992) The complete primary structure of the boar spermadhesin AQN-1, a carbohydrate-bidning protein involved in fertilization. *Eur. J. Biochem.,* **205**, 645–652.

Sabeur, K., Edwards, D.P. and Meizel, S. (1996) Human sperm plasma membrane progesterone receptor(s) and the acrosome reaction. *Biol. Reprod.*, **54**, 993–1001.

Sachdeva, G., Shah, C.A., Kholkute, S.D., Puri, C.P. (2000) Detection of progesterone receptor transcript in human spermatozoa. *Biol. Reprod.*, **62**, 1610–1614.

Schaefer, M., Habenicht, U.F., Bräutigam, M. and Gudermann, T. (2000a) Steroidal sigma receptor ligands affect signaling pathways in human spermatozoa. *Biol. Reprod.,* **63**, 57–63.

Schaefer, M., Hofmann, T., Schultz, G. and Gudermann, T. (1998) A new prostaglandin E receptor mediats calcium influx and acrosome reaction in human spermatozoa. *Proc. Natl. Acad. Sci. USA*, **95**, 3008–3013.

Schaefer, AM., Plant, T.D., Obukhov, A.G., Hofmann, T., Gudermann, T. and Schultz, G. (2000b) Receptor-mediated regulation of the nonselective cation channels TRPC4 and TRPC5. *J. Biol. chem.*, **275**, 17517–17526.

Shimizu, Y., Yorimitsu, A., Maruyama, Y., Kubota, T., Aso, T. and Bronson, R. (1998) Prostaglandins induce calcium influx in human spermatozoa. *Mol. Hum. Reprod.*, **4**, 555–561.

Sinclair, M.L., Wang, X.Y., Mattia, M., Conti, M., Buck, J., Wolgemuth, D.J. and Levin, L.R. (2000) Specific expression of soluble adenylyl cyclase in male germ cells. *Mol. Reprod. Dev.* **56,** 6–11.

Su, T.P., London, E.D. and Jaffe, J.H. (1988) Steroid binding at σ receptors suggest a link between endocrine, nervous and immune systems. *Science*, **240**, 219–221.

Sunahara, R.K., Dessauer, C.W. and Gilman, A.G. (1996) Complexity and diversity of mammalian adenylyl cyclases. *Annu. Rev. Pharmacol. Toxicol.* **36**, 461–480.

Tesarik, J., Moos, J. and Mendoza, C. (1993) Stimulation of protein tyrosine phosphorylation by a progesterone receptor on the cell surface of human sperm. *Endocrinlogy*, **133**, 328–335.

Thomas, P. and Meizel, S. (1989) Phosphatidylinositol 4,5-bisphosphate hydrolysis in human sperm stimulated with follicular fluid or progesterone is dependent upon Ca^{2+} influx. *Biochem. J.*, **264**, 539–546.

Turner, R.M., Johnson, L.R., Haig-Ladewig, L., Gerton, G.L. and Moss, S.B. (1998) An X-linked gene encodes a major human sperm fibrous sheath protein, hADAP82. Genomic organization protein kinase A-RI binding, and distribution of the precursor in the sperm tail. *J. Biol. Chem.*, **273**, 32135–32141.

Vannier, B., Peyton, M., Boulay, G., Brown, D., Quin, N., Jiang, M., Zhu, X. and Birnbaumer, L. (1999) Mouse trp2, the homologue of the human trpc2 pseudogene, encodes mTRP2, a store depletion-activated capacitative Ca^{2+} enry channel. *Proc. Natl. Acad. Sci. USA*, **96**, 1060–2064.

Visconti, P.E. and Kopf, G.S. (1998) Regulation of protein phosphorylation during sperm capacitation. *Biol. Reprod.*, **59**, 1–6.

Ward, C.R., Storey, B.T. and Kopf, G.S. (1992) Activation of a G_i protein in mouse sperm membranes by solubilized proteins of the zona pellucida, the egg's extracellular matrix. *J. Biol. Chem.*, **267**, 14061–14067.

Ward, C.R., Storey, B.T. and Kopf, G.S. (1994) Selective activation of G_{i1} and G_{i2} in mouse sperm by the zona pellucida, the egg's extracellular matrix. *J. Biol. Chem.*, **269**, 13254–13258.

Wassarman, P.M. (1999) Mammalian fertilization: Molecular aspects of gamete adhesion, exocytosis, and fusion. *Cell*, **96**, 175–183.

Weyand, I., Godde, M., Frings, S., Weiner, J., Müller, F., Altenhofen, W., Hatt, J. and Kaupp, U.B. (1994) Cloning and functional expression of a cyclic-nucleotide-gated channel from mammalian sperm. *Nature*, **368**, 859–863.

Wiesner, B., Weiner, J., Middendorf, R., Hagen, V. and Kaupp U.B. (1998) Cyclic nucleotide-gated channels on the flagellum control Ca^{2+} entry into sperm. *J. Cell Biol.*, **142**, 473–484.

Wissenbach, U., Schroth, G., Philipp, S. and Flockerzi, V. (1998) Structure and mRNA expression of a bovine trp homologue related to mammalian trp2 transcripts. *FEBS Lett.*, **429**, 61–66.

Wolfe Jr., S.A., Culp, S.G. and De Souza, E.B. (1989) σ-Receptors in endocrine organs: identification characterization and autoradiographic localization in rat pituitary, adrenal, testis, and ovary. *Endocrinology*, **124**, 1160–1172.

Yanagimachi, R. (1994) Mammalian fertilization. In *The Physiology of Reproduction*, eds. Knobil, E. and Neill, J.D. (Raven Press, New York), pp. 189–317.

Follicular Growth, Ovulation and Fertilization: Molecular and Clinical Basis
Anand Kumar and Amal K. Mukhopadhyay (Eds.)
Narosa Publishing House, New Delhi, India, 2001

24

Methodological Problems and Clinical Advantage of Cryopreservation of Ovarian Tissue

Frank Nawroth[1,3], Mohammad Kouselhar[1], Torsten Schmidt[1], Thomas Romer[1] and Roland Sudik[2]

[1]Department of Obstetrics and Gynaecology, University of Cologne, [2]Department of Obstetrics and Gynaecology, Municipal Hospital of Neubrandenburg, Germany

[3]Universitäts-Frauenklinik Köln, Kerpener Str. 34, 50931 Köln, Germany

Introduction

The premature cessation of the ovarian reproductive function can have its cause in a premature menopause of varying geneses. However, with an incidence of less than 1% (Coulam *et al.*, 1986) it is rather seldom. Malignant diseases and their fertility limiting therapy, on the other hand, play an increasingly frequent role in the premature loss of fertility. Particularly hematologic-oncologic diseases often affect patients in their reproductive and childhood years. A reduced rate in recurrence and an increased likelihood of survival, owing to measures for early diagnosis and more aggressive therapies, have abetted the search for possibilities of preserving the germinative tissue in prepubescent patients and in women wishing to have children (Apperley and Reddy, 1995).

In the case of Morbus Hodgkin more than 40% of patients become sterile, that is, experience premature menopause as a result of the therapy (Meirow *et al.*, 1998). The only possibility for these women to fulfil their desire for offspring is through an oocyte donation (Nugent *et al.*, 1997), which is prohibited in Germany.

A way out of this dilemma may be found in the pretherapeutical extraction of mature oocytes and their cryopreservation in an unfertilized or (in the case of a present partner) fertilized state. Another possibility is the extraction and cryopreservation of ovarian tissue for post-therapeutical reimplantation (Opsahl *et al.*, 1997). The cryopreservation of ovarian tissue is therefore an alternative or supplement to cryopreservation of embryos or oocytes for patients with a risk of premature menopause (Wood *et al.*, 1997).

The studies towards the attainment of this goal are at the highest level of activity, yet in some aspects still far from practical application. The first successful application of cryopreservation/autotransplantation of human ovarian tissue was recently published (Oktay *et al.*, 1999). The goal of this essay is to provide an overview of the current state of development and clinical applicability of the methods for cryopreservation of ovarian tissue.

The Follicle Reservoir

Mammals have almost without exception a capacity for oocyte production limited to fetal or early post-natal life. They are born with a limited germ cell pool, the primordial follicle of the ovarian cortex and, over time, lose their population of stem cells of the oogenesis. The recruited primordial follicles for growth or atresia are therefore irreplaceable, so that already in the first years of life the functional ovarian reservoir declines (Gosden, 1992a). In female fetuses the greatest number of germ-cells, 7 million, may be found in the 20th week. their number decreases constantly through atresia (Faddy and Gosden, 1995). Only about one tenth of them survive until birth. This portion is reduced to approximately one fourth by puberty (Baker, 1963). At the onset of the reproductive phase, the follicular reserve, normally sufficient for roughly 30–40 years of reproductive life, is reduced to approximately 250,000. The germinative function of the ovaries ceases at a cut-off of roughly 1000 primordial follicles (Faddy and Gosden, 1995). Hartshorne *et al.* (1999) could show that the human oocytes cultured in human fetal ovarian tissue survive and that progression through prophase I continues.

Sensitivity of Germinal Tissue to Radiation and Chemotherapy, and Possibilities of Protection

The ovaries are sensitive to a cytotoxic therapy (Apperley and Reddy, 1995), although in most cases they are dormant primordial follicles. Tests on animals have shown that primordial follicles are even more sensitive to ionizing rays and alkylating substances than maturing follicles (Baker, 1971). The estimated radiation dose in which the number of follicles reduced by half (LD_{50}) in humans is 4 Gy (Wallace *et al.*, 1989). The extent of gonadal toxicity through chemotherapy and radiation depends on the dosage and time between treatments, the duration of therapy, the type of medication, as well as the patient's age (Baker, 1971, Horning *et al.*, 1981, Kreuser *et al.*, 1990).

An attempt was made to protect the ovaries from damage caused by radiation or chemotherapy by reducing the gonadotrophins with a Gonadotrophin-Releasing-Hormone-Analog (GnRHa). Blumenfeld *et al.* (1996) found in lymphoma patients after a GnRHa treatment parallel to chemotherapy a 93.7% rate of spontaneous ovulation 3–8 months following therapy, as opposed to 39.0% without Down-Regulation. Gosden *et al.*, (1998) do not hold these results for convincing, as there is no definitive proof that FSH triggers the growth of primordial follicles and that the follicle reservoir can be preserved by means of suppressed gonadotrophin secretion (Gougeon, 1996). Small follicles are sensitive to gonadotrophins, but require neither FSH nor LH for development until preantral stages. (Dufour *et al.*, 1979, Wang and Greenwald, 1993, Oktay *et al.*, 1998b).

Experimental studies could not prove any protective effect of a hypogonadotrophism on sensitivity to radiation (Gosden *et al.*, 1997a). It remains questionable whether or not the survival of germ-cells that have been exposed to DNA-damaging substances is at all worthwhile (Gosden *et al.*, 1997a), as the spontaneous abortion rate after therapy is high (Sanders *et al.*, 1996).

A pretherapeutical oophoropexy for removal from the field of radiation is in isolated cases suitable for protection against radiation, but ineffectual in a systemic therapy. Furthermore, the ovaries may absorb scattered rays (Ray *et al.*, 1970, Hunter *et al.*, 1980). Morice *et al.*, (1998) maintained the ovarian function in 15 of 19 patients (79%) through a bilateral laparoscopic transposition of the ovaries prior to operation/radiation. Three of the 18 patients with preserved uteruses became pregnant (18%).

The application of individual schemata (monotherapies, reduction of cumultive doses) is possible with systemic chemotherapeutics, but because the doses often do not reach the therapeutic level, they cannot be used in every case.

Cryopreservation of Ovarian Tissue

Three main problems limit cryopreservation of isolated mature oocytes (Metaphase II-Oocytes), which may be obtained quite easily through vaginal follicular puncture after ovarian stimulation:

1. The survival rate of unfertilized oocytes after cryopreservation and thawing is low at 18–52% (Trounson, 1986, Mandelbaum *et al.*, 1988).
2. The essential ovarian stimulation delays the beginning of therapy (the operation as well as chemotherapy) and possibly exercises a stimulating effect on hormone-dependent tumours.
3. In prepubescent patients ovarian stimualtion is not yet possible.

Therefore freezing of ovarian tissue insures the survival of germ-cells, just as the Cryo-TESE (Testicular Sperm Extraction) concept in azoospermic men already shows that cryopreserved sperm in testicular tissue remain active and especially after ICSI (Intracytoplasmmatic Sperm Injection), may lead to pregnancy (Devroey *et al.*, 1994).

Cryoprotectants and Cryopreservation of the Slices of Ovarian Tissue

Substances known as cryoprotectants, which limit the lethal effects of ice crystal development as well as the extracellular build-up of salt, are necessary for successful cryopreservation. Numerous agents are in use today which may be distinguished as permeating and non-permeating cryoprotectants.

Cryopreservation is relatively easy in isolated cells. However, it becomes problematic in larger tissues or organs (Mazur, 1972). A permeating cryoprotectant penetrates quickly into isolated cells, so that little time is required to reach the required concentration intracellularly. The necessary exposure time of the tissues is, on the other hand, incomparably longer, so that the surface cells risk being exposed to toxins. It is therefore important to consider the ratio of equilibration time to the negative influence of over exposure of the cryoprotectant prior to cryopreservation and during its removal (Nash, 1962, Pegg and Diaper, 1988).

Glycerol, successfully used as early as 1961 by Smith for the storage of spermatozoa at –80°C, has long been replaced by other substances. These include dimethylsulfoxide (DMSO), ethylene glycol (EG) and propylene glycol, which are highly water-soluble and can therefore penetrate quickly) (Gosden and Aubard, 1996).

Hovatta *et al.*, (1996) found comparable results (follicle maturity stages, atresia, tissue necrosis) after freezing/thawing 0,2–3 mm sized slices of ovarian cortex with 1.5 M DMSO compared to 1,2-propandiol/saccharose as a cryoprotectant. Newton *et al.*, (1998) compared the permeation of EG, DMSO, propylene glycol and glycerol in 2 × 2 mm sized ovarian biopsy samples. The use of 1. M EG or DMSO over 30 min. at 4°C provided the best results, after which 80% of the tissue was equilibrated. The addition of saccharose as a nonpermeating cryoprotectant did not provide a significant protective effect against freezer damage.

The currently used protocols for cryopreservation of ovarian tissue orient towards those used for mature oocytes (Newton *et al.*, 1998). With tissue, however, there occurs the additional problem of cellular heterogeniety with different diffusion rates of the cryoprotectant (Pegg *et al.*, 1979). The optimal protocol with regard to the suitable cryoprotectant and freezing methods has obviously not yet been found (Newton *et al.*, 1998). "Slow cooling—rapid thawing" is the favoured method. (Gosden, 1998).

Preliminary attempts at cryopreservation of ovarian tissue were published by Parkes and Smith, (1953), in which they froze and thawed ovarian tissue from rats, and transplanted them in oophorectomized animal recipients. The ovarian tissue survived partially and produced estrogens that led to a cornification of the vaginal epithelium.

Pregnancies occurred in the mice after replantation of frozen/thawed ovarian tissue (Parrott, 1960). With a storage temperature of – 79°C, glycerol was used concurrently as a cryoprotectant. Gunasena *et al.* (1997) described the successful autologic transplantation of cryopreserved mice ovaries under the periovarian peritoneum.

From the prognostically favourable freezing of small bands of tissue or cells came the consideration of freezing only pieces of the ovary (cortex) or the primordial follicles that have been isolated from them.

At present, ovarian tissue can be cultivated for only a few weeks. However, the maturing of human primordial follicles requires more than 3 months (Gougeon, 1986). Moreover, these follicles grow considerably larger than those in mice (Oktay *et al.*, 1998).

For this reason the morphological proof of follicle growth up to the antral stage was possible only after transplantation of ovarian tissue in cats and sheep (Gosden *et al.*, 1994), as well as in human ovarian tissue under the kidney capsule of mice (xenografting) (Oktay *et al.*, 1998). The mice in this case were SCID (severe combined immunodeficient) with a mutation of the chromosome 16. The latter led to a agammaglobulinaemia and a lack of mature B and T cells, so that xenografts were tolerated over a longer period of time (Bosma *et al.*, 1983).

10% of the human primordial follicles survived the cryopreservation and the following xenografting under the kidney capsule of SCID-mice with the use of glycerol. With EG and DMSO it was 84% and 74% (Newton *et al.*, 1996). Similar results were achieved by Candy *et al.*, (1995) with fresh and cryopreserved ovarian tissue of primates transplanted under the kidney capsule of immune incompetent mice.

Cox *et al.*, (1996) were able to prove for the first time that the fertility of ovarectomized adult mice can be regained by transplantation of fresh as well as cryopreserved fetal mice ovaries. This is the case for orthotopic (under the periovarian capsule) as well as the heterotopic transplantation (under the kidney capsule). At the same time it was shown that maturation and pregnancy were only accelerated when the recipients had been bilaterally oophorectomized. The increase in gonadotrophins in mice after a bilateral ovariectomy might stimulate the accelerated maturation of transplanted fetal ovaries (Naik *et al.*, 1984). When the recipients possessed one or both ovaries, the tissue either did not develop or developed only partially (Cox *et al.*, 1996). Possible explanations for this might be, for example, secreted factors from the intact adult ovaries which directly or indirectly impede the development of the fetal ovaries.

The transplantation of slices of ovarian cortex into the ovaropedicle of the removed ovary led to ovulation and births in sheep (Gosden *et al.*, 1994). Animal-experiment studies aimed at protecting the germinal epithelium in connection with radiation exposure showed an intact follicle population in unilaterally ovariectomized, contralaterally radiated rats with frozen/thawed ovarian tissue from the opposite side transplanted under the periovarian capsule (Aubard *et al.*, 1998). However, pregnancies occurred both in this group as well as in a merely unilaterally ovariectomized and contralaterally radiated control group. Due to the obviously incomplete sterilization of the rats the authors could not unequivocally conclude that the cryopreserved transplants maintained fertility.

In sheep the autotransplantation of frozen/thawed hemiovaries after an ovariectomy led to progesterone synthesis and endometrial maturity (Salle *et al.*, 1999). Meirow *et al.* (1999) described a laparoscopic technique for obtaining cortical ovarian tissue for cryopreservation.

Cryopreservation of Isolated Primordial Follicles

Isolated primordial follicles may be obtained and frozen from ovarian tissue through enzymatic digestion (Carroll and Gosden, 1993).

That fertility may be restored this way was already demonstrated by

Gosden in 1990. He transferred primordial follicles from infant mice to the ovaries of adult animals sterilized through radiation or in their periovarian capsule after an ovariectomy. Ovulation and pregnancies occurred.

Oktay *et al.* (1997) isolated for the first time human primordial follicles from fresh and cryopreserved ovarian tissue. The isolation took place after a partial disaggregation with type 1A-collagenase and a following microdissection. The tissue blocks, with a surface of 4 mm^2 and a thickness of 1 mm, received during the histological examination on average 38 primordial follicles from which 16–18 (56–58%) could be isolated from fresh and frozen/thawed blocks. In both groups no significant difference as to viability of the follicles and oocytes could be found. In the electron microscopy the majority of oocytes were without ultrastructural signs of lesion following isolation and cryopreservation.

Previously it was only possible to isolate human preantral follicles with collagenase from ovarian tissue, which, however, necessitated a 36 hour digestion time for enzyme penetration (Roy and treacy, 1993). At least for primordial follicles, this long time span does not appear necessary (Oktay *et al.*, 1997). This could reduce possible undesired effects of the enzyme action (Eppig, 1994). The more mature oocytes from preantral follicles are more sensitive to damage by cryopreservation (Baka *et al.*, 1995). That could speak for the use of primordial follicles. If one compares the culture of primordial or even primary and secondary follicles in small tissue blocks and after partial isolation (enzymatic or mechanical), the isolation improves neither the survival nor the further development of the follicle (Hovatta *et al.*, 1999).

Ovarian Transplantation

The transplantation of ovaries has been practised in animals for a long time (Krohn, 1977, Gosden 1992b). The use of immature organs produced better results, as they possess a larger follicle population with less connective tissue, and are smaller, whereby a rapid revascularization is promoted. Most follicles are destroyed during the postoperative ischaemia, but up to 50% of the primordial follicles can survive as a result of their peripheral location and their low metabolism (Gosden, 1992b).

Possibly more follicles could survive after the induction of vascular anastomosis. However, studies have not yet been able to verify this (Winston and Browne, 1974). Nonetheless, the observation that a heterotopic autotransplant of ovarian tissue without vascular pedicle in syngenic Lewis rats is successful for $\geq$ 6 months has been disproved.

Animal experiment studies were undertaken to clarify the question as to whether or not a unilateral ovariectomy for obtaining ovarian tissue for cryopreservation affects the cycle and the folliculogenesis (Gosden *et al.*, 1989). These studies showed that in a time-period of 2 months changes in the number of growing as well as atretic follicles occurred after a unilateral ovariectomy in mice. A growing number of large preantral and antral follicles in the remaining ovary compared with ovaries in a control group, as well as a diminishing number of atretic follicles, indicate that in an act of compensation, follicles actually destined for atresia survived. In mice, the transplanted organs can no longer be morphologically distinguished from non-transplanted ones after less than a week after the operation (Felicio *et al.*, 1984).

It could be demonstrated in rats that the subcutaneous autotransplantation (of the whole ovary as well as of ovarian fragments) after a bilateral ovariectomy led to a maintenance of the hormone synthesis (Von Eye Corleta *et al.*, 1998). A vessel anastomosis, as for example described by Harrison *et al.*, (1979), does not appear to be necessary. Transplanted ovaries are capable of synthesizing substances which support the angiogenesis. The rise in gonadotrophin secretion after the transplantation led in rats to a revascularization through an increased gene expression of the angiogenesis factors VEGF and $TGF_{\beta 1}$ (Dissen *et al.*, 1994). Aubard *et al.*, (1999) found that the ortho- as well as the heterotopic transplantation in sheep was more disadvantageous to the follicle than the cryopreservation itself. Merely 5% of primordial follicles survived the grafting both after cryopreservation and without freezing. They concluded that a heterotopic may be an alternative to an orthotopic transplantation when the follicle survival is generally improved. A hypothetical transference of these studies to humans opens for the future new possibilities of sparing, with a relatively simple method, prophylactically bilateral oophorectomized women the necessity of constant exogenous hormone substitution (Von Eye Corleta *et al.*, 1998). Oktay *et al.*, (1999) reported for the first time on a patient whose tissue was cryopreserved during a extirpation of the second ovary and later subperitoneally autotransplanted. Initially, an ovarian reaction occurred after gonadotrophin stimulation, and later, spontaneously.

Risk of Transmission of Malignant Cells with Transplanted Ovarian Tissue

If the ovarian tissue contains malignant cells at the time of extraction, which survive the cryopreservation and transplantation, the possibility exists of their transmission to the recipient. Shaw *et al.*, (1996) established this in mice with lymphomas in both fresh as well as cryopreserved ovarian tissue.

The risk of transmission depends on the histology, stage and activity of the malignant tumour (Meirow *et al.*, 1998). Meirow *et al.*, (1998) were unable to establish the presence of malignant cells in the ovarian cortex in 7 Hodgkin patients in advanced stages. They consider the cryopreservation in such patients safe when the ovaries appear sonographically and laparoscopically inconspicuous.

With the aid of sensitive PCR, the remaining activity of lymphomas and leukaemic diseases may be analysed more precisely, and the time of tissue extraction thus decided (Kurokawa *et al.*, 1996).

In the case of tumors with frequent ovarian metastasizing, such as in mammary carcinomas or leukaemic diseases with a possible presence of malignant cells in the ovarian blood, the tissue extraction could take place after the first courses of chemotherapy if ovarian hormonal activity can still be proven (Meirow *et al.*, 1998).

Definite safety can be established only after the maturation of the extracted oocytes in vitro, fertilization and the following intrauterine transfer through the missing tumour cell transplantation. The primordial follicles could first grow to Graafian follicles as xenotransplants, for example, in SCID-mice (Gosden *et al.*, 1997b).

The replantation of ovarian tissue in selected patients appears to be applicable in the future (Gosden *et al.*, 1997b), but 100% safety of this procedure is controversial (Shaw and Trounson, 1997) and reserved for further preclinical studies.

References

Apperley, J.F. and Reddy, N. (1995) Mechanism and management of treatment related gonadal failure in recipients of high dose chemoradiotherapy. *Blood Rev.*, **9**, 93–116.

Aubard, Y., Newton, H., Scheffer, G. *et al.*, (1998) Conservation of the follicular population in irradiated rats by the cryopreservation and orthotopic autografting of ovarian tissue. *Eur. J. Obstet. Gynecol. Reprod. Biol.*, **79**, 83–87.

Aubard, Y., Piver, P., Cognie, Y. *et al.* (1999) Orthotopic and heterotopic autografts of frozen-thawed ovarian cortex in sheep. *Hum. Reprod.*, **14**, 2149–2154.

Baka, S.G., Toth, T.L., Veeck, L.L. *et al.* (1995) Evaluation of spindle apparatus of in vitro matured oocytes following cryopreservation. *Hum. Reprod.*, **7**, 1816–1820.

Baker, T.G. (1963) A quantitative and cytological study of germ cells in human ovaries. *Proc. Roy. Soc. Lond.*, **158**, 414–433.

Baker, T.G. (1971) Radiosensitivity of mammalian oocytes with particular reference to the human female. *Am. J. Obstet. Gynecol.*, **110**, 746–761.

Blumenfeld, Z., Avivi, I., Linn, S. *et al.* (1996) Prevention of irreversible chemotherapy-induced ovarian damage in young women with lymphoma by gonadotrophin releasing hormone agonist in parallel to chemotherapy. *Hum. Reprod.*, **11**, 1620–1626.

Bosma, G.C., Custer, R.P., Bosma, M.J. (1983) A severe combined immunodeficiency mutation in the mouse. *Nature*, **301**, 527–530.

Callejo, J., Jauregui, M.T., Valls, C. *et al.*, (1999) Heterotopic ovarian transplantation without vascular pedicle in syngeneic Lewis rats: six-month control of estradiol and follicle-stimulating hormone concentrations after intraperitoneal and subcutaneous implants. *Fertil. Steril.*, **72**, 513–517.

Candy, C.J., Wood, M.J., Whittingham, D.G. (1995) Follicular development in cryopreserved marmoset ovarian tissue after transplantation. *Hum. Reprod.*, **10**, 2334–2338.

Carroll, J. and Gosden R.G. (1993) Transplantation of frozen-thawed mouse primordial follicles. *Hum. Reprod.*, **8**, 1163–1167.

Coulam, C.B., Adamson, S.C., Annegers, J.F. (1986) Incidence of premature ovarian failure. *Obstet. Gynecol.*, **67**, 604–606.

Cox, S.L., Shaw, J., Jenkin, G. (1996) Transplantation of cryopreserved fetal ovarian tissue to adult recipients in mice. *J. Reprod. Fertil.*, **107**, 315–322.

Devroey, P., Liu, J., Nagy, Z. *et al.* (1994) Normal fertilization of human oocytes after testicular sperm extraction and intracytoplasmic sperm injection. *Fertil. Steril.*, **62**, 639–641.

Dissen, G.A., Lara, H.E., Fahrenbach, W.H. *et al.* (1994) Immature rat ovaries become revascularized rapidly after autotransplantation and show a gonadotropin-dependent increase in angiogenic factor gene expression. *Endocrinology,* **134**, 1146–1154.

Dufour, J., Cahill, L.P., Mauleon, P. (1979) Short- and long-term effects of hypophysectomy and unilateral ovariectomy on ovarian follicular population in sheep. *J. Reprod. Fertil.*, **57**, 301–309.

Eppig, J.J. (1994) Further reflections on culture systems for the growth of oocytes in vitro. *Hum. Reprod.,* **9**, 974–976.

Faddy, M.J. and Gosden, R.G. (1995) A mathematical model for follicular dynamics in human ovaries. *Hum Reprod.,* **10**, 770–775.

Felicio, L.S., Nelson, J.F., Gosden, R.G. *et al.* (1984) Restoration of ovulatory cycles by young ovarian grafts in aging mice: potentiation by long-term ovariectomy decreases with age. *Proc. Natl. Acad. Sci. USA*, **80**, 6076–6080.

Gosden, R.G., Telfer, E., Faddy, M.J. *et al.* (1989) Ovarian cyclicity and follicular recruitment in unilaterally ovariectomized mice. *J Reprod Fertil,* **87**, 257–264.

Gosden, R.G. (1990) Restitution of fertility in sterilized mice by transferring primordial follicles. *Hum. Reprod.*, **5**, 499–504.

Gosden, R.G. (1992a) Transplantation of fetal germ cells. *J. Assist. Reprod. Genet.*, **9**, 118–123.

Gosden, R.G. (1992b) Transplantation of ovaries and testes. In Edwards, R.G. (ed) Fetal Tissue Transplants in Medicine. Cambridge University Press. Cambridge, pp. 253–279.

Gosden, R.G., Baird, D.T., Wade, J.C. *et al.* (1994) Restoration of fertility to oophorectomized sheep by ovarian autografts stored at –196°C. *Hum Reprod.,* **9**, 597–603.

Gosden, R.G. and Aubard, Y. (1996) *Transplantation of ovarian and testicular tissues.* Austin TX: RG landes Co.

Gosden, R.G., Wade, J.C., Fraser, H.M. *et al.*, (1997a) Impact of congenital or experimental hypogonadotrophism on the radiation sensitivity of the mouse ovary. *Hum. Reprod.,* **12**, 2483–2488.

Gosden, R.G. Rutherford, A., Norfolk, D.R. (1997b) Transmission of malignant cells in ovarian grafts, *Hum Reprod,* **12**, 403.

Gosden, R.G., Newton, H., Kim, S.S. (1998) The cryopreservation of human ovarian tissue. In Kempers, R.D., Cohen, J., Haney, A.F., Younger JB (eds) *Fertility and Reprodutive Medicine*. Elsevier Science B.V. pp. 615–620.

Gougeon, A. (1986) Dynamics of follicular growth in the human: a model from preliminary results. *Hum. Reprod.,* **1**, 81–87.

Gougeon, A. (1996) Regulation of ovarian follicular development in primates: facts and hypotheses. *Endocr. Rev.*, **17**, 121–155.

Gunasena, K.T., Villnes, P.M., Crister, E.S. *et al.*, (1997) Live birth after autologous transplant of cryopreserved mouse ovaries. *Hum. Reprod.,* **12**, 101–106.

Harrison, F.A., Chambers, S.G., Green, E.A. (1979) Autotransplantation of the ovary to the neck in the sow: normal cyclic activity and plasma hormone levels. *J. Endocrinol.*, **83**, 46.

Hartshorne, G.M., Barlow, A.L., Child, T.J., *et al.* (1999) Immunocytogenetic detection of normal and abnormal oocytes in human fetal ovarian tissue in culture. *Hum. Reprod.*, **14**, 172–182.

Horning, S.J., Hoppe, R.T., Kaplan, H.S. *et al.* (1981) Female reproductive potential after treatment for Hodgkin's disease. *N. Eng. J. Med.*, **304**, 1377–1382.

Hunter, M.C.H., Gless, J.P., Gazet, J.C. (1980) Oophoropexy and ovarian function in the treatment of Hodgkin=B4s disease. *Clin. Radiol* **31**: 21–26.

Hovatta, O., Wright, C., Krausz, T. *et al.*, (1999) Human primordial, primary and secondary follicle in long-term cutlure: effect of partial isolation. *Hum Reprod.*, **14**: 2519–2524.

Hovatta, O., Silye, R., Krausz, T., Abir, R., Margara, R., Trew, G., Lass, A., Winston, and propanediol-sucrose as cryprotectants. *Hum. Reprod.* **11**: 1268–1272.

Kreuser, B.D., Hetzel, W.D., Billia, D.O. *et al.* (1990) Gonadal toxicity following cancer therapy in adults: significance, diagnosis, prevention and treatment. *Cancer Treat. Rev.*, **17**, 169–175.

Krohn, P.L. (1977) Transplantation of the ovary. In Lord Zuckermann and Weir, B.J. (eds) *The Ovary*, 2nd edn, vol. II. Academic Press. New York, pp. 101–128.

Kurokawa, T., Kinoshita, T., Ito, T. *et al.* (1996) Detection of minimal residual disease B cell lymphoma by PCR mediated RNase protection assay. *Leukemia,* **10**, 1222–1231.

Mandelbaum J., Junca AM, Plachot M. *et al.*, (1988) Cryopreservation of human embryos and oocytes. *Hum. Reprod.,* **3**, 117–119.

Mazur, P., Leibo, S.P., Chu, E.H.Y. (1972) A two-factor hypothesis of freezing injury. *Exp. Cell. Res.,* **71**, 345–351.

Meirow, D., Yehuda, D.B., prus, D. *et al.* (1998) Ovarian tissue banking in patients with Hodgkin's disease: Is it safe? *Fertil. Steril.,* **69**, 996–998.

Meirow, D., Fasouliotis, S.J., Nugent, D. *et al.* (1999) A laparoscopic technique for obtaining ovarian cortical biopsy specimens for fertility conservation in patients with cancer. *Fertil Steril.*, **71**, 948–951.

Morice, P., Castaigne, D., Haie-Meder, C. *et al.* (1998) Laproscopic ovarian transposition for pelvic malignancies: indications and functional outcomes. *Fertil. Steril.*, **70**, 956–960.

Naik, S.I., Young, L.S., Charlton, H.M. *et al.* (1984) Pituitary gonadotropin-releasing hormone receptor regulation in mice. II: *Females Endocrinology,* **115**, 114–120.

Nash, T. (1962) The chemical consitution of compounds which protect erythrocytes against freezing damage. *J. Gen. Physiol.*, **46**, 167–175.

Newton, H., Aubard, Y., Rutherford, A. *et al.* (1996) Low temperature storage and grafting of human ovarian tissue. *Hum. Reprod.*, **11**, 1487–1491.

Newton, H., Fisher, J., Arnold, J.R.P. *et al.*, (1998) Permeation of human ovarian tissue with cryoprotective agents in preparation for cryopreservation. *Hum. Reprod.,* **13**, 376–380.

Nugent, D., Meirow, D., Brook, P.F. *et al.* (1997) Transplantation in reproductive medicine: previous experience, present knowledge and future prospects. *Hum. Reprod. Update,* **2**, 103–117.

Oktay, K., Nugent, D., Newton, H *et al.*, (1997) Isolation and characterization of primordial follicles from fresh and cryopreserved human ovarian tissue. *Fetil. Steril.,* **67**, 481–486.

Oktay, K., Newton, H., Mullan, J. *et al.* (1998) Development of human primordial follicles to antral stages in SCID/hpg mice stimulated with follicle stimulating hormone. *Hum. Reprod.*, **13**, 1133–1138.

Oktay, K., Karlikaya, R., Gosden. R. *et al.*, (1999) Ovarian function after autologous transplantation of frozen-banked human ovarian tissue. Fertil. Steril., 70 (Suppl. 3), 21.

Opsahl, M.S., Fugger, E.F., Sherins, R.J. *et al.*, (1997) Preservation of reproductive function before therapy for cancer: new options involving sperm and ovary cryopreservation. *Cancer. J. Sci. Am.,* **3**, 189–191.

Parkes, A.S. and Smith, A.U. (1953) Regeneration of rat ovarian tissue grafted after exposure to low temperatures. *Proc. Roy. Soc.,* **140,** 455–467.

Parrott, D.M.V. (1960) The fertility of mice with orthotopic ovarian grafts derived from frozen tissue. *J. Reprod. Fertil.*, **1**, 230–241.

Pegg, D.E., Jacobsen, I.A., Armitage, W.J. *et al.*, (1979) Mechanisms of cryoinjury in organs. In Pegg, D.E. and Jacobsen, I.A. (eds) Organ Preservation, 2. churchill Livingstone, Edinbourgh, pp. 132–144.

Pegg, D.E., and Diaper, M.P. (1988) On the mechanism of injury to slowly frozen erythrocytes frozen. *Biophys. J.*, **54**, 471–488.

Ray, G.R., Trueblood, H.V., Enwright, L.P. *et al.*, (1970) Oophoropexy: a means of preserving ovarian function following pelvic megavoltage radiotherapy for Hodgkin's disease. *Radiology*, **96**, 180–195.

Roy, S.K., and Treacy, B.J. (1993) Isolation and long-term culture of human preantral follicles. *Fertil Steril.*, **59**, 783–790.

Salle, B., Lorgange, J., Demirci, B. *et al.*, (1999) Restoration of ovarian steroid secretion and histologic assessment after freezing, thawing, and autograft of a hemi-ovary in sheep. *Fertil. Steril.*, **72**, 366–370.

Sanders, J.F., Hawley, J., Levy, W. *et al.* (1996) Pregnancies following high-dose cyclophosphamide with or without high-dose busulphan or total-body irradiation and bone marrow transplantation. *Blood*, **87**, 3045–3052.

Shaw, J.M. and Bowles, J. Koopman, P. *et al.*, (1996) Fresh and cryopreserved ovarian tissue samples from donors with lymphoma transmit the cancer to graft recipients. *Hum. Reprod.*, **11**, 1668-1673.

Shaw, J. and Trounson, A. (1997) Oncological implications in the replacement of ovarian tissue. *Hum. Reprod.,* **12**, 403–405.

Smith, A.U. (1961) Biological effects of freezing and supercooling. London: Edward Arnold.

Trounson, A. (1986) Preservation of human eggs and embryos. *Fertil. Steril.*, **46**, 1–11.

Von Eye Corleta, H, Corleta, O, Capp, E. *et al.*, (1998) Subcutaneous autologous ovarian transplantation in Wistar rats maintains hormone secretion. *Fertil. Steril.,* **70**, 16–19.

Wallace, W.H.B., Shalet, S.M., Hendry, J.M. (1989) Ovarian failure following abdominal irradination in childhood: the radiosensitivity of the human oocyte. *Br. J. Radiol.,* **62**, 995–998.

Wang, X.N. and Greenwald, G.S. (1993) Hypophysetomy of the cyclic mouse. I. Effects on folliculogenesis, oocye growth and FSH-and hCG-receptors. *Biol. Reprod.*, **48**, 585–594.

Winston, R.M. and Browne, C. (1974) Pregnancy following autograft transplantation of fallopian tube and ovary in the rabbit. *Lancet,* **2**, 494–495.

Wood, C.E., Shaw, J.M., Trounson, A.O. (1997) Cryopreservation of ovarian tissue. *Med. J. Australia*, **166**, 366–369.

Follicular Growth, Ovulation and Fertilization: Molecular and Clinical Basis
Anand Kumar and Amal K. Mukhopadhyay (Eds.)
Narosa Publishing House, New Delhi, India, 2001

25

The Use of a 1.48 μm Diode Laser in Assisted Reproduction

Markus Montag and Hans van der Ven
Department of Endocrinology and Reproductive Medicine,
University of Bonn, 53105 Bonn, Germany

Introduction

The creation of artificial openings into the zona pellucida of oocytes or embryos is essential for numerous applications in ART like polar body or blastomere biopsy and assisted hatching. Several methods for opening the zona pellucida have been proposed. Mechanical zona opening involves the use of a sharp glass capillary to cut or dissect the zona (Zona cutting, Tsunoda *et al.*, 1986; Partial zona dissection, Malter and Cohen, 1989). Chemical opening of the zona is achieved by acidic tyrode's solution (Gordon and Talanky, 1987). The relevance of these techniques has been described in detail in the literature (Depypere *et al.*, 1988; Cohen, 1992).

Several groups investigated the use of lasers for the creation of openings into the zona pellucida by contact and non-contact laser systems. Contact lasers require additional micromanipulation equipment, because the laser radiation has to be delivered directly to the zona with microcapillaries. Due to this disadvantage, these lasers are no longer used in ART.

Non-contact lasers allow for direct delivery of the laser beam through the optical system of the microscope. The first non-contact laser emitted light in the ultraviolet. However, using a UV laser one can not exclude the possibility of potential mutagenic and cytotoxic effects (Kochevar, 1989). Secondly, the UV-wavelength requires immersion objectives and special culure dishes and therefore the use of this type of laser is not very practicable and cost-effetive in routine laboratory work.

More recently, an indium-gallium-arsenic-phosphorus (InGaAsP) 1.48 μm diode laser was introduced (Rink *et al.*, 1994). This laser can be fitted to every inverted microscope and the unique absorption characteristics of the diode laser by water and culture medium allow its use in combination with standard culture dishes (Rink *et al.*, 1996). Zona drilling can be performed with a single laser irradiation using a pulse and the size of a drilled opening is directly controlled by variation of the irradiation time. The safety and efficacy of the system was first demonstrated in mouse experiments (Germond *et al.*, 1995; 1996). This laser was applied in human *in-vitro* fertilization for assisted hatching and in 1995 the first baby was born in Switzerland. Since then, a variety of applications has been developed for the diode laser technique and these will be reviewed in this paper.

Assisted Hatching

Prior to implantation the embryo must escape the surounding zona pellucida. It has been discussed, that an impairment of the hatching process may be responsible for failed implantation in human IVF (Cohen, 1991). Several methods have been proposed to assist the hatching process (Tsunoda *et al.*, 1986; Gordon and Talansky, 1987; Malter and Cohen, 1989) and in the past a large number of studies were undertaken to investigate the possible benefit of assisted hatching. The results of these studies, prospective as well as retrospective, were highly controversial. Some authors found a significant rise in the pregnancy rate for selected patient groups (Cohen *et al.*, 1992; Liu *et al.*, 1993; Schoolcraft *et al.*, 1994; Stein *et al.*, 1995), but others reported on the contrary (Hellebaut *et al.*, 1996; Tucker *et al.*, 1996; Bider *et al.*, 1997; Lanzendorf *et al.*, 1998; Magli *et al.*, 1998). The majority of these studies employed acidic tyrode's solution for zona drilling, although the use of acidic tyrode's solution does not allow for reliable drilling of equally sized openings. However, the size of an artificial opening in the zona is of great importance for hatching efficiency (Cohen and Feldberg, 1991; Montag and van der Ven, 1999).

The introduction of the 1.48 μm diode laser system allowed for the first time to drill openings into the zona pellucida which are identical in size. The ease of laser application and the proven efficiency and safety of this system made it an ideal tool for assisted hatching (Fig. 1). All available data on laser-assited hatching published so far confirm that it is a successful treatment therapy for selected patients. This includes patients with a maternal age over 35–37 years and patients with previous implantation failure (Germond *et al.*, 1998; Montag and van der Ven, 1999).

Blastomere Biopsy

A prerequisite for preimplantation genetic diagnosis is the penetration of the zona pellucida with microcapillaries for biopsy of polar bodies or blastomeres (Grifo, 1992). Usually embryo biopsy is performed in combination with zona drilling by acidic tyrode's solution (Handyside *et al.*, 1990). This technique requires immediate washing of embryos due to embryo-toxicity of the acidic solution. Additional steps during embryo handling enhance the potential risk of contamination and this has to be considered if further molecular genetic analysis needs to be performed by polymerase chain reaction (PCR).

Obviously, the diode laser is a valuable tool for any biopsy technique. We have developed a laser-based biopsy procedure, which allows for fast and efficient removal of polar bodies (Montag *et al.*, 1998) and the same technique was applied to blastomere biopsy by *Boada et al., (1998)*. The zygote or

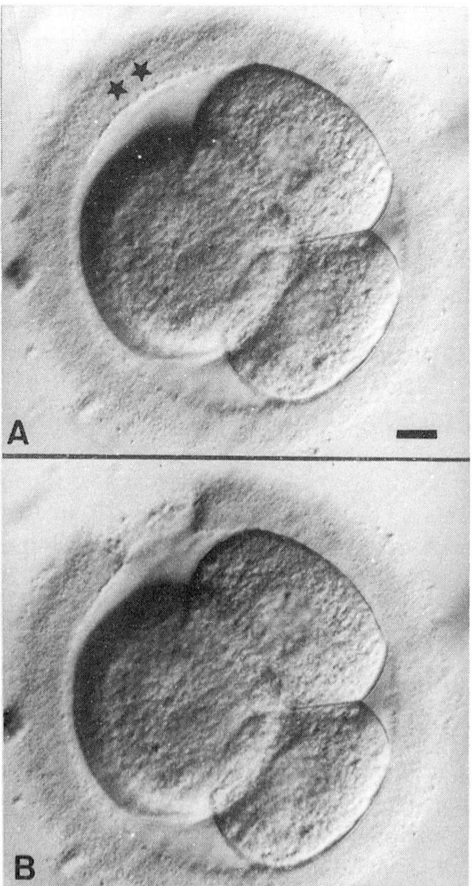

Fig 1 Human four-cell stage embryo before (a) and after (b) laser-drilling of an opening into the zona pellucida for assisted hatching. The opening was drilled with a single laser irradiation at 18 ms pulse length. The sharp border of the opening can be clearly seen (arrows). Immediately after laser-drilling, the embryo was transferred back into the uterus. Bar = 12 µm (reprinted by permission from Springer-Verlag GmbH & Co. KG: Montag *et al.*, Reproduktionsmedizin 1999; 15: 45–54).

embryo is affixed to the holding capillary (Fig. 2). The zona pellucida is opened with a single laser irradiation and through this opening a blunt-ended, flame-polished aspiration capillary can be inserted into the perivitelline space. One or more blastomeres can be gently aspirated and used for cytogenetic analysis. The laser technique allows for a one step procedure. Further, the size of the opening required for polar body or blastomere biopsy can be easily adapted online to the diameter of the biopsy material or the aspiration capillary.

Another advantage of laser biopsy is the use of blunt-ended capillaries which greatly reduce possible damage of the oocyte or the biopsy material. Experimental studies in the mouse have shown, that the procedure does not affect further embryonic development. We have combined this technique with

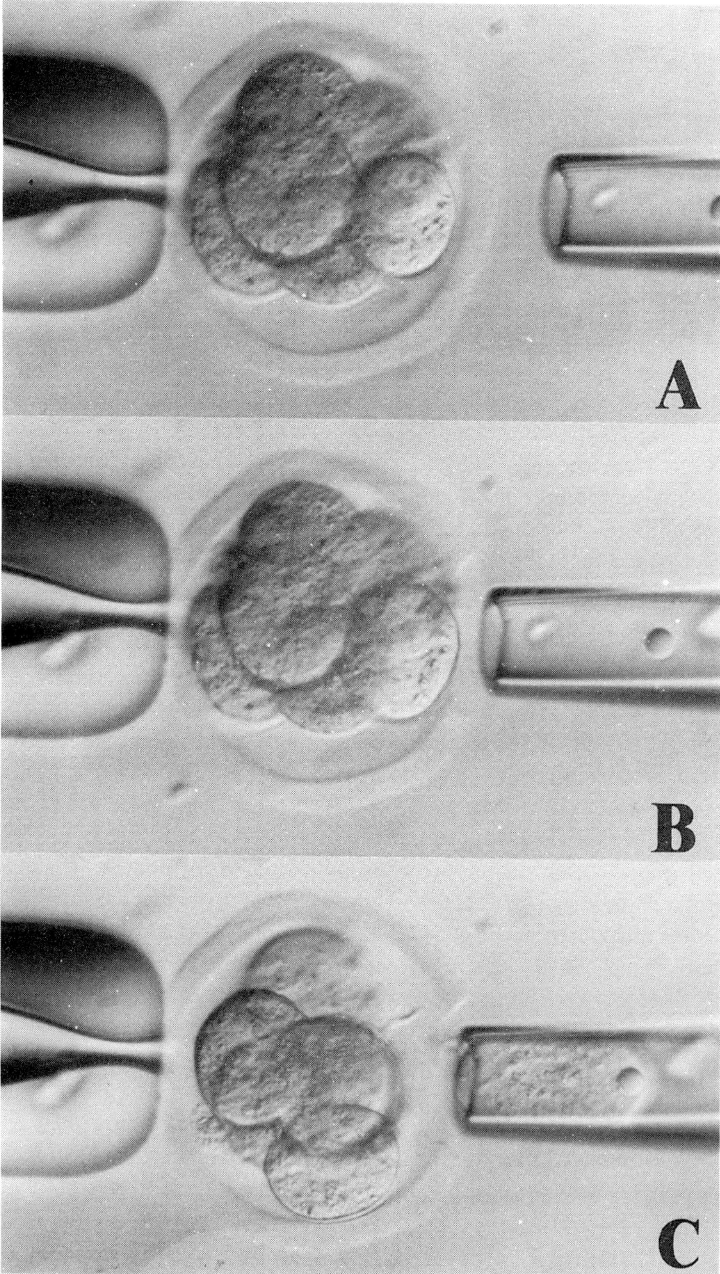

Fig 2 Laser-assisted blastomere biopsy of a mouse 8 cell embryo. The embryo is positioned properly (A) and a laser drilled opening is created in the zona pellucida (B). One or more blastomeres can be easily aspirated with the aspiration capillary (C).

cytogenetic analysis by primed in situ hybridisation and were able to get a diagnosis within less than 2 hours (Montag *et al.*, 1997).

Cryopreservation of Single Spermatozoa in Empty Zona Pellucida

Cohen *et al.*, (1997) proposed the cryopreservation of single spermatozoa in empty zona pellucida. We have improved this method by using the laser for production of empty zonae pellucidae and for cryopreservation of laser-immobilised spermatozoa (Montag *et al.*, 1999a). In our approach, the ooplasm is aspirated completely through a laser-drilled opening to generate an empty zona. Through the same opening up to 10 spermatozoa can be introduced and the opening of the zona can be closed with an oil droplet expelled out of the capillary during its withdrawal. The oil droplet serves to close the opening, but also to allow for easy detection of empty zonae during subsequent freezing and thawing procedures. This treatment may allow for improved cryopreservation of low numbers of spermatozoa e.g. after testicular biopsy or in patients who have occasionally a few spermatozoa in the ejaculate.

Production of Hemizonae

The hemizona assay is a valuable tool for sperm function testing, e.g. sperm binding to human zona pellucida (Burkman *et al.*, 1988). The accuracy of this assay is dependent on the availability of equally sized hemizonae. So far the production of hemizonae was performed by a microblade affixed to a micromanipulator (Burkman *et al.*, 1988) or by manual bisection (Sanchez *et al.*, 1995). We have used the diode laser to cut hemizonae by complete removal of the outer rim of the zona pellucida of an oocyte. The oocyte is affixed to a holding capillary and consecutive openings are drilled beside each other into the zona (Fig. 3). Finally, the oocyte is turned by 90° and the two hemizonae can be separated with two holding capillaries, one attached to each side. We have compared this method of zona production to the manual bisection method and found no difference in the binding capacity of spermatozoa from control patients (Montag *et al.*, 2000c). In our hands, laser-based bisection was much more practicable than manual bisection, especially because corresponding hemizonae were identical in size and all hemizonae could be used in the hemizona assay. This is of great advantage, as the number of human oocytes which are available for hemizona studies is usually limited.

Immobilization of Spermatozoa and Permeabilization of the Sperm Membrane

Besides the creation of openings in the zona pellucida, the laser can be applied for immobilization of spermatozoa (Montag *et al.*, 1999b, 2000d). We found that it is possible to immobilise spermatozoa by applying a single laser shot (Fig. 4). If this is aimed directly to the sperm tail, immobilization is

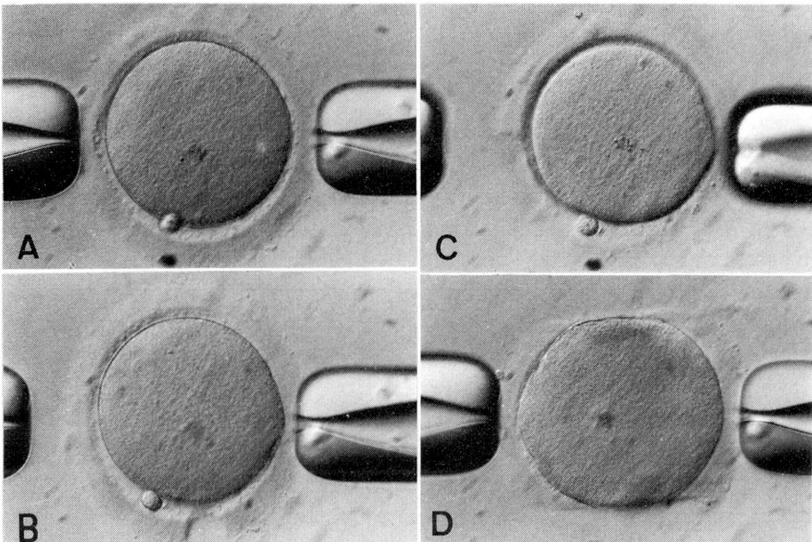

Fig 3 A human oocyte affixed to two holding capillaries (A). Note the sharp focus of the equatorial zona pellucida, which is subsequently removed by a series of single laser shots (Arrowhead in B; C). After complete removal of the equatorial part of the zona, the oocyte is turned by 90° and the two hemizonae are pulled away from the remaining oocyte (D). (reprinted by permission from Blackwell Wissenschafts-Verlag GmbH: Montag *et al.*, Andrologia 2000; **32**, 179–180).

permanent due to permeabilization of the sperm plasma membrane. A single laser shot aimed next to the sperm tail can also cause sperm immobilization. Dependent on the amount of laser energy applied and on the progressive motility of the spermatozoa, this approach will either lead to temporary or to permanent sperm immobilization. The immobilization of fast moving sperm requires a higher energy compared to slower ones. We have further developed this method by the invention of a double laser shot technique, where a first laser shot aimed aside the sperm tail is used for immobilization and a second laser shot aimed directly at the tail is used for reliable permeabilisation (Montag *et al.*, 2000d). This technique allows for the treatment of spermatozoa prior to ICSI and sperm immobilization can be easily performed in culture medium. This approach may help to avoid the use of polyvinylpyrrolidone

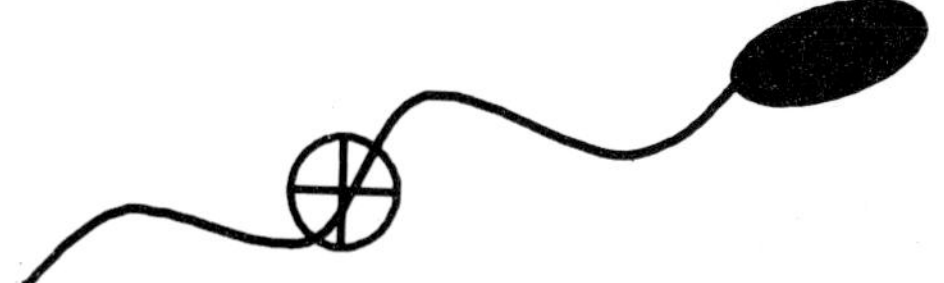

Fig 4 Schematic illustration of the laser-immobilization of human spermatozoa by laser either by shooting aside the middle of the sperm tail or directly at the sperm tail. If the laser shot is applied aside the sperm tail, immobilization is either temporary or permanent dependent on the energy. A direct shot at the tail causes permanent immobilization due to permeabilization of the sperm membrane.

(PVP) for ICSI, as PVP probably has some toxic side-effects (Feichtinger *et al.*, 1995; Jean *et al.*, 1996, 1997; McDermott and Ray, 1996; Strehler *et al.*, 1998).

Measuring of Zona Hardening

The size of a laser-drilled opening in the zona pellucida is an indirect measure of zona hardness. Using constant conditions (temperature of the medium and the heated microscope stage, culture dishes, laser energy, optical setup, culture medium), several drilled openings in one zona will have the smae size, because always the some energy is deposited in zona. Therefore differences in the size of drilled openings in one zona compared to another indicate a difference in zona-structure or hardness. In their first publication on laser assisted hatching, Germond *et al.*, (1995) reported that the size of a laser drilled opening in the zona of an oocyte differed from that of zygote. This finding confirmed fertilization induced zona hardening which has been detected by chemical means already as early as 1976 (Inoue and Wolf, 1976).

We investigated the size of laser-drilled openings in zonae pellucidae from mouse embryos grown in vitro or in vivo (Montag *et al.*, 2000a). In early blastocysts, the size of a laser-drilled opening in embryos grown in vitro was significantly smaller compared to that of in vivo grown embryos. This data show, that in vitro zona hardening occurs presumably due to blastocyst expansion. As blastocyst expansion does not occur in vivo the zona of in vivo grown embryos is softer. This softness may even be more pronounced by the action of uterine lysins which promote global zona lysis in vivo prior to implantation (Montag *et al.*, 2000b).

Conclusions

The 1.48 mm diode laser system is an efficient tool to drill openings into the zona pellucida. The laser technique allows to perform all manipulations which require the instantaneous opening of the zona pellucida, such as assisted hatching, polar body and blastomere biopsy. In addition, new applications include production of hemizonae, generation of empty zona pellucida for cryopreservation of single spermatozoa, immobilization of spermatozoa and assessment of zona pellucida hardness. The precise control of the laser, the ease of application and its safety will stimulate further research and development of new laser-based techniques in assisted reproduction.

Acknowledgements

The authors thank the staff of the IVF unit at the University of Bonn and Mrs. Przybilka for photographic art work.

References

Bider, D., Livshits, A., Yonish, M. *et al.,* (1997) Assisted hatching by zona drilling of human embryos in women of advanced age. *Hum. Reprod.,* **12**, 317–320.

Boada, M., Carera, M., De La Iglesia, C. *et al.* (1998) Successful use of a laser for human embryo biopsy in preimplantation genetic diagnosis: report of two cases. *J. Assist. Reprod. Genet.,* **15**, 302–307.

Burkman, L.J., Coddington, C.C., Franken, D.R. *et. al.* (1988) The hemizona assay (HZA): development of a diagnostic test for the binding of human spermatozoa to the human hemizona pellucida to predict fertilization potential. *Fertil. Steril.* **49**, 688–697.

Cohen, J. (1991) Assisted hatching of human embryos. *J. In Vitro Fertil. Embryo Transfer*, **8**, 179–190.

Cohen, J. and Feldberg, D. (1991) Effects of the size and number of zona pellucida openings on hatching and trophoblast outgrowth in the mouse embryo. *Mol. Reprod. Dev.,* **30**, 70–78.

Cohen, J. (1992) Zona pellucida micromanipulation and consequences for embryonic development and implantation. In Cohen, J., Malter, H.E., Talansky, B.E. and Grifo, J. (eds), *Micromanipulation of human gametes and embryos.* Raven Press, New York, pp. 191–222.

Cohen, J., Alikani, M., Trowbridge, J. and Rosenwaks, Z. (1992). Implantation enhancement by selective assisted hatching using zona drilling of human embryos with poor prognosis. *Hum. Reprod.*, **7**, 685–691.

Cohen, J., Garrisi, G.J., Congedo-Ferrara, T.A. *et al.*, (1997) Cryopresservation of single human spermatozoa. *Hum. Reprod.*, **12**, 994–1001.

Depypere, H.A.T., McLaughlin, K.J., Seamark, R.F. *et al.* (1988) Zona cutting and zona drilling for assisted fertilization in the mouse. *J. Reprdo. Fert.,* **84**, 205–211.

Feichtinger, W., Obruca, A. and Brunner, M. (1995) Sex chromosomal abnormalities and intracytoplasmic injection. *Lancet*, **346**, 1566.

Germond, M., Nocera, D., Senn, A. *et al.* (1995) Microdissection of mouse and human zona pellucida using a 1.48 μm diode laser beam: efficacy and safety of the procedure. *Fertil. Steril.*, **25**, 604–611.

Germond, M., Nocera, D., Senn, A. *et al.*, (1996) Improved fertilization and implantation rates after non-touch zona pellucida microdrilling of mouse oocytes with a 1.48 μm diode laser beam. *Hum. Reprod.*, **11**, 1043–1048.

Germond, M., Primi, M.P., Senn, A. *et al.* (1998) Diode laser for assisted hatching: preliminary results of a multicentric prospective randomized study. *Hum. Reprod.*, **13 Abstract Book 1,** 84–85.

Gordon, J.W. and Talansky, B.E. (1987) Assisted fertilization by zona drillng: a mouse model for correction of oligospermia. *J. Exp. Zool.,* **239,** 347–381.

Grifo, J. (1992) Preconception and preimplanation geneic diagnosis: polar body, blastomere, and trophectodeerm biopsy. In Cohen, J., Malter, H.E., Talansky, B.E. and Grifo, J. (eds), *Micromanipulation of human gametes and emryos.* Raven Press, New York, pp. 191–222.

Handyside, A., Kontogianni, E., Hardy, K. and Winston, R.M.L. (1990) Pregnancies from biopsied human preimplantation embryos sexed with a Y-specific DNA amplification. *Nature,* **344,** 768–780.

Hellebaut, S., De Sutter, P., Dozortsev, D. et. al. (1996) Does assisted hatching improve implantation rates after in vitro fertilization or intracytoplasmic sperm injection in all patients? A prospective randomized study. *J. Assist. Reprod. Genet.,* **13,** 19–22.

Inoue, M. and Wolf, D.P. (1976) Comparative solubility properties of the zonae pellucidae of unfertilized and fertilized mouse ova. *Biol. Reprod.,* **11,** 558-565.

Jean, M., Barriere, P. and Mirallie, S. (1996) Intracytoplasmic sperm injection without polyvinylpyrrolidone: an essential precaution? *Hum. Reprod.,* **11,** 2332.

Jean, M., Barriere, P. and Mirallie, S. (1997) Development of a successful ICSI programme without the use of PVP. *Hum. Reprod.*, **12,** 1115-1116.

Kochevar, I.E. (1989) Cytotoxicity and mutagenicity of excimer laser radation. Lasers Surg. Med., 9, 440-445.

Lanzendorf, S.E., Nehchiri, F., Mayer, J.F. *et al.* (1998) A prospective, randomized, double-blind study for the evaluation of assisted hatching in patients with advanced maternal age. *Hum. Reprod.*, **13,** 409–413.

Liu, H.C., Cohen, J., Alikani, M. et. al. (1993) Assisted hatching facilitates earlier implantation. *Fertil. Steril.,* **60,** 871–875.

Magli, M.C., Gianaroli, L., Ferraretti, A.P. *et al.* (1998) Rescue of implantation potential in embryos with poor prognosis by assisted hatching. *Hum. Reprod.,* **13,** 1331–1335.

Malter, H.E. and Cohen, J. (1989) Partial zona dissection of the human oocytes: a non-traumatic method using a micromanipulation to assist zona pellucida penetration. *Fertil. Steril.,* **51,** 139–148.

McDermott, A. and Ray, B. (1996) Intracytoplasmic sperm injection without polyvinylpyrrolidone: an essential precaution? *Hum. Reprod.,* **11,** 2332.

Montag, M. and van der Ven, H. (1999) Laser-assisted hatching in assisted reproduction. *Croat. Med. J.* **40.**, 398–403.

Montag, M., van der Ven, K., Delacretaz, G. *et al.* (1997) Efficient preimplantation genetic diagnosis using laser assisted microdissection of the zona pellucida for polar body biopsy followed by primed in situ labeling (PRINS). *J. Assist. Reprod. Genet.,* **14,** 476.

Montag, M., van der Ven, K., Delacretaz, G. *et al.* (1998) Laser-assisted microdissection of zona pellucida facilitates polar body biopsy. *Fertil. Steril.,* **69,** 539–542.

Montag, M., Dieckmann, U., Rink, K, *et al.* (1999a) Laser-assisted cryopreservation of single human spermatozoa in cell-free zona pellucida. *Andrologia,* **32,** 49–53.

Montag, M., Rink, K., Descloux, L. *et al.* (1999b) The use of a 1.48 µm diode laser system in assisted reproduction: laser-drilling of the zona pellucida and laser-assisted immobilization of spermatozoa. *Ass. Reprod.,* **9,** 205–213.

Montag, M., Koll, B and van der Ven, H. (2000a) Use of a laser to evaluate zona pellucida hardness at different stages of mouse embryonic development in vitro and in vivo. *J. Assist. Reprod. Genet.,* **17**, 178–181.

Montag. M., Koll, B., Holmes, P. and van der Ven, H. (2000b) The significance of the number of embryonic cells and the state of the zona pellucida for hatching of mouse blastocysts in vitro versus in vivo. *Biol. Reprod.,* **62**, 1738–1744.

Montag, M., Lemola, R and van der Ven, H. (2000c) A new method to produce equally sized hemizonae pellucidae for the hemizona assay. *Andrologia,* **32**, 179–180

Montag, M., Rink, K., Delacretaz, G. and van der Ven, H. (2000d) Laser-induced immobilization and plasma membrne permeabilization in human spermatozoa. *Hum. Reprod.,* **15**, 846–852

Rink, K., Delacretaz, G., Salathe, R.P. *et al.* (1994) 1.48 µm diode laser microdissection of the zona pellucida of mouse zygotes. *Proceedings SPIE.,* **213A**, 412–422.

Rink, K., Delacretaz, G., Salathe, R.P. *et al.* (1996) Non-contact microdrilling of mouse zona pellucida with an objective-delivered 1.48 µm diode laser. *Lasers Surg. Med.,* **18,** 52–62.

Sanchez, R., Finkenzeller, C., Schill, W.B. and Miska, W. (1995) Comparison of two methods to obtain hemizonae pellucidae for sperm function tests. *Hum. Reprod.,* **10,** 2945–2947.

Schoolcraft, W.B., Schlenker, T., Gee, M. *et al.* (1994) Assisted hatching in the treatment of poor prognosis in vitro fertilization candidates. *Fertil. Steril.,* **62**, 551–554.

Stein, A., Rufas, O., Amit, S. *et al.* (1995) Assisted hatching by partial zona dissection of human pre-embryos in patients with recurrent implantation failure after in-vitro fertilization. *Fertil. Steril.*, **63,** 838–841.

Strehler, E., Baccetti, B., Sterzik, K. *et al.* (1998) Detrimental effects of polyvinylpyrrolidone on the ultrastructure of spermatozoa (Notulae seminologicae 13). *Hum. Reprod.*, **13,** 120–123.

Tsunoda, Y., Yasiu, T., Nakamura, K. *et al.* (1986) Effect of cutting the zona pellucida on the pronuclear transplantation in the mouse. *J. Exp. Zool.*, **240,** 119–125.

Tucker, M.J., Morton, P.C., Wright, G. *et al.* (1996) Enhancement of outcome from intracytoplasmic sperm injection: does co-culture or assisted hatching improve implantation rates? *Hum. Reprod.* **11**, 2434–2437.

Oviduct

Follicular Growth, Ovulation and Fertilization: Molecular and Clinical Basis
Anand Kumar and Amal K. Mukhopadhyay (Eds.)
Narosa Publishing House, New Delhi, India, 2001

26

Phospholipase A_2 (PLA_2) and PLA_2 Inhibitor Activities in Ewe Reproductive Tracts

G.C. Upreti, D. Koppens, J.E. Oliver, G.R. Murray, R.M. McDonald and J.F. Smith
AgResearch, Ruakura Research Centre Private Bag 3123, Hamilton, New Zealand

Introduction

The spermatozoal population in the female tract of farm animals is maintained for 6 to 24 hours following artificial insemination before sperm-ovum interactions occur (Hancock, 1962). It has been suggested that the sperm-oviduct epithelial cell interactions are likely to influence sperm competence. Undoubtedly, there are bound to be numerous components of such a complex interaction. As a preliminary attempt to gain insight into this complex interaction, we have focussed our research on ewe tract Phospholipase A_2 (PLA_2). Our approach is based on the observation of Killian *et al* (1989) that reproductive tract fluids are rich in lysolecithin, thereby suggesting involvement of PLA_2 activity.

PLA_2 is a ubiquitous enzyme which catalyses the conversion of lecithin to lysolecithin (a fusogenic lipid) and fatty acid (usually arachidonic acid). The involvement of PLA_2 in the regulation of the reproductive processes has been described both for the sperm PLA_2 (Antaki *et al.*, 1989; Roldan and Fargio, 1993; Roldan and Mollinedo, 1991 and Upreti *et al.*, 1999) and for the PLA_2 activity present in the female tract (Bonney *et al.*, 1987; Morishita *et al.*, 1992; Rajabi and Cybulsky, 1995). There is a contrasting difference between these gender-based studies. Studies on sperm PLA_2 focus on the lysolecithin induced acrosome reaction of the capacitated spermatozoa, which is a pre-requisite of sperm fertilising ability (Fleming and Yanagimachi, 1981; Roldan and Fargio, 1993). In addition, a role for protein inhibitor(s) of PLA_2 (present in seminal plasma and in sperm) in the regulation of sperm competence has also been suggested (Manjunath, 1994; Upreti *et al.*, 1999). In contrast, studies on PLA_{2s} of female reproductive tracts focus on arachidonic acid (a precursor of prostaglandins) for its profound effects on reproduction, particularly corpus luteum function. We report here our preliminary results on the effect of oviductal extracts on ram spermatozoa and on the presence of PLA_2 and PLA_2 inhibitor activities in these extracts.

2 Materials and Methods

2.1 Materials

Porcine pancreatic PLA_2 (pp-PLA_2) was supplied by Sigma Chemical Company, St. Louis, MO, USA. PLA_2 and PLA_2 inhibitor activities from reproductive tract tissues were extracted in our laboratory. The fluorescent labelled substrate of PLA_2—Pyrene PM, [1-hexadecanoyl-2 (1-pyrene decanoyl) - sn - glycerate -3- phosphomethanol, sodium salt] was supplied by Molecular Probes, USA. All other chemicals were of reagent grade quality and were obtained either from BDH, New Zealand or from Sigma Chemical Company, USA.

2.2 Methods

2.2.1 Preparation of Tissue Extracts

Entire reproductive tracts were removed within fifteen minutes of slaughter from non-synchronised, non-pregnant ewes between late April and early May. The tracts were transported from the local abattoir to the laboratory in plastic bags on ice. Oviducts were dissected from the reproductive tract and any with structural abnormalities were discarded. To facilitate injection of perfusion buffer 1–2 cm of the uterine horn was also included with the oviduct dissection. Excess connective tissue surounding the oviducts was also removed. The oviducts were then tied at the infundibulum end with surgical thread and filled with perfusion buffer.

The perfusion buffer described by Ibrahim *et al.* (1994) was modified by replacing EDTA with EGTA and by including proteolytic inhibitors. The modified perfusion buffer contained: 5mM Trizma base (Tris), 5 mM EGTA, 3.4 M NaCl, 1 mM benzamidine and 1 mM phenylmethylsulfonylfluoride (PMSF), adjusted to pH 8.5. Oviducts filled with perfusion buffer were placed in 0.9% NaCl on ice for 1 to 2.5 hours.

Fluid and epithelial layers were extruded by initially squeezing the oviduct gently between forceps then compressing between two glass microscope slides from the isthmus to infundibulum (Chian and Sirard, 1995). Extract was then diluted 1:1 v/v with a $MgSO_4$ solution (30 mM $MgSO_4$, 5 mM Tris, 1 mM benzamidine and 1 mM PMSF, pH 8.5) to neutralise the EGTA. The extract-$MgSO_4$ solution was placed on ice and sonicated for ten seconds with a sonicating probe.

The sonicated extract was then centrifuged at 12,000 for 10 minutes at 4°C. The supernatant (containing PLA_2 in a soluble form) was diluted 1:1 v/v with 2X storage solution [200 mM NaCl, 100 mM Tris, 40% glycerol, 0.5 mM $CaCl_2$, 0.5% defatted-BSA (d-BSA), 2 mM benzamidine, 2 mM PMSF, pH 8.5]. The pellet (containing PLA_2 in a bound form) was resuspended in 9 ml of 1X storage solution (2X solution diluted 1:1 v/v with H_2O). Both samples were assayed for protein concentration and then stored at 4°C (to be used within a week) or at – 20°C for long term storage. Protein was assayed by a modified Lowry's method (Upreti *et al.*, 1988 a, b). BSA was defatted by acid washed charcoal (Upreti, 1984) and its properties have been described previously (Upreti *et al.*, 1988).

2.2.2 *Phospholipase A_2 Assay*

The fluorescence-based PLA_2 assay previously described in detail (Upreti *et al.*, 1999) was used in this study. The only additional modification included in this study was that the enzyme (or tissue extract) was diluted in a buffer containing 0.5% d-BSA: (final composition 50 mM Tris, 100 mM NaCl, 20% glycerol, 0.5% d-BSA, 0.25 mM $CaCl_2$: pH8.5) to minimise non-specific adsorption of the enzyme to the test tube.

Different assay buffers were used for different PLA_{2S}. ppPLA_2 activity was assayed in 50 mM Tris, 100 mM NaCl and 0.25 mM $CaCl_2$, pH 8.5. Soluble and tissue bound oviductal PLA_{2S} were assayed in MES (pH range 5.5 to 6.7) and CAPS (pH range 9.0 to 11.0) buffers, respectively, at 50 mM concentration at stated pH in the presence of 100 mM NaCl and 0.25 mM $CaCl_2$. A typical assay contained 2 ml of appropriate buffer, 30 µl of pyrene-PM, 10 µl of 3.5% d-BSA and 5 µl of PLA_2. Changes in fluorescence (Excitation–345 nm, Emission–395 nm) were recorded and the enzyme activity was calculated from the Relative Increase (RI) in fluorescence intensity. One unit of PLA_2 activity was defined as RI/min/mg protein.

2.2.3 *PLA_2 Inhibitor Activity*

Inhibitor activity in the tissue extract was measured after heat inactivation (70°C for 30 min) of the endogenous PLA_2 activity of the extract. Percent inhibition was calculated from the measured activities of the added PLA_2 in the presence and absence of the putative inhibitor as described previously (Upreti *et al.*, 1999).

2.2.4 *Effect of Oviductal Extracts on Ram Spermatozoa*

Semen was collected by artificial vagina from 5 Romney rams and each sample was processed independently. Semen was transported to the biochemistry laboratory in a Thermus maintained at 35°C where samples were evaluated for the spermatozoal concentration and semen motility. The semen was diluted to 800×10^6 spermatozoa per ml in RSD-1 (a chemically-defined ram semen diluent; Upreti *et al.*, 1995) containing soluble (1.0 mg per ml) and bound (1.8 mg per ml) fractions of epithelial extract or no extract. The diluted semen was gradually cooled to 15°C over a 3 hour period and stored at this temperature under nitrogen atmosphere for up to 6 days.

Aliquots of the diluted and cooled semen were processed on day 0 (D0) and/or day 6 (D6). The semen (100 µl) was mixed with 700 µl of RSD-1 containing soluble and bound fractions (maintained at 15°C) as described in Tables 1 and 2. The diluted semen, containing 100×10^6 spermatozoa per ml, was divided in two portions. The first portion was transferred to 38°C and qualitative assessements of visual motility were made at 1 and 24 hours on D0 for the five semen samples. The second portion was further diluted by a

Table 1 Effect of oviduct epithelial extract fractions on motility, viability and acrosomal status of ram spermatozoa[a]

Fraction[b]	*Incubation(h)/ storage (D)[c]*	*Visual motility*		*%Live*		*Acrosomal status[d]*					
						R		*C*		*I*	
		D0	*D6*	*D0*	*D6*	*D0*	*D6*	*D0*	*D6*	*D0*	*D6*
RSD-1	1 h	75	80	69	48	11	17	23	37	66	46
RSD-1	24 h	85	60	60	44	15	24	15	58	70	18
Soluble	1 h	70	70	78	39	5	6	23	45	72	49
Soluble	24 h	75	20	56	13	10	20	18	55	72	25
Bound	1 h	70	50	85	49	11	1	33	35	56	64
Bound	24 h	20	20	33	23	70	83	22	15	8	2

[a]Three drops from each treatment were smeared on slides and duplicate counts were made on each drop. The data is the mean of 6 determinations.

[b]RSD-1 is the control. Soluble and bound fractions contained ~ 1.8 and 1.0 mg protein per ml respectively.

[c]Samples were stored at 15°C for up to 6 days (D0 and D6) and subsequently incubated at 38°C for 1 and 24 hours.

[d]Percent sperm at each acrosomal status; R—reacted, C—capacitated and I—Intact non-capacitated.

Table 2 Effect of oviductal extract fractions on sperm motility parameters

Fraction	*Motility parameters at 38°C[a]*									
	VCL		*VSL*		*MAD*		*BCF*		*LIN*	
	1h	*24h*	*1h*	*24h*	*1h*	*24h*	*1h*	*24h*	*1h*	*24h*
RSD-1	176.6	119.9	124.2	54.9	26.5	44.3	9.81	6.88	69.8	45.3
+ Soluble	153.3	118.2	112.6	61.8	26.7	38.5	8.16	7.48	72.8	52.0
+ Bound	173.4	93.2	125.6	38.1	26.8	38.0	8.59	5.09	71.2	32.8
s.e.d.	5.64	13.05	5.5	7.19	1.23	4.26	0.32	0.86	1.84	5.35
Δ Storage (15°C; 6 days)	–16	–17.3	–25	–3.4	+5	–10.9	+0.1	–0.45	–9.64	–4.1

[a]VCL—Curvilinear velocity, VSL—Straight line velocity, MAD—Mean angular displacement and LIN—Linearity. Values are means of 40 observations (8 measures on each of 5 semen samples from 5 rams).

factor of 10 fold in the presence of the corresponding oviductal fraction and incubated at 38°C on D0 and D6. Aliquots were taken at 1 and 24 hours for the determination of viability (by propidium iodide and SYBR-14 staining) and acrosomal status [by Chlortetracycline (CTC) staining; Smith and Murray, 1997] after scoring visual motility. These measurements were made for one semen sample (randomly selected) only. Spermatozoal motility parameters were made for the five semen samples using the Hobson Sperm Tracker (Briggs *et al.*, 1996). Measurements for each treatment were made on four drops using two fields of view. The data (mean of 8 measurements) was statistically analysed using ANOVA.

3 Results and Discussion

3.1 Phospholipase A_2 Activity

Oviductal homogenates prepared in NaCl Tris-EGTA buffer containing protease inhibitors contained 3.4 (n=2) units of PLA_2 activity. In comparison, oviductal homogenates prepared by Ibrahim *et al* (1994) protocol, using Tris-EDTA buffer gave 1.25 $\pm$ 0.25 units of PLA_2 activity. The differences in PLA_2 activity could be attributed to the differences in the binding affinities of EDTA (stronger binding) and EGTA towards calcium, which is a co-factor for PLA_2 activity (Upreti and Jain, 1980).

The specific activity of PLA_2 in the bound form was ~4 times more than in the soluble form (Soluble 11.5 $\pm$ 2.3; Bound 46.3: n=2). Further characterization of both forms was carried out to determine if both types of PLA_2 activities represent different isoforms or a manifestation of the differential release of the soluble enzyme activity from its bound form.

The effects of assay temperature, pH and the presence of calcium ions were evaluated for both the soluble and the bond forms (Fig. 1 A, B and C). Enzyme activities were higher for the bound form than the soluble form at all temperatures up to 60°C. An increase in enzyme activity was observed for the soluble form, which reached a saturation point at 30°C. A biphasic response was observed for the enzyme in the bound form, with activity peaking at ~40°C. Contrasting differences were observed between the soluble and the bound form in regard to pH response. The soluble enzyme had a pH optima of 6.0 with a rapid decline above pH 8.0. In contrast, the bound form showed only residual PLA_2 activity in the pH range of 6–8.0. A steep increase in activity was observed at pH > 9.0. Subsequent studies with bound form were carried out at pH 10.5, which was maintained using CAPS buffer. Like pH, the response to changes in calcium concentration, beween soluble and bound forms, were different. The soluble enzyme required calcium for its activity, which plateaued between 0.1 and 0.2 mM calcium. The PLA_2 activity in the bound enzyme, was highest in the absence of calcium (+15 mM EGTA) and was inhibited with increasing calcium. Similar results have also been reported for a membrane bound isoform of human 88 kDa calcium independent PLA_2 (Larsson *et al.*, 1999). PLA_2 activities in the soluble and bound forms also differed in their heat inactivation (70°C for 30 min; see below). In conclusion, the above studies indicate that PLA_2 activities in the soluble and bound fractions of the ewe oviducts are likely to be isoforms. Multiple isoforms of PLA_2 have recently been reported in a variety of tissues e.g. rat brain (Molloy *et al.*, 1998), corpus luteum of pregnant rat (Kurusu *et al.*, 1999) and mouse thymus (Valentin *et al.*, 1999). Development of assays that distinguish various

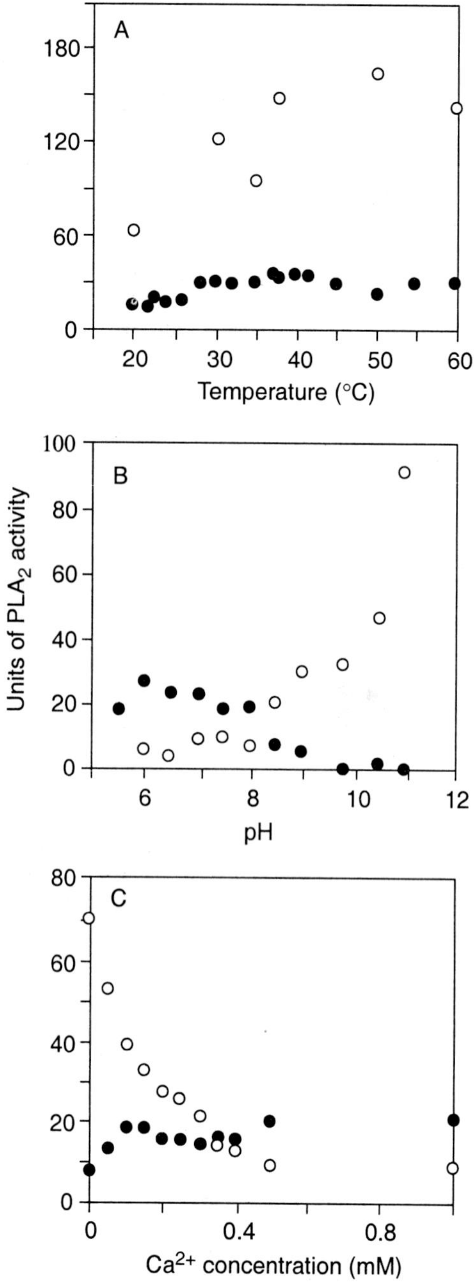

Fig. 1 PLA$_2$ activity in oviduct epithelial fractions(Soluble—•; Bound—o) at various temperatures, pHs and in the presence of varying Ca^{2+} concentrations.

forms of PLA_2 activities in biological samples would advance the understanding of the role of PLA_2 in physiological processes (Yang *et al.*, 1999).

3.2 Phospholipase A_2 Inhibitor Activity

It is necessary to heat inactivate the endogenous PLA_2 activity before measuring the inhibitor activity against exogenous PLA_2 (Upreti *et al.*, 1999). The endogenous PLA_2 activity of the soluble fraction was heat labile, but the activity present in the bound form was heat resistant. Both pp-PLA_2 and bound form of the oviductal PLA_2 were inhibited to 67.3 and 72.1 percent respectivley by the heat inactivated soluble fraction. The inhibitor activity in the bound form could not be assessed, as the inherent PLA_2 activity of this fraction was stable at 70°C for 30 min. The presence of natural inhibitors of PLA_2 in biological systems (Manjunath, 1994; Rice *et al.*, 1987; Upreti *et al.*, 1999) generates additional possibilities for the in vivo regulation of PLA_2 activity.

3.3 Effect of Oviductal Extracts on Ram Spermatozoa

Qualitative evaluation of ram sperm motility on D0 in the presence of soluble (1.8 mg protein per ml) and bound (1.0 mg protein per ml) fractions revealed that the soluble fraction did not influence the sperm motility over a 24 hour incubation at 38°C. Motility was reduced in the presence of bound fractions. The motility pattern changed from linear to circular within 1 hour and spermatozoal head to head clumping appeared prominent after 24 hours incubation. The qualitative motility patterns were similar for the five semen samples. The different amounts of soluble and bound fractions were used in these experiments to compensate the inherent differences in the PLA_2 activities.

The overall effects on motility and viability were also quantitated by staining sperm samples for viability and acrosomal status by specific stains and by measuring spermatozoal motility using a Hobson Sperm Tracker. Spermatozoa which were cooled to 15°C (but not stored at this temperature; D0) maintained their motility, viability (% live) and acrosomal intactness during 24 hour incubation at 38°C for both the control (RSD-1) and in the presence of the soluble fraction. These parameters declined in the presence of the bound fraction for D0 samples. Storage of spermatozoa (D6) had only marginal effects on motility, but both the viability and the acrosomal intactness declined. In general, approximately 50% of sperm were capacitated by D6 for both the control and in the presence of the soluble fraction. Sperm motility decline became prominent upon incubation at 38°C for 24 hours. The bound fraction induced acrosomal reaction with concomitant decline in motility and viability for D6 samples (Table 1). Similar results on the effect of human tubal fluid on spermatozoa have been reported (Mukherjee and Lippes, 1972).

All motility parameters measured with the Hobson Sperm Tracker (Table 2) declined over the 24 hour incubation at 38°C, excepting MAD, which increased by a factor of ~1.5. This is an expected behaviour of a slow moving spermatozoal population. Differences in the motility parameters between control (RSD-1) and treatments (Soluble and Bound) were only marginal for the soluble form at 1 hour. Even this marginal reduction did not persist over 24 hour incubation. There were significant reductions in VSL, LIN and BCF in the presence of the bound form after 24 hour incubation. There were no interactions with the storage of semen for 6 days at 15°C. A gradual decline in most motility parameters was observed.

In conclusion, extracts from ewe oviductal epithelia had a significant impact on sperm viablity, motility and acrosomal status of ram spermatozoa. It is therefore likely, that factors which modulate the activities of PLA_2 and PLA_2 inhibitor, would play a key role in the maintenance of spermatozoal competence in oviducts, prior to sperm-ovum interactions. The involvement of PLA_2 from the female reproductive tract in pregnancy has been suggested (Bonney *et al.*, 1987; Morishita *et al*, 1992; Rajabi and Cybulsky, 1995). Ours is the first report we are aware of which positively implicates sperm-mediated effects of the female tract PLA_2 on the outcome of pregnancy.

Acknowledgements

We thank Mr J. Parr and C. Harris for their technical suppot and Dr. H.R. Tervit for his critical reading and suggestions. This work was supported by grants from Meat New Zealand (91 MT 66/3.2) and the New Zealand Lottery Board.

References

Antaki, P., Guerette, P. and Roberts, K.D. (1989) Detection of PLA_2 in human spermatozoa. *Biol. Reprod.* **41**, 241–246.

Bonney, R.C., Qizilbash, S.T. and Franks, S. (1987) Endometrial phospholipase A_2 enzymes and their regulation by steroid hormones. *J. Steroid Biochem,* **27**, 1057–1064.

Briggs, R.M., Smith, J.F. and Duganzich, D.M. (1996) Optimisation of a Hobson Sperm Tracker for ram sperm assessment. *Proc. International Congress Anim. Reprod. (Sydney)* **13(3)**, 24–28.

Chian R. and Sirard, M. (1995) Fertilizing ability of bovine spematozoa with oviduct epithelial cells. *Biol. Reprod.* **52**, 156–162.

Fleming, A.D. and Yanagimachi, R. (1981) Effects of various lipids on the acrosome reaction and fertilizing capacity of guinea pig spermatozoa with special reference to the possible involvement of lysophospholipids in the acrosome reaction. *Gamete Res.* **4**, 253–273.

Hancock, J.L. (1962) Fertilization in farm animals. *Animal Breeding Abstracts* **30**: 285–310.

Ibrahim, M et al., (1994). Calpain from rat intestinal epithelial cells: age dependent dynamics during cell differentiation. *Mol. Cell. Biochem.* **131**, 49–59.

Killian, G.J. *et al.*, (1989) Changes in phospholipids, cholesterol and protein content of oviduct fluid of cows during the oestrus cycle. *J. Reprod. Fertil.* **86**, 419–426.

Kurusu, S. *et al.* (1999) Luteal phospholipase A_2 activity increases during functional and structural luteolysis in pregnant rats. *FEBS Letters* **454(3)**, 225–228.

Larsson, F.P.K.A., Kennedy, B.P., Claesson, H.E. (1999) The human calcium-independent phospholipase A2 gene: Multiple enzymes with distinct properties from a single gene. *Eur. J. Biochem.* **262(2)**, 575–585.

Manjunath, P. *et al.* (1994) Major proteins of bovine (SP) inhibit phopholipase A_2. *Biochem. J.* **303**, 121–128.

Molloy, G.Y., Rattray, M. and Williams, R.J. (1998) Genes encoding multiple forms of phospholipase A_2 are expressed in rat brain. *Neuroscience Letters* **258(3)**, 139–142.

Morishita, T. *et al.* (1992) Regional differences of phospholipase A_2 activity in the rabbit oviductal epithelium. *Prostaglandins, Leukotrienes and Essential Fatty Acids.* **47**, 199–202.

Mukherjee, A.B. and Lippes, J. (1972) Effect of human follicular and tubal fluids on human, mouse and rat spermatozoa in vitro. *Can. J. Genet. Cytol.* **14**, 167–174.

Rajabi, M.R. and Cybulsky, A.V. (1995) Phospholipase A_2 activity is increased in guinea pig uterine cervix in late pregnancy and at parturition. *Am. J. Physiol.* **269**(5 part I), E940–E947.

Rice, G.E., Wong, M.H. and Thorburn, G.D. (1987) Ovine myometrial cytosol inhibits phospholipase A_2 activity, in vitro. *Prostaglandins* **34**, 563–573.

Roldan, E.R.S. and Fragio, C. (1993) Phospholipase A_2 activity and exocytosis of the ram sperm acrosome: Regulation by bivalent cations. *Biochem. Biophys. Acta* **1168(1)**, 108–114.

Roldan, E.R.S. and Mollinedo, F. (1991) Diacylglycerol stimulates the Ca^{2+} dependent PLA_2 of ram spermatozoa. *Biochem. Biophys. Res. Commun.* **176**, 294–300.

Smith, J.F. and Murray, G.R. (1997) Evaluation of different staining techniques for determination of membrane status in spermatozoa. *New Zealand Soc. Anim. Prod.* **57**, 246–250.

Upreti, G.C. (1984) Colorimetric estimation of inorganic phosphate in colored and/or turbid biological samples: Assay of phosphohydrolases. *Anal. Biochem.* **137**, 485–492.

Upreti, G.C. and Jain, M.K. (1980) Action of phospholipase A_2 on unmodified phosphotidylcholine bilayers: Organizational defects are preferred sites of action. *J. Membrane Biol.* **55**, 113–121.

Upreti, G.C., Ratcliff, R.A. and Riches, P.C. (1988) Protein estimation in tissues containing high levels of lipid: modifications to Lowry's method of protein determination. *Anal. Biochem.* **168**, 421–427.

Upreti, G.C., Riches, P.C. and Johnson, L.A. (1988) Attempted sexing of bovine spermatozoa by fractionation on a Percoll density gradient. *Gamete Research* **20**, 83–92.

Upreti, G.C. *et al.*(1995) Development of a chemically defined ram semen diluent (RSD-1). *Anim. Reprod. Sci.* **41**, 143–157.

Upreti, G.C. *et al.*, (1999) Studies on the measurement of phospholipase A_2 (PLA_2) and PLA_2 inhibitor activities in ram semen. *Anim. Reprod. Sci.* **56**, 107–121.

Valentin, E. *et al.* (1999) Cloning and recombinant expression of a novel mouse secreted phospholipase A_2. *J. Biol. Chem.* **274(27)**, 19152–19160.

Yang, H.C. *et al.* (1999) Group-specific assays that distinguish between the four major types of mammalian phospholipase A_2. *Analytical Biochemistry* **269**, 278–288.

Placenta

Follicular Growth, Ovulation and Fertilization: Molecular and Clinical Basis
Anand Kumar and Amal K. Mukhopadhyay (Eds.)
Narosa Publishing House, New Delhi, India, 2001

27

Cloning and Characterisation of a Novel Apoptotic Gene in the Human Placenta

M. Rekha Rao[1], A.M. Dharmarajan[2], Robert R. Friis[3] and A. Jagannadha Rao[1]

[1]Department of Biochemistry, Indian Institute of Science, Bangalore, India
[2]Department of Anatomy and Human Biology, The University of Western Australia, Nedlands, Perth, Australia
[3]Department of Clinical Research, Faculty of Medicine, University of Bern, Switzerland

Introduction

The evolution of viviparity necessitated the simultaneous evolution of placenta, a transient feto-meternal association that develops during all mammalian pregnancies, primarily involved in fulfilling the nutritional, respiratory, excretory and endocrine requirements of the fetus. In addition it also serves as an effective immunologiocal barrier and prevents the rejection of the fetus by the maternal immune system.

The first-trimester human placenta and term placenta represent two distinct stages of differentiation, both of which differ not only in their function but also in their ability to synthesize and secrete a variety of proteins, peptides and steroids (Villee, C.A., 1997). Moreover, the cell populations that make up the placental villi comprise primarily of the cytotrophoblasts in the first trimester which continuously fuse and differentiate into syncytiotrophoblasts which are a unique and multinucleated cell type (Carter, J.E., 1964). The cytotrophoblasts are the predominant cell type during the first and second trimester of pregnancy. Both cytotrophoblast and syncytiotrophoblasts differ in their protein and steroid production profiles; the syncytiotrophoblasts being primarily steroidogenic in nature (Petraglia, F., Florio, P., *et al* 1996). Interestingly the incidence of apoptosis is also known to increase at term (Smith, S.C., Philip, N., *et al* 1997a). All these factors make the human placenta an excellent model for the study of tissue-specific and stage-specific gene expression.

We employed the technique DD (Differential display)-RT PCR in order to gain an insight into the kind of differentially expressed genes between the human first-trimester and term placental villi. Using this strategy we have been able to identify several cDNAs as being differentially expressed and have been able to partially charcterise a few of them. In this study we present the cloning and characterisation of a novel cDNA from the term placental display and provide some evidence suggesting the involvement of this gene product in apoptosis (Rekha, M.R., Dharmarajan, A., M., 2000).

Apoptosis is a widespread physiologic process that has steadily gained recognition due to its now clearly established roles in development, tissue homeostasis and disease. Since the initial report describing the presence of internucleosomal DNA activity in rat granulosa and luteal cells (Zeleznik, A.J., Ihrig, L.L., *et al* 1989) a role for apoptosis in several reproductive tissues including the human placenta has been demonstrated (Smith, S.C., Philip, N., *et al* 1997). Furthermore, the expression of several pro- and anti-apoptotic markers like Fas, FasL and Bcl-2 have been demonstrated in the human placenta. The expression of FasL by the syncytiotrophoblast has been proposed to contribute to the immuneprevilaged status in the placenta during pregnancy and prevent the immunological response of maternal lymphocytes against the semi-allogenic fetus (Uckan, D., Steele, A., 1997).

Although the cytotrophoblasts are known to express the full length Fas and the villous cytotrophoblasts and the syncytiotrohoblasts are known to express FasL. The Fas induced signaling leading to the destruction of the cytotrophoblasts is blocked in what appears to be a novel mechanism in the placenta (Payne, S.G., Smith, S.C., *et al* 1999). The high levels of Bcl-2 in the syncytiotrophoblasts have been proposed to play a role in preventing or delaying the death of these terminally differentiated cells (Sakuragi, N., Matsuo, H., *et al* 1994). In another study, p53 and Bax were found to be expressed in the cytotrophoblast indicating that the proliferative capacity of the cytotrophoblast may be under the control of p53 tumor suppresor gene (Qiao, S., Nagasaka, T., *et al* 1998). More recently signaling molecules like TRAIL (TNF-related-apoptosis inducing ligand) and its receptor TRAIL-R expression in the placenta has been assessed and its implications in the establishment of placental immune privilege has been proposed (Philips, T. A., Ni, J., *et al* 1999). There is now a growing list of factors that are being identified in the human placenta that are involved in apoptosis.

Of direct relevance to the present studies, recent investigation of functional and regressing human placenta demonstrates that placental apoptosis increase significantly as pregnancy progresses. Furthermore, the study clearly indicates that apoptosis may play a role in the normal development and aging of the placenta (Smith, S.C., Philip, N., *et al* 1997b). It has also been proposed that apoptosis in the human placenta maybe a mechanism of maintaining tissue homeostasis (Chan, C.C., Lao, T.T., *et al* 1998). Todate, however, all the molecular mechanisms underlying the complex process of placental apoptosis have not been studied exhaustively.

In previous studies of mammary gland involution following forced weaning, vental prostate involution as a consequence of castration, and corpus luteum regression at the end of pregnancy, differential display has been employed to isolate genes up-regulated specifically in association with apoptosis (Guo, K., Wolf. V., 1998). In the present investigation we attempted to document lines of similarity with the incidence of apoptosis in human placenta so as to identify genes that are up-regulated in the term placenta, a stage of gestation when apoptosis is known to be more extensive compared to the first trimester tissue. We also describe the establishment of an in vitro model system for analysis of the apoptotic machinery in the human placenta under defined experimental conditions.

Differential Display Cloning of T-18

A differential display RT-PCR was performed with first trimester and term placental villi, using T_{12}AG and arbitrary 10-mer AGCCAGCGAA as decribed previously (Rekha, M.R., Dharmarajan, A.M., *et al* 2000). The differentially expressed cDNAs identified from the display were cloned and sequenced. Sequence comparison of one of the cDNAs differentially expressed in the term placental tissue, with the databases revealed that it is a novel gene with absolutely no homology to any known sequence (Insert size 221bp, GenBank accession AF089811). This gene will be referred to as T-18 in this article. The differential expression of this novel gene was confirmed by Northern Blot analysis.

Since placenta is very intimately associated with the uterine tissue, even traces of uterine contamination would result is displays that would be a gross mis-representation of the genes expressed in the placenta. Hence, all samples used for DD-RT PCR analysis were first checked for the absence of uterine contamination. A sensitive RT-PCR for PP14, an abundantly expressed endometrial transcript, was used as a marker for the checking for the presence of uterine contamination. Only those placental samples that failed to show an amplification using PP14 specific primers were used for DD-RT PCR analysis.

In vivo Expression of T-18 in the First Trimester and Term Placenta

Transcripts of the sizes corresponding to 4.4, 3.0 and 1.4 Kb were present only in the term placenta and no expression was observed during the first trimester. Previous studies have also reported increased levels of apoptosis in syncytiotrophoblast during later stages of pregnancy. This finding together with our finding of increased expression of T-18 at this time suggest a possible role for this gene in placental apoptosis.

Validation of the in vitro Placental Culture System

Since our initial experiments suggested a potential role for T-18 in placental apoptosis, an attempt was made to establish a model for analysis of apoptosis in the placenta under defined experimental conditions. As recent studies have shown that organ culture of intact ovarian follicles and corpus luteum provide a useful model to study the hormonal regulation of atresia (Tilly, J.L., Tilly, K.I., *et al* 1995) and luteolysis (Dharmarajan, A.M., Goodman S.B. *et al.*, 1994), we utilized a similar organ culture system to incubate healthy placental villi as a model to study the regulation of apoptosis in human placenta.

Incubation of placental villi from first trimester without tropic support caused a time-dependent onset of apoptosis in vitro (Fig. 1). DNA breakdown as assessed by ethidium bromide staining was detected as early as 1 hr.

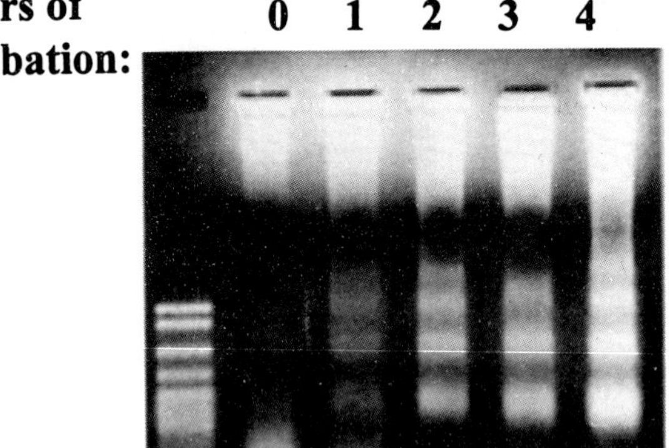

Fig. 1 Biochemical analysis of DNA extracted from first trimester placental villi and snap frozen immediately (0 hour) or following serum-free organ culture for increasing lengths of time (1, 2, 3 and 4 hours).

Evidence that Superoxide Dismutase (SOD) Promotes Placental Villi Survival

Based on previous studies which demonstrated an inhibitory effect of SOD in corpus luteum apoptosis (Dharmarajan, A.M., Hisheh, S., *et al* 1999), we next tested the ability of this free radical scavenger to alter the onset of apoptosis in the human placenta following 4 hrs of incubation. The time 4 hrs was chosen as this length of incubation caused half-maximal DNA fragmentation. Incubation of both first-trimester and term placental villi for 4 hrs with SOD caused an extensive reduction in the internucleosomal DNA cleavage associated with apoptosis.

T-18 mRNA Levels in the Incubated Placental Villi Undergoing Spontaneous Apoptosis and Following SOD Treatment

Northern blot analysis to total RNA extracted from first trimester placental villi following incubations at 0, 2, 4, 6 and 8 hr revealed an induction of T-18 transcripts upon incubation of villi under serum-free conditions, reaching maximal expression at 6hr. T-18 mRNA levels following a 4-hr treatment with SOD in vitro significantly decreased compared to control levels. The 4.4. Kb transcript showed a 33% reduction while the 3.0 Kb transcript reduced by as much as 48% (Rekha, M.R., Dharmarajan, A.M., *et al* 2000). We did not see any significant changes in the 1.4 Kb transcript under these conditions.

Discussion

Although differential display was originally developed to identify differences in gene expression in two specific types of cells (Liang, P. and Pardee, A.B., 1992), it is a molecular protocol that is ideally suited for detecting genes that are newly expressed in endocrine tissues in response to tropic hormones and/or those that are turned on during the process of tissue differentiation. This technique has been successfully employed to study a variety of phenomena like differentiation, senescence, control of cell-cycle, gene expression upon hormone stimulation, etc.

In the present study, we sought to evaluate the temporal relationship between the process of placental differentiation and the incidence of apoptosis, as well as to begin analysis of those genes that are actively transcribed during apoptosis in the placenta. Using internucleososmal DNA cleavage as a marker for apoptotic cells, biochemical analysis of DNA integrity revealed that placental cells are dying by apoptosis in our *in vitro* model for spontaneous induction of apoptosis. This DNA laddering was not detected in snap-frozen healthy placenta, supporting the results of previous findings that only a small population of placental cells are dying by apoptosis *in vivo*. Hence it agrees well with our finding that we donot detect any laddering using ethidium bromide staining in unincubated snap frozen samples (0 hour sample in Fig. 1).

Studies with extragonadal tissues have used *in vitro* models, such as primary cell cultures, to investigate the hormonal regulation of cell death; however, these approaches fail to recreate the three dimensional architecture and cell-cell interactions present within any given tissue. As recent studies have demonstrated that organ/explant culture of intact ovarian follicles and corpus luteum provide a useful model to study the regulation of apoptosis. In the present study, we report the establishment of a similar in vitro model based on serum-free culture of placental villi isolated from first trimester and full term placenta. Placental cell culture and villi culture experiments suggest the presence of an active apoptotic machinery in term placental cells as assessed by comparing nuclei positive for TUNEL (Cirelli, N., Moens, A., *et al* 1999). Furthermore, the acute sensitivity of the first trimester placental syncytiotrophoblasts to oxygen-mediated damage has also been documented when cultured *in vitro* (Watson, A.L., Skepper, J.N., *et al* 1998).

Using the explant culture model, our data have indicated that healthy placental villi collected from first trimester rapidly undergo spontaneous apoptosis when incubated without tropic hormone support *in vitro* and the onset of apoptosis is suppressed by treatment of SOD, a free radical scavenger enzyme. The survival of placental villi in the presence of SOD is consistent with previous reports in rat follicle and rabbit corpus luteum (Tilly, J.L., Tilly, K.I., 1995). Interestingly, a time dependent increase in expression of T-18 was observed when the first trimester placental villi was cultured in serum-free conditions. It is significant to note that first-trimester placental villi donot show expression of T-18. However, the transcripts corresponding to T-18 appear upon induction of apoptosis or upon incubating the villi under serum-free conditions. This data together with the decreased expression of T-18 when the DNA fragmentation was suppressed by SOD further strengthens the role for T-18 in placental apoptosis.

As discussed above, placental villi exhibited extensive apoptosis after 4 hours of incubation and this extensive breakdown was blocked by treating the placental villi with 75U/ml SOD, suggesting accumulation of superoxide anion and/or its downstream reactive metabolites in tropic factor-deprived placenta is involved in triggering placental cell death. This finding for the first time indicates that an inhibition of oxidative stress using SOD inhibits apoptosis in placental cells in vitro. Considering the fact that the generation of superoxide anion could be an important source of oxidative stress in pre-eclampsia (Wang, Y., and Walsh, S.W., 1998) and that pre-eclampsia is associated with widespread apoptosis of the cytotrophoblast that invade the uterus (DiFederico, E. and Genbacev, O., 1999) there is a need to study the regulation of SOD expression in greater detail. Our model hence provides a convenient handle to simulate and study the mechanistic aspects of SOD action.

In summary, biochemical analysis of DNA integrity has demonstrated that human placental villi cells undergo apoptosis. Moreover, the data provide evidence that healthy placental villi isolated from first trimester placenta exhibit a spontaneous onset of apoptosis when incubated without hormonal support *in vitro*, and that appoptosis is prevented by SOD, a free radical scavenging enzyme. Collectively, these findings suggest that an important role exists for T-18 in inducing apoptosis within the human placenta or T-18 is an apoptosis associated gene product, and that serum-free placental culture provides an useful model to further investigate the molecular and intracellular signaling events which control placental cell apoptosis. Currently studies on the expression of several apoptotic markers in the placenta, both in our in vivo and in vitro models and their regulation in vitro are being carried out.

Acknowledgements

This study was supported by the Council for Scientific and Industrial Research, Government of India (AJR, RMR), Department of Biotechnology (AJR), Department of Science and technology (AJR), Indian Council for Medical Research (AJR), Rockefeller Foundation (AJR), National Health and Medical Research Council of Australia (AMD), Australian Research Council (AMD), and Raine Foundation (AMD).

The authors are grateful to Mr. Ricky Lareu and Mr. Markus Lacher for their useful scientific discussions.

References

Carter J.E., (1964) Morphologic evidence of syncytial formation from cytotrophoblast cells. *Obstet Gynecol.*, **23**, 647.

Chan C.C., Lao T.T. et al., (1998) Apoptosis in human placenta. *Am J Obstet Gynecol.* **179**(5), 1377–1378.

Cirelli, N., Moens, A, *et al.* (1999) Apoptosis in human placenta is not increased during labor but can be massively induced *in vitro. Biol Reprod.,* **61(2)**, 458–463.

Dharmarajan A.M, Goodman S.B., *et al* (1994) Apoptosis during functional corpus luteum regression: evidence of a role for chorionic gonadotropin in promoting luteal cell survival. *Endocrine J (Endocrine)* **2**, 295–303.

Dharmarajan A. M., Hisheh, S., *et al* (1999) Anti-oxidants minic the ability of chorionic gonadotropin to suppress apoptosis in the rabbit corpus luteum in vitro: A novel role for superoxide dismutase in regulating bax expression. *Endocrinol.,* **140(6)**, 2555–2561

DiFederico, E., Genbacev, O., *et al* (1999) Preeclampsia is associated with widespread apoptosisi of Placental cytotrophoblasts within the uterine wall. *Am J Pathol.*, **155 (1):** 293–301

Guo, K., Wolf, V., *et al* (1998) Apoptosis-associated gene expression in the corpus luteum of the rat. *Biol Reprod*, **58**, 739–746

Payne, S.G., Smith, S.C., *et al* (1999) Death receptor Fas/Apo-1/CD95 expressed by human placental cytotrophoblasts does not mediate apoptosis. *Biol Reprod.*, **60(5)**, 1144–1150.

Petraglia, F., Florio, P., *et al* (1996) peptide signaling in human placenta and membranes: Autocrine, paracrine, and endocrine mechanisms. *Endocrine Reviews*, **17**, 156–186.

Phillips, T.A., Ni, J., (1999) TRAIL (Apo-2L) and TRAIL receptors in human placentas: implications for immune privilege. *J. Immunol.,* **162(10)**, 6053–6059.

Qiao, S., Nagasaka, T., *et al* (1998) p53, Bax and bcl-2 expression, and apoptosis in gestational trophoblast of complete hydatidiform mole. *Placenta,* **19**, 361–369.

Rekha, M.R., Dharmarajan, A.M., *et al* (2000) Cloning and characterisation of an apoptotsis-associated gene in the human placenta. *Apoptosis*, **5**, 49–56.

Sakuragi, N., Matsuo, H., *et al* (1994) Differentiation-dependent expression of the Bcl-2 proto-oncogene in the human trophoblast lineage. *J Soc. Gynecol Invest.*, **1**, 164–172.

Smith, S.C., Philip, N., *et al* (1997a) Placental apoptosis in normal human pregnancy. *Am J. Obstet & Gynecol.,* **177**, 57–65.

Smith, S.C., Philip, N., *et al* (1997b) Increased placental apoptosis in intrauterine growth restriction. *Am J. Obstet & Gynecol.*, **177**, 1395–1401.

Tilly, J.L. & Tilly K.I. (1995) Inhibitors of oxidative stress mimic the ability of follicle-stimulating hormone to suppress apoptosis in cultured rat ovarian follicles. *Endocrinology,* **136**, 242–252.

Tilly, J.L., Tilly, K.I., *et al* (1995) Expression of members of the bcl-2 gene family in the immature rat ovary:equine chorionic gonadotropin-mediated inhibition of granulosa cell apoptosis is associated with decreased bax and constitutive bcl-2 and bcl-x messenger ribonucleic acid levels. *Endocrinol.,* **136**, 232–241.

Uckan, D., Steele, A., *et al* (1997) Trophoblst express Fas ligand: a proposed mechanism for immune privilege in plaenta and maternal invasion. *Molecular Human Reproduction,* **3**, 655–662.

Villee, C.A. (1977) Synthesis of proteins in placenta. *Gynecol Invest.*, **8,** 145–161.

Wang, Y., Walsh, S.W., *et al* (1998) Placental mitochondria as a source of oxidative stress in pre-eclampsia. *Placenta.* **19(8)**, 581–586.

Watson, A.L., Skepper, J.N., *et al* (1998) Susceptibility of human placental syncytiotrophoblastic mitochondria to oxygen-mediated damage in relation to gestational age. *J Clin Endocrinol Metab.*, **83(5)**, 1697–1705.

Zeleznik, A.J., Ihrig, L.L., *et al.* (1989) Developmental expression of Ca/Mg dependent endonuclease activity in rat granulosa and luteal cells. *Endocrinol,* **125**, 2218–2220.

Author Index

Subject Index